Industrial Applications of Oil and Gas Resources

Industrial Applications of Oil and Gas Resources

Edited by **Oliver Haghi**

SYRAWOOD
PUBLISHING HOUSE

New York

Published by Syrawood Publishing House,
750 Third Avenue, 9th Floor,
New York, NY 10017, USA
www.syrawoodpublishinghouse.com

Industrial Applications of Oil and Gas Resources
Edited by Oliver Haghi

© 2016 Syrawood Publishing House

International Standard Book Number: 978-1-68286-140-0 (Hardback)

Contents

Preface   VII

Chapter 1   **Permeability of EVOH Barrier Material Used in Automotive Applications:**
             Metrology Development for Model Fuel Mixtures   1
             Jing Zhao, Charbel Kanaan, Robert Clément, Benoît Brulé, Henri Lenda and
             Anne Jonquières

Chapter 2   **Control-Oriented Models for Real-Time Simulation of Automotive Transmission**
             Systems   15
             N. Cavina, E. Corti, F. Marcigliano, D. Olivi and L. Poggio

Chapter 3   **Energy Management of Hybrid Electric Vehicles: 15 Years of Development at the**
             Ohio State University   39
             Giorgio Rizzoni and Simona Onori

Chapter 4   **Evaluation of Long Term Behaviour of Polymers for Offshore Oil and**
             Gas Applications   53
             P.-Y. Le Gac, P. Davies and D. Choqueuse

Chapter 5   **Energy Management Strategies for Diesel Hybrid Electric Vehicle**   64
             Olivier Grondin, Laurent Thibault and Carole Quérel

Chapter 6   **Investigation of Cycle-to-Cycle Variability of NO in Homogeneous Combustion** 81
             A. Karvountzis-Kontakiotis and L. Ntziachristos

Chapter 7   **Combustion Noise and Pollutants Prediction for Injection Pattern and Exhaust**
             Gas Recirculation Tuning in an Automotive Common-Rail Diesel Engine   94
             Ivan Arsie, Rocco Di Leo, Cesare Pianese and Matteo De Cesare

Chapter 8   **Development of Look-Ahead Controller Concepts for a Wheel Loader Application** 113
             Tomas Nilsson, Anders Fröberg and Jan Åslund

Chapter 9   **Shale Gas Pseudo Two-Dimensional Unsteady Seepage Pressure Simulation**
             Analysis in Capillary Model   133
             Lai Fengpeng, Li Zhiping, Li Zhifeng, Yang Zhihao and Fu Yingkun

Chapter 10   **Integrated Energy and Emission Management for Diesel Engines with Waste Heat**
             Recovery Using Dynamic Models   142
             Frank Willems, Frank Kupper, George Rascanu and Emanuel Feru

Chapter 11   **Design Methodology of Camshaft Driven Charge Valves for Pneumatic Engine Starts**   **158**
Michael M. Moser, Christoph Voser, Christopher H. Onder and Lino Guzzella

Chapter 12   **Development of Reactive Barrier Polymers against Corrosion for the Oil and Gas Industry: From Formulation to Qualification through the Development of Predictive Multiphysics Modeling**   **174**
X. Lefebvre, D. Pasquier, S. Gonzalez, T. Epsztein, M. Chirat and F. Demanze

Chapter 13   **Biofuels Barrier Properties of Polyamide 6 and High Density Polyethylene**   **187**
L.-A. Fillot, S. Ghiringhelli, C. Prebet and S. Rossi

Permissions

List of Contributors

Preface

This book aims to highlight the current researches and provides a platform to further the scope of innovations in this area. This book is a product of the combined efforts of many researchers and scientists from different parts of the world. The objective of this book is to provide the readers with the latest information in the field.

Oil and gas act as both raw materials and fuels to the industrial sector across the globe. This book discusses the tools and techniques employed for exploration and production of oil and gas. It also elucidates the process for refining of petroleum. Some emerging sources for alternative energy have also been covered herein. It strives to provide a fair idea about this subject to students, and also helps academicians and researchers to help develop a better understanding of the latest technological advances within this field.

I would like to express my sincere thanks to the authors for their dedicated efforts in the completion of this book. I acknowledge the efforts of the publisher for providing constant support. Lastly, I would like to thank my family for their support in all academic endeavors.

Editor

Permeability of EVOH Barrier Material Used in Automotive Applications: Metrology Development for Model Fuel Mixtures

Jing Zhao[1], Charbel Kanaan[1], Robert Clément[1], Benoît Brulé[2],
Henri Lenda[1] and Anne Jonquières[1]*

[1] Laboratory of Macromolecular Physical Chemistry, LCPM FRE CNRS-UL 3564, ENSIC, Université de Lorraine,
1 rue Grandville, BP 20451, 54001 Nancy Cedex - France
[2] Arkema – CERDATO, rue du Grand Hamel, 27470 Serquigny - France
e-mail: jing.zhao@total.com - charbel.kanaan@fr.michelin.com - clement.robert@free.fr - benoit.brule@arkema.com
henri.lenda@univ-lorraine.fr - anne.jonquieres@univ-lorraine.fr

* Corresponding author

Abstract — *Permeability of EVOH Barrier Material Used in Automotive Applications: Metrology Development for Model Fuel Mixtures* — *EVOH (Ethylene-Vinyl Alcohol) materials are widely used in automotive applications in multi-layer fuel lines and tanks owing to their excellent barrier properties to aromatic and aliphatic hydrocarbons. These barrier materials are essential to limit environmental fuel emissions and comply with the challenging requirements of fast changing international regulations. Nevertheless, the measurement of EVOH permeability to model fuel mixtures or to their individual components is particularly difficult due to the complexity of these systems and their very low permeability, which can vary by several orders of magnitude depending on the permeating species and their relative concentrations. This paper describes the development of a new automated permeameter capable of taking up the challenge of measuring minute quantities as low as $1\ mg/(m^2.day)$ for partial fluxes for model fuel mixtures containing ethanol, i-octane and toluene at 50°C. The permeability results are discussed as a function of the model fuel composition and the importance of EVOH preconditioning is emphasized for accurate permeability measurements. The last part focuses on the influence of EVOH conditioning on its mechanical properties and its microstructure, and further illustrates the specific behavior of EVOH in presence of ethanol oxygenated fuels. The new metrology developed in this work offers a new insight in the permeability properties of a leading barrier material and will help prevent the consequences of (bio)ethanol addition in fuels on environmental emissions through fuel lines and tanks.*

Résumé — **Perméabilité d'un matériau barrière EVOH utilisé dans des applications automobiles : développement métrologique pour des mélanges modèles de carburants** — Les matériaux EVOH (*Ethylene-Vinyl Alcohol*) sont largement utilisés dans des tubes et réservoirs multicouches pour les carburants car ils présentent d'excellentes propriétés barrières aux composés hydrocarbonés aromatiques et aliphatiques. Ces matériaux barrières sont essentiels pour limiter les fuites de carburants dans l'environnement et satisfaire aux normes environnementales internationales de plus en plus sévères. Néanmoins, la mesure de la perméabilité de ces matériaux à des mélanges modèles de carburants, ou à leurs composants individuels, est particulièrement difficile en raison de la complexité de ces systèmes et de leurs très faibles perméabilités, qui peuvent cependant varier de plusieurs ordres de grandeur selon la nature des espèces transférées et leurs concentrations respectives. Cet article décrit le développement d'un nouveau perméamètre automatisé capable de relever le défi de mesurer des flux partiels aussi faibles que 1 mg/(m^2.jour) pour des mélanges modèles de carburants contenant de l'éthanol, de l'*i*-octane et du toluène. Les résultats de perméabilité sont discutés en fonction de la composition des mélanges modèles et l'importance d'un pré-conditionnement des films d'EVOH est soulignée pour obtenir des mesures fiables. La dernière partie se concentre sur l'influence du conditionnement de l'EVOH sur ses propriétés mécaniques et sa microstructure et illustre de façon complémentaire le comportement spécifique de l'EVOH en présence de mélanges modèles de carburants oxygénés. La métrologie développée apporte de nouveaux éléments de compréhension de la perméabilité d'un matériau barrière leader et devrait contribuer à prévenir les conséquences de l'addition de (bio)éthanol dans les carburants sur les émissions environnementales à travers les tubes et réservoirs polymères multicouches.

INTRODUCTION

Ethylene-vinyl alcohol (EVOH) copolymers are amongst the best commercial barrier materials used worldwide. They are mainly used in packaging applications for their excellent barrier properties to oxygen [1]. The chemical physical properties of EVOH copolymers are highly dependent upon their vinyl alcohol content and the best barrier properties are obtained with high contents in vinyl alcohol ($\geq$ 68%). Nevertheless, the content in vinyl alcohol does usually not exceed 80% because higher contents are responsible for an increased sensitivity to water. The data reported in Table 1 show the very low oxygen permeability of EVOH copolymers compared to other polymer materials commonly used in packaging applications. Nevertheless, the EVOH outstanding barrier performances are highly sensitive to humidity and temperature. Several papers have for instance discussed the influence of humidity on sorption [2, 3], glass transition temperature [4],

morphology [5] and permeability properties for packaging applications [2, 6, 7]. Typically, EVOH barrier properties decrease considerably above 75% relative humidity. Therefore, in packaging applications, the EVOH barrier material is usually co-extruded with hydrophobic polyolefin layers which protect the barrier layer from humidity.

EVOH copolymers are also used in automotive applications in fuel lines and tanks owing to their excellent barrier properties to aromatic and aliphatic hydrocarbons. Polymer fuel lines and tanks usually also involve complex multi-layer structures to limit environmental fuel emissions and comply with the challenging requirements of fast changing international regulations. EVOH copolymers are in the middle of the multi-layer structure and play a key role on the barrier properties. The measurement of EVOH permeability to model fuel mixtures or to their individual components is particularly difficult due to the complexity of these systems and their very low permeability, which can nevertheless vary by several

TABLE 1

Comparison of the barrier properties of EVOH copolymers to oxygen and model fuels with other common barrier materials [9, 16]

	Oxygen (20°C)[*] (cm^3.cm/cm^3.s.cmHg.10^{13})		Hydrocarbons (40°C) (g.mm/m^2.day)
EVOH 29	0.1		0.005 (Fuel C)
EVOH 38	0.4		0.02 (Fuel C + 10% ethanol)
LDPE	2 400	EVOH 32	4.5-5.5 (Fuel C + 15% methanol)
HDPE	1 100		0.004 (toluene)
Polyamide	23	HDPE	62.6 (Fuel C)
PVC	40	PA6	5.06 (Fuel C)

[*] Oxygen permeability for dry state.
Note: EVOH X corresponds to an EVOH copolymer containing X mol% of ethylene and Fuel C is an equivolumic mixture of toluene and *i*-octane.

orders of magnitude depending on the permeating species and their relative concentrations.

Related permeability investigations for EVOH copolymers and model fuels (or their individual components) remain really scarce in the literature [8-11]. Nevertheless, the results reported in Table 1 show the outstanding barrier properties of EVOH copolymers to Fuel C which is a model fuel commonly used for assessing the permeability properties of fuel lines and tanks in the automotive industry. Fuel C is an equivolumic mixture of toluene and i-octane, both components being representative of the aromatic and aliphatic hydrocarbons in real fuels. When EVOH films were tested in absence of humidity, the fluxes of toluene and i-octane could not be detected by Nulman et al. [9] at the Ford Motor Company equipped with a permeation cell connected to a carbon adsorbent trapping and desorbing circuit. Lagaron et al. [10] in collaboration with BP Chemicals Limited succeeded in measuring a very low permeability to toluene for Fuel C but the permeability to i-octane was again not detected on the basis of weight loss measurements for semi-closed permeation cells.

Nevertheless, a few former works have already reported that the EVOH outstanding barrier properties to aromatics and aliphatics were strongly decreased in presence of water and oxygenated species like methanol and ethanol [8-11]. This issue is particularly critical for the automotive industry where increasing amounts of oxygenated species are being added to fuels to minimize the fuel environmental impact and take advantage of the ethanol biofuel. Samus and Rossi [8] have thoroughly analyzed the complex behavior of EVOH with respect to methanol permeability on the basis of sorption, diffusion and dynamic mechanical experiments. They have reported a strong linear decrease of EVOH glass transition temperature as function of methanol weight fraction absorbed in EVOH. The plasticization induced by methanol had an important impact on its permeability which strongly increased above glass transition temperature. Nulman et al. [9] have also reported interesting permeability results for EVOH and a model fuel methanol/i-octane/toluene (15/42.5/42.5 in vol% – Fuel CM15). The methanol flux obtained with Fuel CM15 was almost 20% of that of pure methanol. The authors mentioned that the corresponding fluxes of toluene and i-octane were less than 0.01 g mm/(m^2.day) at 40°C. Even if the partial fluxes of toluene and i-octane could not be quantified for Fuel CM15 at 40°C, the latter work showed qualitatively that the addition of methanol to Fuel C strongly increased the toluene flux and decreased the overall barrier performance of EVOH. The quantitative

permeability data reported by Lagaron et al. [10] further confirmed this observation and showed that the total flux of EVOH was increased by three orders of magnitude by adding 15% of methanol to Fuel C. Furthermore, when 10% of ethanol was added to Fuel C, the total flux increased by a factor 4 only, which showed that ethanol had less impact on the EVOH barrier properties than methanol.

Later on, Gagnard et al. [11] in collaboration with the worldwide company Arkema developed an automated permeameter combining a semi-closed permeation cell, a desorption trap and on-line GC-analysis and reported permeability data for several barrier materials used in fuel tank applications. For the first time for an EVOH film, the fluxes of each permeating species (partial fluxes) were determined for methanol/toluene mixtures over the whole composition range. For all the methanol/toluene mixtures, the methanol flux was much higher than the toluene flux. The data also revealed the strong enhancement of the toluene flux over a wide range of compositions in presence of methanol at 60°C. The maximum toluene flux was obtained for 70% of methanol and was 2 500 times higher than that of pure toluene in the same conditions.

Following our former work for assessing related sorption properties [12], this new paper in collaboration with Arkema describes a new metrology development for measuring the permeability properties of barrier materials used in fuel tank applications. Our first objective was to characterize accurately the partial fluxes for an EVOH copolymer and complex model fuel mixtures containing ethanol, toluene and i-octane over a wide composition range typical of the targeted application. As mentioned above, the addition of ethanol in Fuel C leads to a smaller increase in fuel permeation than the addition of methanol. From an experimental point of view, the subject of this paper is more challenging than the measurements formerly performed at Arkema with methanol based fuels. The first part of the paper describes the development of a new automated permeameter capable of taking up the challenge of measuring minute quantities as low as 1 mg/(m^2.day) for the partial fluxes. The second part of the paper discusses the permeability results as a function of the model fuel composition and emphasizes the importance of EVOH preconditioning for accurate permeability measurements. The last part focuses on the influence of EVOH conditioning on its mechanical properties and its microstructure, and further illustrates the specific behavior of EVOH in presence of ethanol oxygenated fuels. The new metrology developed in this work offers a new insight in the permeability properties of one of the leading barrier polymers for fuel lines and tanks.

1 MATERIALS AND METHODS

1.1 Solvents and EVOH Copolymer

Solvents of maximal purity were purchased for this investigation because the presence of impurity can be detrimental to the permeability analysis. Toluene (*Aldrich*, ACS reagent, 99.5 + %, boiling point 110.6°C), *i*-octane (*Aldrich*, ACS reagent, 99 + %, boiling point 98.5°C) and ethanol (*Aldrich*, absolute, 99.8 + %, boiling point 78.5°C) were used without further purification.

The EVOH copolymer was a high barrier material commonly used in multi-layer fuel tanks. This copolymer was an EVOH DT Soarnol® containing 29 mol% of ethylene. Table 2 reports its main physical characteristics. In this study, we focused on the permeability properties of EVOH monolayer films, the corresponding data being necessary for a future modeling of their multicomponent permeability. The EVOH films used for the permeability experiments were extruded films produced by *Arkema*. The film thicknesses were optimized with respect to permeability in order to obtain good accuracy of the partial fluxes for the different compositions of the model fuels.

1.2 Physical Characterization of EVOH Material

Differential Scanning Calorimetry (DSC) experiments were carried out with a DSC 2920 *TA Instruments* with liquid nitrogen cooling. Sample weights were ca. 10 mg. EVOH films of 10 μm were first heated from −20°C to 250°C at a heating rate of 20°C/min then cooled down from 250°C to −20°C at the same rate. A second heating was then performed in the same conditions. The data were recorded over the whole cycle with EVOH films before and after preconditioning. During preconditioning for DSC analysis, the EVOH films were immersed in a model fuel mixture ethanol/*i*-octane/ toluene (85/7.5/7.5 – Fuel CE85) at 40°C until saturation

was reached and then dried at ambient temperature in air.

Wide Angle X-ray Scattering (WAXS) experiments completed the DSC analysis for assessing the influence of EVOH conditioning on copolymer crystallinity. The spectra were recorded in transmission with EVOH films before and after preconditioning like for the DSC experiments. The spectra were recorded on an *INEL* equipment with a wavelength defined by the main ray of *K*α1 of copper (1.54 Angström), an incident energy of 32 kV and a counting time of 3 000 seconds. Eight EVOH films of 10 μm had to be stacked together to obtain a sufficient signal.

Original Dynamic Mechanical Analysis (DMA) experiments were performed on a DMA 242C *Netzsch* while the EVOH films were immersed in pure solvents or model fuel mixtures to follow the consequences of conditioning on the mechanical properties in real time. The thickness of the EVOH films had to be increased to 60 μm to obtain good DMA sensitivity in these conditions. The experiments were carried out at a frequency of 1.6 Hz at 50°C, which corresponded to the temperature used for the permeability experiments. Dynamic Mechanical Thermal Analysis (DMTA) experiments were also carried out to assess the influence of model fuel composition on the temperature of the α transition. In DMTA experiments, the loss factor tan δ was recorded at the same frequency of 1.6 Hz for temperatures increasing from −40°C to 80°C with a heating rate of 2°C/min.

1.3 Permeability Experiments

The permeability experiments were performed with EVOH films of 8.4 cm^2 at 50°C with an automated permeameter specifically developed for this purpose in our laboratory (*Fig. 1*). The thickness of the EVOH films was chosen according to the total permeability in order to optimize the permeability measurements. This permeameter enabled to analyze permeation in a continuous way in a fully closed system by the pervaporation membrane technique. In this membrane process, a model fuel mixture was circulated on top of the EVOH film by a pump from a stainless-steel reservoir at constant temperature. The temperature of 50°C was checked in the thermally insulated permeation cell. The downstream side of the EVOH film was maintained under low pressure (typically less than 0.5 mm Hg) by an *Adixon* vacuum pump. In these conditions, the boundary conditions for mass transfer across the EVOH film were well defined and easily reproducible. The permeate was collected in a sampling loop cooled by a liquid nitrogen trap, which was mounted on a pneumatic jack for automation (*Fig. 2*).

TABLE 2
Main characteristics of the EVOH investigated in this work

	EVOH DT Soarnol® 29 mol% ethylene
Density (g/cm^3)	1.21
Melting temperature (°C)	188
Crystallisation temperature (°C)	163
Glass transition temperature* (°C)	62

* DSC measurement on dry state.

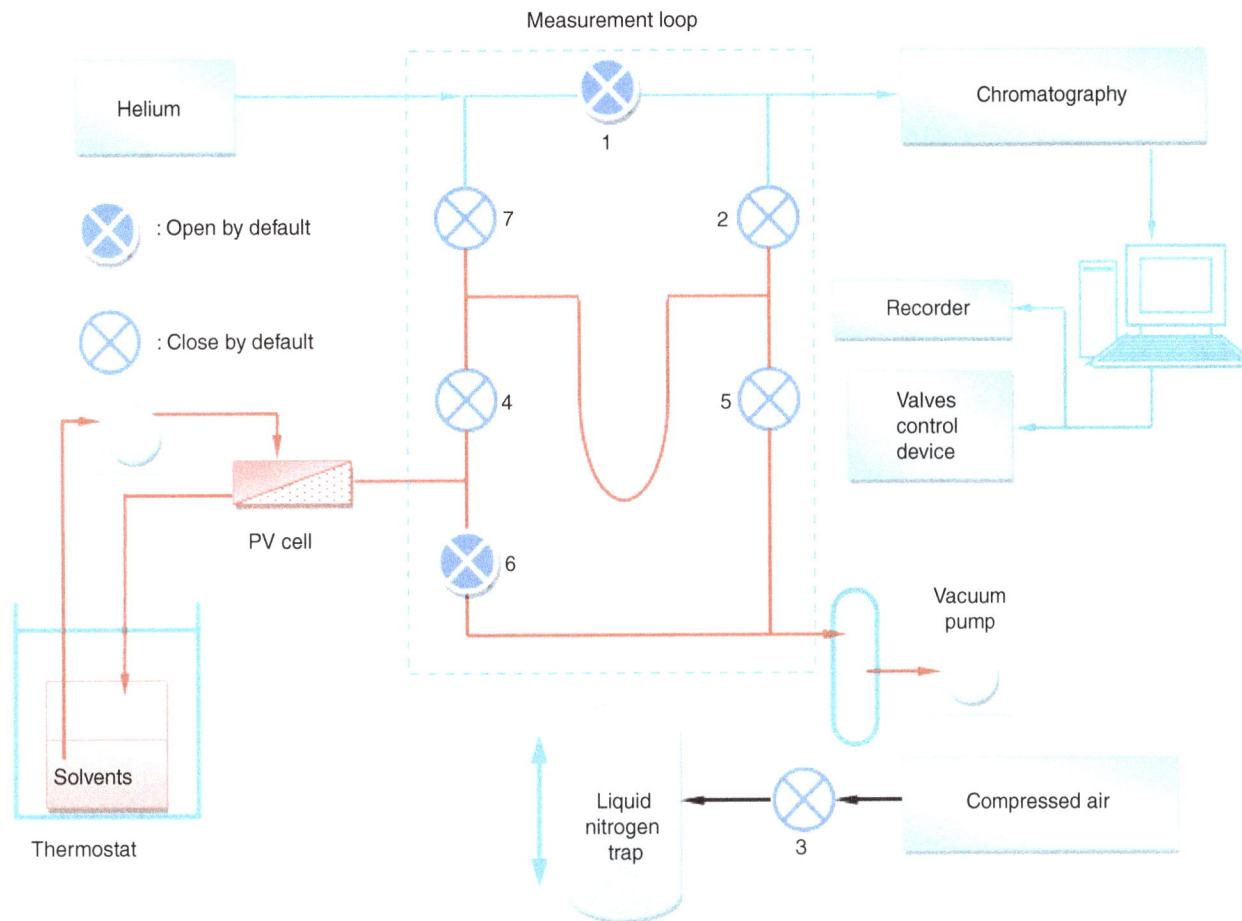

Figure 1

Scheme of the automated experimental set-up developed for permeability measurements with barrier polymer materials and multi-component fuel model mixtures.

The sampling loop was equipped with on-line gas chromatography (*Shimadzu* GC 8A/ PDMS SE30 column). The automation of the whole system was performed *via* solenoid valves (*Bacosol* inox-teflon valves) which were controlled by a personal computer through an appropriate electronic high sensitivity interface (*Keithley* KPCI 3107 Data Acquisition Board). The same controller was also used to drive the automated pneumatic jack necessary for the cooling of the measurement loop by liquid nitrogen.

In a typical permeation experiment, the permeate was collected over a limited period of time which depended on the total permeation rate. Each permeation experiment was carried out in several cycles consisting of condensing the permeate vapors in liquid nitrogen, heating the permeate to room temperature followed by on-line

chromatography analysis. The sampling time was automatically adjusted from one cycle to another to avoid saturating the GC FID detector and to remain in its linearity range.

A *Visual Basic* interface was developed to follow the progress of each experiment. Figure 3 gives an example for the computer screen displayed during an experiment. This computer screen showed the state for all the solenoid valves and a plot of the last chromatogram. The weights of each permeated species were obtained from the area of each peak by a preliminary GC absolute calibration. Even if it could have theoretically been done and saved some time, the absolute GC calibration was not performed directly with specific and well chosen fuel mixtures because it would have been quite difficult to find polymer films with moderate permeability for all

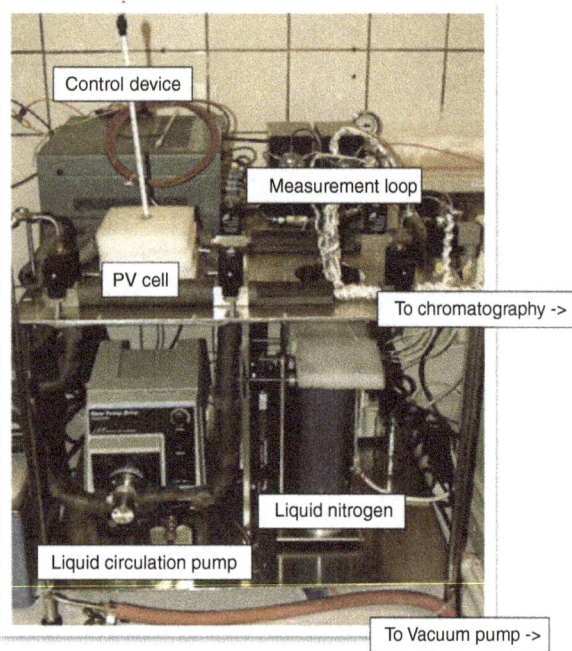

Figure 2

Photograph of the automated permeameter.

---

of the different components of the fuel mixtures. Therefore, the calibration was made with polymer films having known and moderate permeability for each pure species (*i.e.* ethanol, *i*-octane and toluene) to minimize the errors made in estimating each calibration coefficient. Knowing the weights of each permeated species w_i, the sampling time Δt and the active membrane area A_{mem}, the partial flux of each species i were calculated by Equation (1):

$$J_i = \frac{w_i}{\Delta t}\frac{1}{A_{mem}} \qquad (1)$$

To optimize the measurements of the different partial fluxes which could differ by several orders of magnitude for the same experiment, the sensitivity range of the gas chromatograph was adapted to the injected weights. With delays of 0.1 second for valve commutation and sampling times of ca. 1 minute, the measurements were repeatable in stationary regime. Furthermore, the permeameter was also equipped with a secondary measurement loop, which enabled on-line analysis of the model fuel composition. With the small membrane area of 8.4 cm^2 of the EVOH films used in this work, the permeate sampling did not induce any significant change in the composition of the model fuel mixture with a minimum circulating feed volume of 150 mL.

With this automated permeater, each measurement cycle for the partial fluxes took between 15 to 20 minutes, depending on the sampling time which was a function of the total permeation rate but never exceeded 300 seconds. Typically, a few hours were generally necessary for obtaining steady-state permeability data for each composition of the model fuel mixture and partial fluxes as low as 1 mg/(m^2.day) were measured quantitatively.

2 RESULTS AND DISCUSSION

2.1 Permeability Results Obtained for Virgin and Preconditioned EVOH Materials

The EVOH permeability features were characterized for model fuel mixtures containing ethanol, *i*-octane and toluene. Ethanol (rather than methanol) was chosen in this work because, thanks to different tax policies, this biofuel is being more and more used in European fuels. The permeability data were obtained for two important tie lines of the ternary diagram ethanol/*i*-octane/toluene with respect to the permeation properties (*Fig. 4*). The first tie line (called ethanol tie line) corresponded to equivolumic mixtures of toluene and *i*-octane (*i.e.* Fuel C) with increasing contents in ethanol. Ethanol contents were limited in the range of 50 to 100%, which corresponded to the working range of an EVOH layer in a multi-layer fuel tanks as shown by our former sorption measurements for HDPE and EVOH materials [12]. The second tie line (called toluene tie line) corresponded to equivolumic mixtures of ethanol and *i*-octane with increasing contents in toluene. Ethanol and toluene are key species for mass transfer through EVOH and other polymers used in polymer fuel tanks. The corresponding tie lines were chosen to assess the permeability behavior over a wide range of compositions covering a large domain of the ternary diagram and typical of the targeted application.

During metrology development, the first permeability measurements with virgin EVOH films led to partial fluxes which were slightly evoluting over long periods of time. Figure 5 shows an example for the slow decline of the partial fluxes for ethanol, toluene and *i*-octane, even for a relatively thin EVOH film of 25 μm. The steady-state was very difficult to attain with virgin EVOH films and accurate permeability measurements could not be performed in these conditions no matter the sensitivity of the new automated permeameter. It was also important to maintain vacuum at the downstream side of the EVOH films for whole series of permeability measurements over several weeks. Otherwise,

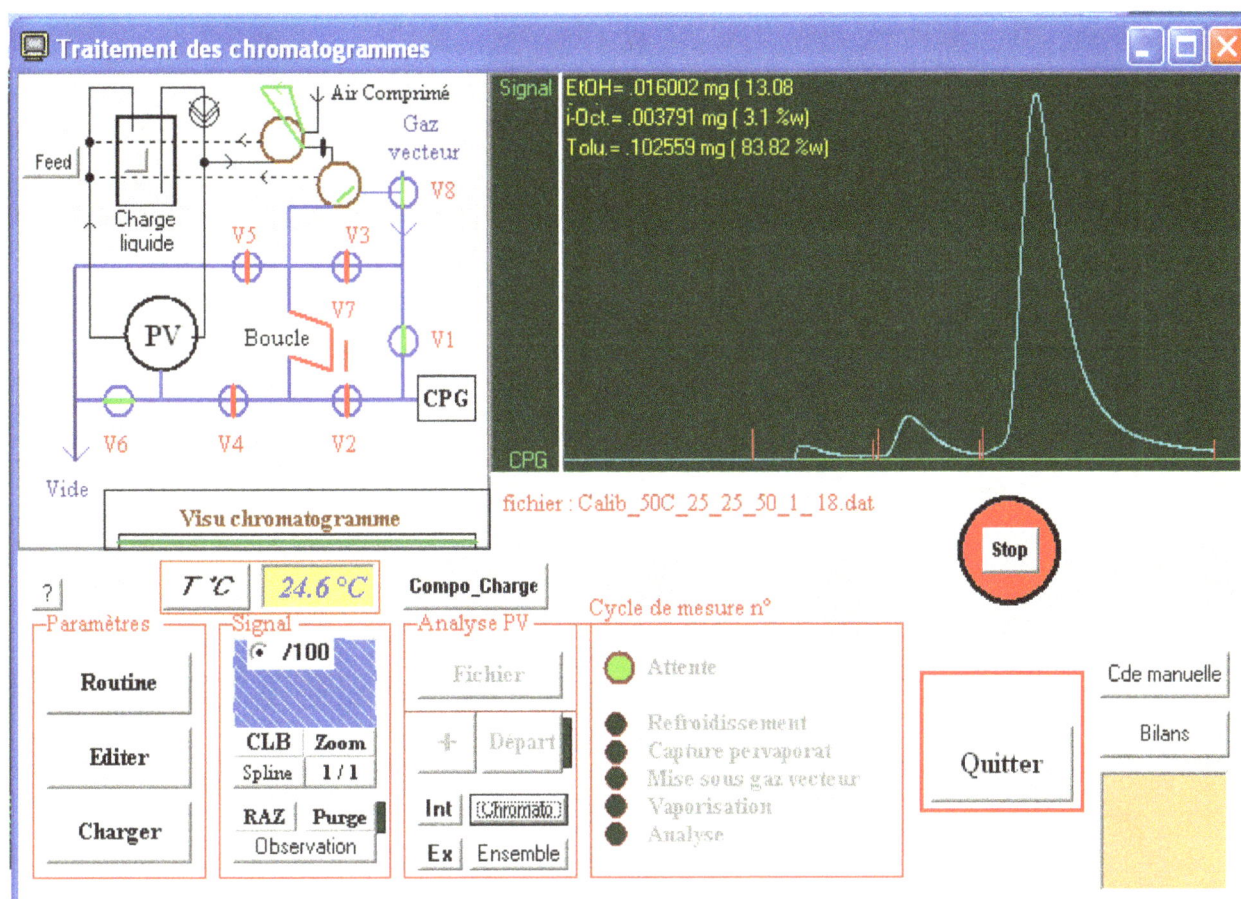

Figure 3

Copy of the computer screen displayed during permeability measurement.

the intermittent swelling of the film downstream side led to significant perturbation in the measured permeability properties (even sometimes to film breaking for the thinnest films) and the come-back to steady-state was again very slow. The preconditioning of the EVOH films in the targeted model fuel mixtures for more than ten days at 50°C considerably accelerated the reaching of the permeation steady-state and it became possible to obtain stable permeation data within a few hours for each composition of the model fuel mixtures.

Figure 6 shows the partial fluxes of ethanol, toluene and i-octane for a preconditioned EVOH film of 9 μm and the ethanol tie line at 50°C. Ethanol partial fluxes increased with feed ethanol content and were about ten times higher than the toluene partial fluxes. The strong interactions of ethanol with EVOH hydroxyl groups were partially responsible for the high EVOH permeability to ethanol. The permeability to both hydrocarbons were much less than the permeability to ethanol, confirming the very good barrier properties of EVOH to apolar

hydrocarbons. These results are in good agreement with those reported by Nulman et al. [9] for Fuel CM15, which also showed the preferential permeation of alcohol ($i.e.$ methanol in that case) through EVOH films. The authors mentioned that the toluene and i-octane fluxes were less than 0.01 g.mm/m^2 24 h, which confirmed that EVOH was acting as a very good barrier towards both hydrocarbon species. Nevertheless, the reported data did not allow any comparison between both hydrocarbon species. The same comment could also be made for the data reported by Lagaron et al. [10] for very closely related systems. The new automated permeameter developed in this work offered some progress for i-octane flux measurements. For the first time, to the best of our knowledge, Figure 6 enabled a quantitative comparison between the partial fluxes of toluene and i-octane. These partial fluxes differed by one order of magnitude over the whole composition range investigated for the ethanol tie line. Although these new data cannot be compared to former related EVOH permeability data, it is well known from

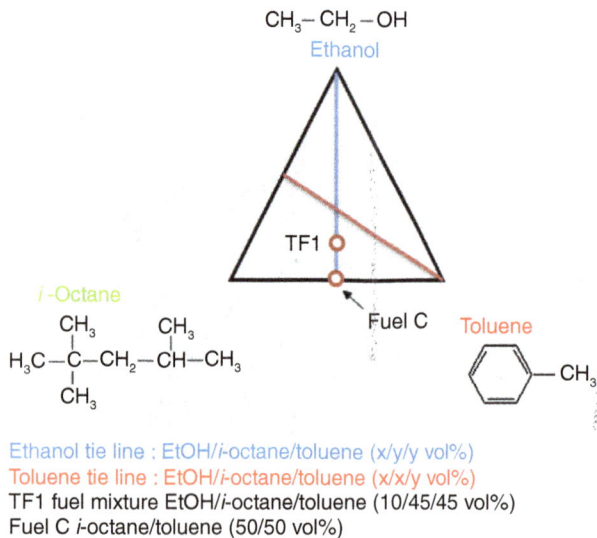

Ethanol tie line : EtOH/*i*-octane/toluene (x/y/y vol%)
Toluene tie line : EtOH/*i*-octane/toluene (x/x/y vol%)
TF1 fuel mixture EtOH/*i*-octane/toluene (10/45/45 vol%)
Fuel C *i*-octane/toluene (50/50 vol%)

Figure 4

The model fuel mixtures investigated in this work in the ternary diagram ethanol/*i*-octane/toluene in comparison with model fuels C and TF1 commonly used in this field.

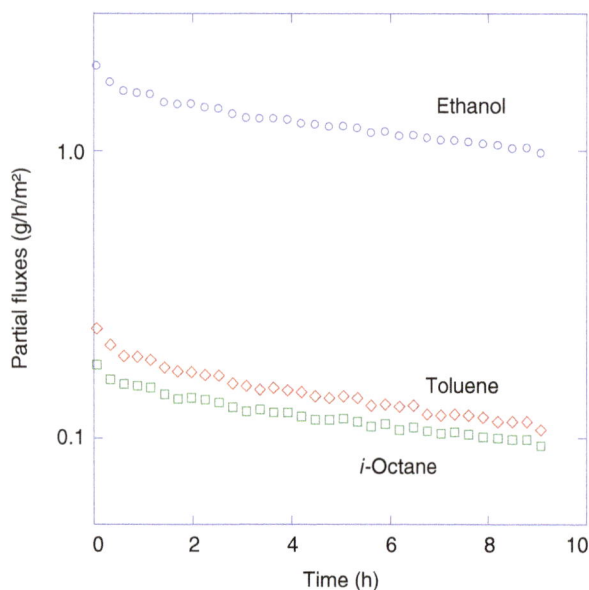

Figure 6

Partial fluxes of a preconditioned EVOH film of 9 μm for model fuel mixtures ethanol/*i*-octane/toluene (*x/y/y* vol%) corresponding to the ethanol tie line at 50°C.

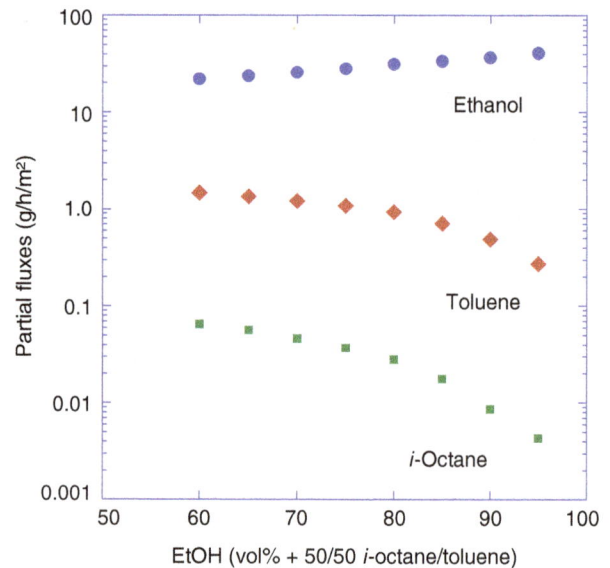

Figure 5

Evolution of the partial fluxes of a virgin EVOH film of 25 μm for a model fuel mixture ethanol/*i*-octane/toluene (12.5/12.5/75 vol%) at 50°C.

the literature on the separation of aromatics/alkanes mixtures by membrane separation processes that polar polymers are usually used for this type of separation [13-15]. Such polymers are indeed capable of interacting with aromatics owing to their strong polarizability.

For the toluene tie line, several attempts were then made to measure the permeability properties of EVOH films of 9 μm, which would have allowed a straightforward comparison with the latter results obtained with the ethanol tie line. Unfortunately, these very thin EVOH films did not withstand continuous exposure to the new model fuel mixtures during permeability measurements and EVOH films of 24 μm were eventually used for the toluene tie line. Figure 7 showed that the progress made with the new permeameter found its limits in these conditions. Above a toluene content of 40%, the *i*-octane fluxes could no more be measured accurately and the same limitations as formerly reported by Nulman *et al.* [9] and Lagaron *et al.* [10] were thus also encountered in this new composition range. The permeability results obtained with the toluene tie line again showed that ethanol permeated at least ten times faster than toluene over the whole composition range. Furthermore, the toluene flux reached a maximum for a toluene content of ca. 30% and then decreased by a factor of almost ten when the toluene content further increased from 30 to 100%. This maximum revealed a strong positive synergy between the ethanol and toluene fluxes. The plasticization of EVOH by ethanol, which will be analyzed in the next section, was responsible for this strong synergy and for the toluene flux maximum. Consequently, the EVOH barrier properties were strongly decreased in presence of ethanol in good agreement with

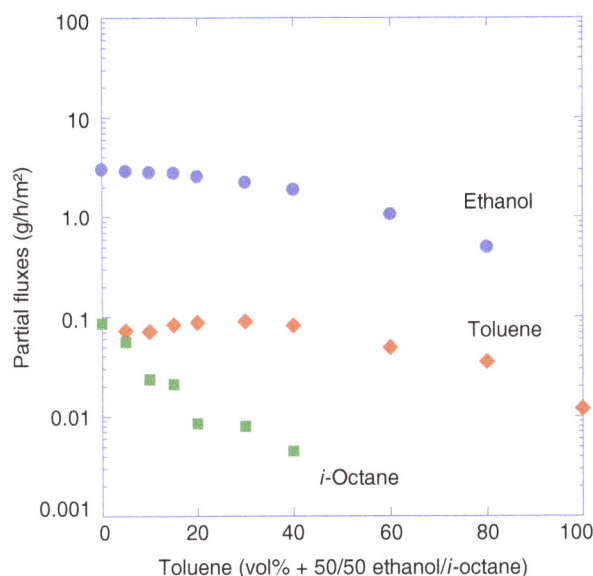

Figure 7

Partial fluxes of a preconditioned EVOH film of 24 μm for model fuel mixtures ethanol/i-octane/toluene (x/x/y vol%) corresponding to the toluene tie line at 50°C.

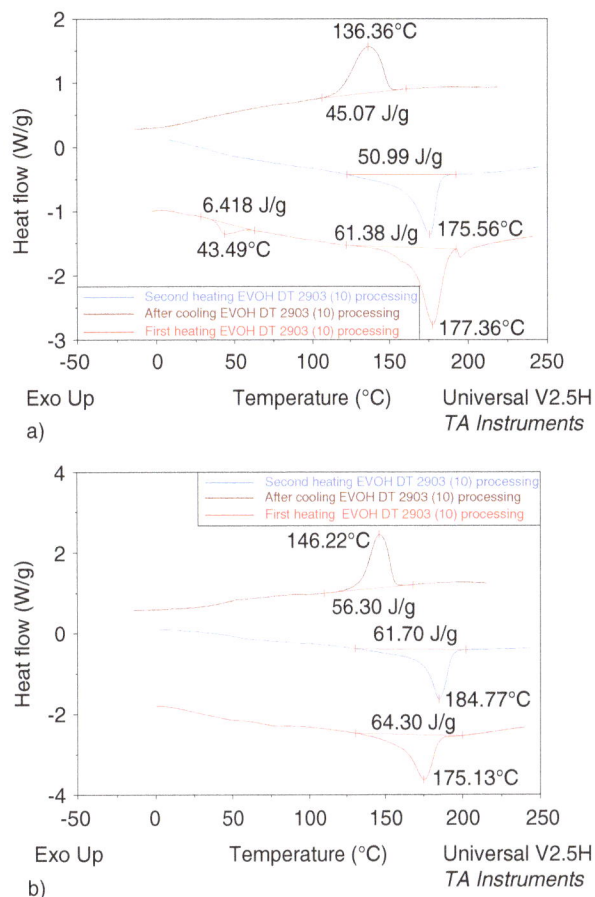

Figure 8

Comparison between the DSC curves obtained for an EVOH film of 10 μm a) before conditioning and b) after conditioning. First heating (red), second heating (blue), after cooling (black).

the related observation reported by Lagaron et al. [10] for the addition of 10% of ethanol to Fuel C and with other data on the related influence of other protic species (water and methanol) on EVOH permeability [8-11].

For both tie lines covering a wide range of the compositions typically found in the targeted application, ethanol was by far the most preferentially permeated species through the EVOH films and the permeate generally contained over 90 wt% of ethanol. The role of ethanol was also important on the toluene flux and the related plasticization decreased the EVOH barrier performance towards aromatics. The addition of increasing contents of (bio)ethanol in fuels could thus have important consequences on the barrier properties of multi-layer fuel tanks.

2.2 Influence of EVOH Conditioning in Model Fuel Mixtures

This section focuses on the influence of EVOH conditioning on its physical properties and its microstructure, and further illustrates the specific behavior of EVOH in presence of ethanol oxygenated fuels with respect to the targeted application. Solvent induced changes were assessed by three complementary techniques DSC, WAXS and DMTA. The DSC and WAXS analyses were performed with films of about 10 μm, which corresponded to the thickness of the polymer film used for the most challenging ethanol tie line (i.e. the most interesting case with respect to recent evolutions in oxygenated fuels). Unfortunately, the sensitivity of DMTA analysis was too low to assess the mechanical property variations for such thin films and much thicker films of 60 μm had to be used to obtain enough sensitivity for this technique. New DSC and WAXS analyses performed for these thicker films could have revealed small differences due to slightly different processing conditions, which were not assessed in this work.

A comparison was first made between the DSC curves obtained for an EVOH film of 10 μm before and after its conditioning in the model fuel mixture ethanol/i-octane/toluene (85/7.5/7.5 – Fuel CE85) at 40°C until saturation (Fig. 8). The temperature for sample conditioning in this part of the work was slightly different from that used for

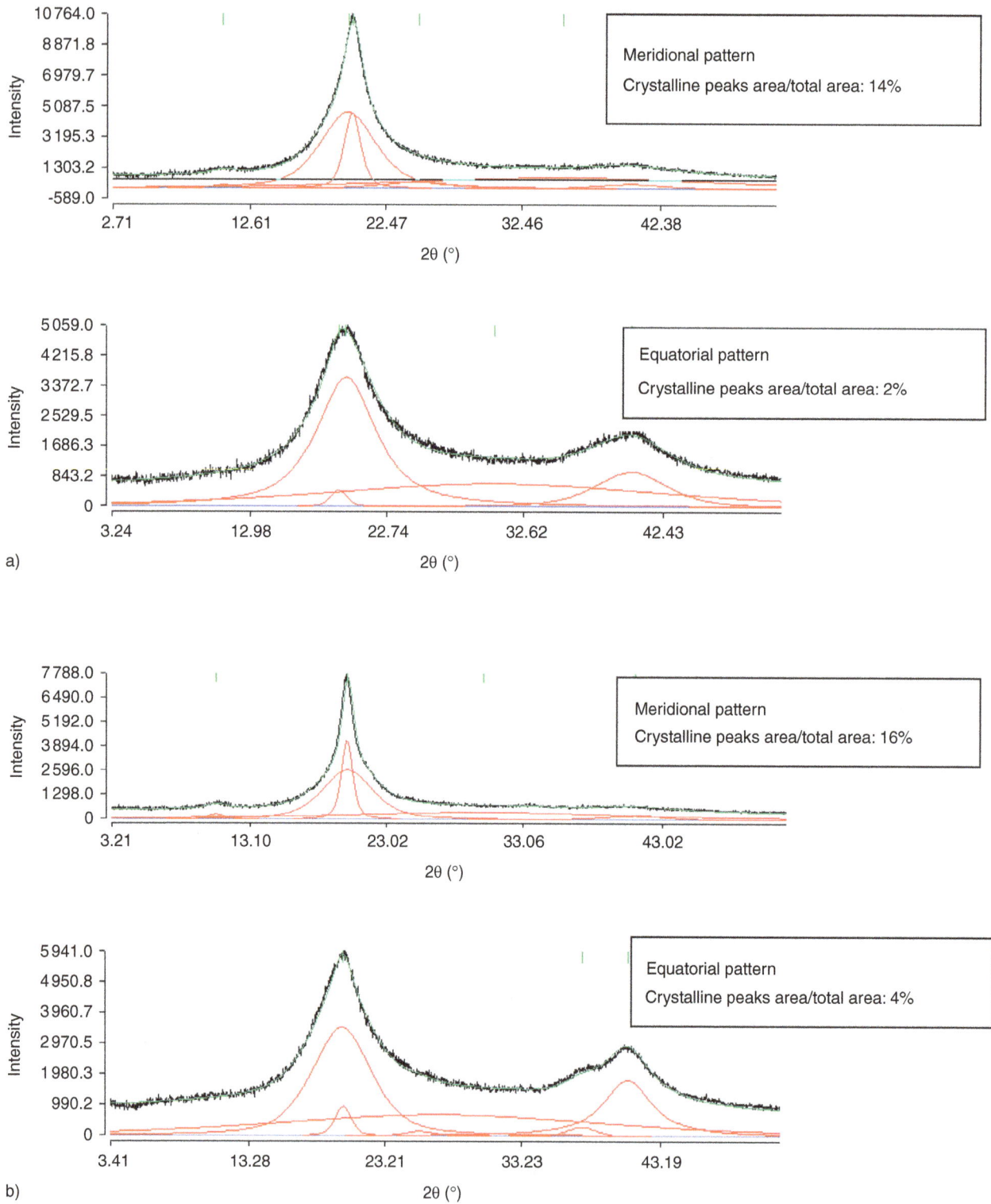

Figure 9

Meridional and equatorial WAXS patterns obtained for an EVOH film of 10 μm a) before conditioning and b) after conditioning.

TABLE 3

WAXS data obtained for an EVOH film of 10 μm before and after conditioning

	Crystalline peak (101)		Amorphous bump		Ratio of the crystalline peaks area to the total area (%)
	Position (Å)	Width	Position (Å)	Width	
Virgin EVOH film (Meridional)	4.42	1.5	4.49	4.9	14
Virgin EVOH film (Equatorial)	4.5	1.5	4.49	5.2	2
EVOH film after conditioning (Meridional)	4.40	1	4.40	4.4	16
EVOH film after conditioning (Equatorial)	4.40	1.47	4.42	5.2	4

the permeability experiments due to practical reasons. This particular mixture was chosen because it corresponded to the model fuel mixture for Superethanol E85, *i.e.* the most challenging ethanol oxygenated fuel used in the European Union (EU) and the United States of America (USA). The ethanol content of E85 varies between 65% and 85% in winter and summer, respectively, and E85 can only be used in so-called "flexible-fuel" cars. The DSC results did not show important differences in terms of crystallinity for EVOH before and after conditioning in Fuel CE85. The thermograms showed similar melting temperatures and a crystallinity of ca. 30% was found for both samples. Nevertheless, a small increase in the melting enthalpy from 51 to 62 J/g (+22%) was observed in the second heating after conditioning. Upon cooling, the crystallization temperature was enhanced by 10°C after conditioning and the corresponding crystallization enthalpy slightly increased (+25%). Furthermore, the small peak at 43.5°C for virgin EVOH was ascribed to slight physical aging during its storage in ambient conditions and disappeared after the first heating, as expected.

WAXS experiments were carried out in transmission mode to complement the latter DSC analysis. EVOH films of 10 μm were analyzed before and after conditioning in the same model mixture Fuel CE85. For the WAXS analysis, eight EVOH films had to be stacked together to obtain a sufficient signal. Figure 9 shows representative meridional and equatorial WAXS patterns before and after conditioning. The strong differences for the meridional and equatorial patterns obtained-proved the strong anisotropy of the EVOH films. The crystallinity degree cannot be calculated from WAXS data when samples are oriented. Nevertheless, deconvolution of the WAXS patterns was performed to assess the

ratio of the crystalline peaks area to the total area (Tab. 3). Even if the limits of the deconvolution method can be debated, deconvolution enabled to compare the results obtained in the same deconvolution conditions and to show that conditioning in Fuel CE85 had only a small influence on EVOH morphology. The results reported in Table 3 confirmed that, if the ratio of the crystalline peaks area to the total area strongly depended upon the observation direction (*ca.* 15% and 4% in meridional and equatorial directions, respectively), it did not vary significantly after conditioning. Whatever the diffraction pattern, the position of the principal peak did not change. The single small change was the decrease of its width after conditioning, which reflected a small increase in the crystallite size. Furthermore, the WAXS results reported in Table 3 also showed that there was no significant change in the amorphous phase after conditioning. Globally, the WAXS data confirmed the DSC results and showed a slight increase in the ratio of the crystalline peaks area to the total area after conditioning. The later results obtained with Fuel CE85 strongly differed from those formerly reported by Lopez-Rubio *et al.* [5] on important morphological alterations of EVOH induced by severe hydro-thermal treatments, which were typical for decontamination procedures used in packaging applications. Importantly for multi-layer fuel tanks, EVOH conditioning in the most challenging model Fuel CE85 had a weak influence on its morphology and only induced a small increase in the ratio of the crystalline peaks area to the total area and in the crystallite size.

DMA and DMTA analyses were then performed to assess the influence of EVOH conditioning on its mechanical and physical properties, following the pioneering work of Samus and Rossi [8] on the influence

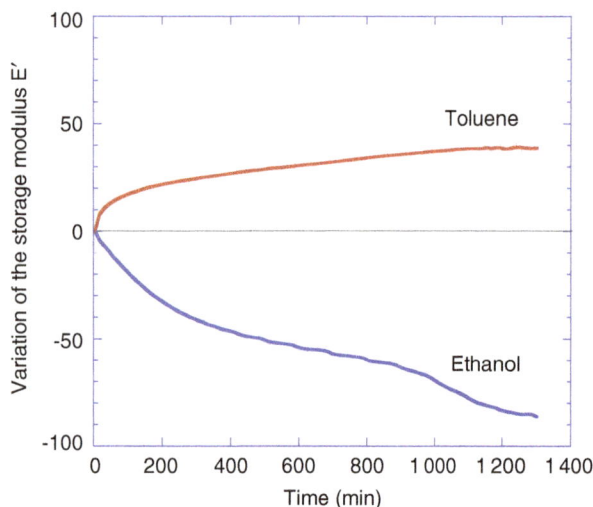

Figure 10

Variation of the storage modulus E' of an EVOH film of 60 µm during its conditioning in pure toluene and pure ethanol.

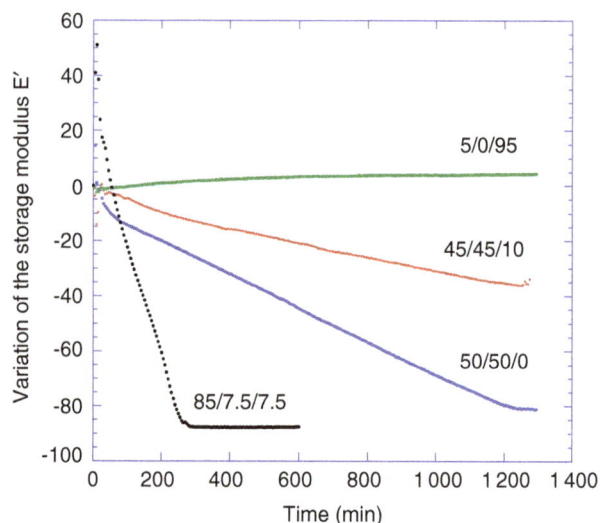

Figure 11

Variation of the storage modulus E' of an EVOH film of 60 µm during its conditioning in model fuel mixtures ethanol/i-octane/toluene with increasing ethanol contents.

of methanol absorption on EVOH mechanical properties and the related changes in methanol diffusion behavior. In the later work, the authors have shown that methanol strongly plasticized EVOH, resulting in a T_g decrease of $-10°C$ per weight percent of absorbed methanol. This strong plasticizing effect was related to a sharp increase in methanol diffusion. In our work, the variation of the storage modulus E' was first followed over time for EVOH films of 60 µm immersed in pure ethanol and toluene at 23°C. This temperature corresponded to ambient temperature chosen for practical reasons. The thickness of the films was chosen to obtain enough sensitivity for the dynamic mechanical analysis at a frequency of 1.6 Hz.

Figure 10 shows that very different trends were found for both pure solvents. In the same way as reported by Samus and Rossi [8] for methanol during static experiments, ethanol induced a strong continuous decrease of the storage modulus (−90%) in ca. 20 hours and thus strongly plasticized the tested films over a very long period of time. The DMA data also showed that the plasticizing effect of EVOH still continued after 10 hours in pure ethanol. This alcohol interacted with the EVOH hydroxyl groups and limited the intermolecular interactions in this material. On the other hand, the storage modulus E' increased during EVOH conditioning in pure toluene and reached a plateau at $+40\%$ after 20 hours. Therefore, pure toluene acted as an efficient anti-plasticizer in EVOH by reinforcing its intermolecular interactions.

Further DMA experiments carried out with model fuel mixtures with increasing ethanol contents showed that the influence of ethanol globally dominated the solvent induced changes at ambient temperature (Fig. 11). Adding 5% of ethanol to toluene (corresponding to the model fuel mixture 5/0/95) was sufficient to almost compensate for the anti-plasticizing effect of toluene and the increase of the storage module E' was very strongly reduced in these conditions. Further increased ethanol contents led to a strong decrease of the storage module over time. Despite the initial artifact of the DMA run in this case, the results obtained for the model Fuel CE85 revealed the strongest influence on EVOH mechanical properties. In this worst case corresponding to the most challenging fuel currently used in the EU and in the USA, the decrease of the storage module was very important and almost similar to that obtained for pure ethanol.

Complementary DMTA experiments enabled to further investigate the influence of EVOH conditioning on its main transition temperature T_α. After conditioning in pure solvents or in model fuel mixtures with different ethanol contents at ambient temperature, the samples were analyzed in presence of the conditioning medium at a frequency of 1.6 Hz with a small heating rate of 2°C/min from $-30°C$ to 80°C. Even if it would have been interesting to investigate temperatures higher than 80°C to enable the full determination of the α relaxation of the neat EVOH film, the upper temperature had to be limited in this work for safety reasons due to the handling of fuel model mixtures containing ethanol.

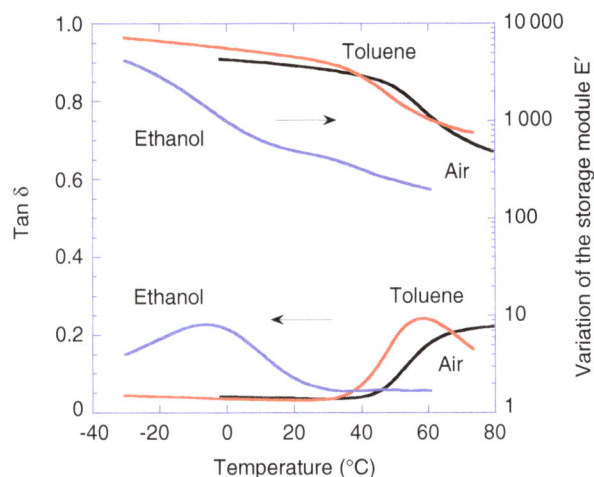

Figure 12

Variation of the loss factor tan δ and of the storage module E' of an EVOH film of 60 μm after its conditioning in pure toluene and pure ethanol.

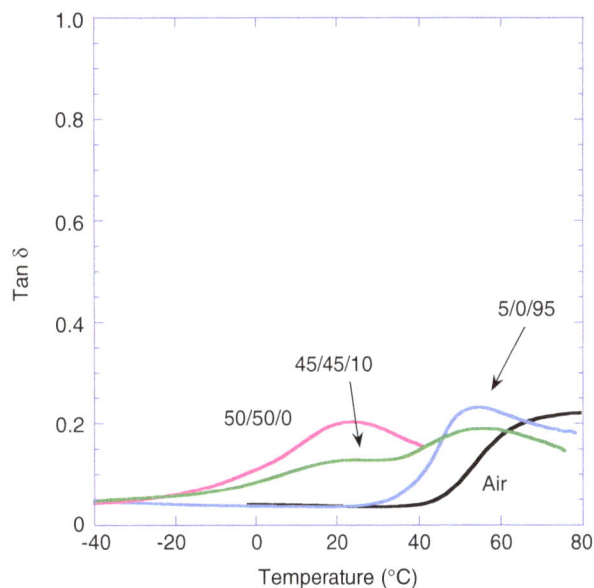

Figure 13

Variation of the loss factor tan δ of an EVOH film of 60 μm after its conditioning in model fuel mixtures ethanol/i-octane/toluene with increasing ethanol contents.

Figure 12 shows the variation of the loss factor tan δ and the storage module E' as a function of temperature for EVOH samples conditioned in pure ethanol and in pure toluene in comparison with virgin EVOH. EVOH conditioning in pure ethanol induced a very strong decrease of its main transition temperature T_α from ca. 75°C to ca. −5°C. This important shift of T_α of ca. −80°C was similar to the corresponding shift reported by Samus and Rossi [8] for an EVOH extruded film after its conditioning in methanol while Lagaron et al. [10] reported an even lower T_g of −25°C for another methanol-saturated EVOH sample. Our new results confirmed the very strong plasticizing effect of ethanol on EVOH, which was also responsible for a strong decrease in the storage modulus E'. Compared to ethanol, toluene led to a much smaller impact on T_α corresponding to a decrease of less than 20°C. This small decrease indicated that short scale motions were slightly improved by toluene absorption. Nevertheless, toluene swelling also improved intermolecular interactions in EVOH, which resulted in a higher storage modulus E' at ambient temperature (Fig. 12). Therefore, a small decrease in T_α co-existed with a significant increase in storage modulus at ambient temperature when EVOH was immersed in pure toluene. This particular behavior confirmed the anti-plasticizing effect of toluene at ambient temperature already observed in Figure 10.

When EVOH conditioning was performed in model fuel mixtures with increasing ethanol contents at ambient temperature, the DMTA patterns were more complex (Fig. 13). Nevertheless, the main transition temperature T_α was considerably lowered in presence of oxygenated model fuels and typically decreased when the ethanol content increased. It is important to note that, for several systems, the main transition temperature T_α was very close to the temperature of 50°C chosen for the permeability measurements. It means that the physical properties of EVOH could vary strongly close to 50°C, depending on the composition of the model fuels. This influence was the strongest for the ethanol tie line, which typically corresponded to the strongest ethanol plasticizing effect and to the highest fluxes.

CONCLUSION

A new automated permeameter was developed for measuring minute quantities as low as 1 mg/(m².day) for partial fluxes with model fuel mixtures containing ethanol, i-octane and toluene at 50°C. With this experimental set-up, it was possible to access to new permeability data and, in some cases, to i-octane partial fluxes which have never been reported for EVOH copolymers so far. However, the obtaining of accurate data necessitated an EVOH preconditioning in the model fuel mixtures to avoid the long drifts in the permeability data over time, which were initially obtained for virgin EVOH samples.

Permeability data were obtained for two important tie lines (*i.e.* ethanol and toluene tie lines) in the ternary diagram ethanol/*i*-octane/toluene. The partial flux of ethanol was systematically ca. ten times higher than that of toluene and ca. hundred times higher than that of *i*-octane. Therefore, the permeate samples generally contained at least 90 wt% of ethanol and this alcohol can thus be considered as a key component for EVOH permeability. Ethanol also played a strong role on EVOH conditioning in the oxygenated model mixtures. Its strong plasticizing effect led to a strong decrease in the EVOH mechanical and physical properties. However, ethanol did not induce any significant change in EVOH crystallinity.

Globally, ethanol induced a strong increase in EVOH total permeability. The role of ethanol was also important on the toluene flux and the related plasticization effect decreased the EVOH barrier performance towards aromatics. The addition of increasing contents of (bio) ethanol in fuels could thus have important consequences on the barrier properties of multi-layer fuel tanks. The new metrology developed in this work offers a new insight in the EVOH permeability properties and will help prevent the consequences of (bio)ethanol addition in fuels on environmental emissions through fuel lines and tanks.

ACKNOWLEDGMENTS

Jing Zhao and his colleagues gratefully acknowledge the support of the chemical company *Arkema*.

REFERENCES

1 Khanah Mokwena K., Tang J. (2012) Ethylene Vinyl Alcohol: A Review of Barrier Properties for Packaging Shelf Stable Foods, *Crit. Rev. Food Sci. Nutrition* **52**, 6405-650.

2 Apicella A., Hopfenberg H.B. (1982) Water-swelling behavior of an ethylene-vinyl alcohol copolymer in presence of sorbed sodium chloride, *J. Appl. Polym. Sci.* **27**, 1139-1148.

3 Aucejo S., Pozo M.J., Gavara R. (1998) Effect of water presence on the sorption of organic compounds in ethylene-vinyl alcohol copolymers, *J. Appl. Polym. Sci.* **70**, 711-716.

4 Zhang Z., Britt I.J., Tung M.A. (1999) Water Absorption in EVOH Films and Its Influence on Glass Transition Temperature, *J. Polym. Sci. Part B Polym. Phys.* **37**, 691-699.

5 Lopez-Rubio A., Lagaron J.M., Giménez E., Cava D., Hernandez-Munoz P., Yamamoto T., Gavara R. (2003) Morphological alterations induced by temperature and humidity in ethylene-vinyl alcohol copolymers, *Macromolecules* **36**, 9467-9476.

6 Zhang Z., Britt I.J., Tung M.A. (2001) Permeation of oxygen and water vapor through EVOH films as influenced by relative humidity, *J. Appl. Polym. Sci.* **82**, 1866-1872.

7 Cava D., Lagaron J.M., Martinez-Gimenez F., Gavara R. (2007) Inverse gas chromatography study on the effect of humidity on the mass transport of alcohols in an ethylene-vinyl alcohol copolymer near the glass transition temperature, *J. Chromatogr. A* **1175**, 267-274.

8 Samus M.A., Rossi G. (1996) Methanol absorption in ethylene-vinyl alcohol copolymers: Relation between solvent diffusion and changes in glass transition temperature in glassy polymeric materials, *Macromolecules* **29**, 2275-2288.

9 Nulman M., Olejnik A., Samus M., Fead E., Rossi G. (1998) Fuel permeation performance of polymeric materials analyzed by gas chromatography and sorption techniques, *SAE Special Publications* **1365**, 41-47.

10 Lagaron J.M., Powell A.K., Bonner G. (2001) Permeation of water, methanol, fuel and alcohol-containing fuels in high-barrier ethylene-vinyl alcohol copolymer, *Polym. Testing* **20**, 569-577.

11 Gagnard C., Germain Y., Keraudren P., Barriere B. (2004) Permeability of semicrystalline polymers to toluene/methanol mixture, *J. Appl. Polym. Sci.* **92**, 676-682.

12 Clément R., Kanaan C., Brulé B., Lenda H., Lochon P., Jonquières A. (2007) An original automated desorption apparatus for measuring multi-component sorption properties of barrier polymer films, *J. Membr. Sci.* **302**, 95-101.

13 Smitha B., Suhanya D., Sridhar S., Ramakrishna M. (2004) Separation of organic-organic mixtures by pervaporation - A review, *J. Membr. Sci.* **241**, 1-21.

14 Garcia Villaluenga J.P., Tabe-Mohammadi A. (2000) A review on the separation of benzene/cyclohexane mixtures by pervaporation processes, *J. Membr. Sci.* **169**, 159-174.

15 Uragami T. (2006) Polymer Membranes for Separation of Organic Liquid Mixtures - Chapter 14 in *Materials Science of Membranes for Gas and Vapor Permeation*, Yampolskii Y., Pinnau I., Freeman B.D. (eds), John Wiley & Sons, New York, pp. 355-371.

16 Iwanami T., Hirai Y. (1983) Ethylene vinyl alcohol resins for gas-barrier material, *Tappi J.* **66**, 85-90.

Control-Oriented Models for Real-Time Simulation of Automotive Transmission Systems

N. Cavina[1]*, E. Corti[1], F. Marcigliano[2], D. Olivi[1] and L. Poggio[2]

[1] DIN, University of Bologna, Viale Risorgimento 2, 40136 Bologna - Italy
[2] Ferrari SpA, Via Abetone Inferiore 4, 41053 Maranello (MO) - Italy
e-mail: nicolo.cavina@unibo.it - enrico.corti2@unibo.it - francesco.marcigliano@ferrari.com - davide.olivi3@unibo.it
luca.poggio@ferrari.com

* Corresponding author

Abstract — *A control-oriented model of a Dual Clutch Transmission (DCT) was developed for real-time Hardware In the Loop (HIL) applications, to support model-based development of the DCT controller and to systematically test its performance. The model is an innovative attempt to reproduce the fast dynamics of the actuation system while maintaining a simulation step size large enough for real-time applications. The model comprehends a detailed physical description of hydraulic circuit, clutches, synchronizers and gears, and simplified vehicle and internal combustion engine submodels. As the oil circulating in the system has a large bulk modulus, the pressure dynamics are very fast, possibly causing instability in a real-time simulation; the same challenge involves the servo valves dynamics, due to the very small masses of the moving elements. Therefore, the hydraulic circuit model has been modified and simplified without losing physical validity, in order to adapt it to the real-time simulation requirements. The results of offline simulations have been compared to on-board measurements to verify the validity of the developed model, which was then implemented in a HIL system and connected to the Transmission Control Unit (TCU). Several tests have been performed on the HIL simulator, to verify the TCU performance: electrical failure tests on sensors and actuators, hydraulic and mechanical failure tests on hydraulic valves, clutches and synchronizers, and application tests comprehending all the main features of the control actions performed by the TCU. Being based on physical laws, in every condition the model simulates a plausible reaction of the system. A test automation procedure has finally been developed to permit the execution of a pattern of tests without the interaction of the user; perfectly repeatable tests can be performed for non-regression verification, allowing the testing of new software releases in fully automatic mode.*

Résumé — **Modélisation orientée-contrôle pour la simulation en temps réel des systèmes de transmission automobile** — Un modèle orienté vers le contrôle d'une transmission à double embrayage (DCT, *Dual Clutch Transmission*) a été développé et implémenté en temps réel dans un système *Hardware In the Loop* (HIL), pour supporter le développement du contrôleur DCT et pour tester systématiquement ses performances. Le modèle représente une tentative innovatrice de reproduire les dynamiques rapides du système d'actionnement tout en conservant une dimension de pas de calcul assez grande pour l'application en temps réel. Le modèle comprend une description physique détaillée du circuit hydraulique, des embrayages, des synchroniseurs et des engrenages, et des sous-modèles simplifiés du véhicule et du moteur à combustion interne. Comme l'huile circulant dans le système est caractérisée par un module de compressibilité élevé, les dynamiques de pression sont très rapides, pouvant provoquer une instabilité dans la simulation en temps réel, et le même risque intéresse la dynamique des servovalves, en raison des très petites masses des éléments mobiles. Par conséquent, le modèle de circuit hydraulique a été modifié et simplifié sans perdre la validité physique. Les résultats des simulations hors ligne ont été comparés à des mesures en ligne pour vérifier la validité du modèle, qui a ensuite été implémenté dans un système HIL et connecté à l'unité de control de la transmission (TCU, *Transmission Control Unit*). Plusieurs tests ont été effectués sur le simulateur HIL, pour vérifier la performance de la TCU : tests de panne électrique sur capteurs et actionneurs, tests de défaillance sur les valves hydrauliques, les embrayages et les synchros, et tests d'applications comprenant toutes les principales actions de contrôle effectuées par la TCU. Étant basé sur des lois physiques, dans toutes les conditions le modèle simule une réaction plausible du système. Une procédure d'automatisation de test a finalement été développée pour permettre l'exécution d'une suite de tests sans interaction de l'utilisateur; les tests parfaitement reproductibles peuvent être effectués pour la vérification de non régression, ce qui permet la validation de nouvelles versions du logiciel en mode entièrement automatique.

NOMENCLATURE

ADC   Analog/Digital Converter
AMT   Automated Manual Transmissions
AT   Automatic Transmission
CVT   Continuously Variable Transmissions
DAC   Digital/Analog Converter
DCT   Dual Clutch Transmission
D/R   Digital/Resistance
ECU   Engine Control Unit
FIU   Failure Injection Unit
HIL   Hardware In the Loop
PWM   Pulse Width Modulation
RTP   Real-Time Processor unit
TCU   Transmission Control Unit

INTRODUCTION

In recent years the need for increased fuel efficiency, driving performance and comfort has driven the development of engine and transmission technology in the automotive industry, and several types of transmissions are currently available in the market trying to meet these needs. The conventional Automatic Transmission (AT), with torque converter and planetary gears, was leading the market of non-manual transmissions, but in recent years it is losing its predominant position, because of the low efficiency of the converter and the overall structure complexity, in favour of other technologies. Continuously Variable Transmissions (CVT) permit avoiding the problem of gear shifting, but are limited in torque capacity and have the disadvantage of a low transmission efficiency due to the high pump losses caused by the large oil flows and pressure values needed. Automated Manual Transmissions (AMT) with dry clutches are the most efficient systems, but they do not meet customer expectations due to torque interruption during gear shift (Zhang *et al.*, 2005). If compared to other transmissions, the DCT technology has the advantage of being suitable both for low revving and high torque Diesel engines, and for high revving engines for sport cars, maintaining a high transmission efficiency, as well as high gear shift performance and comfort (Matthes, 2005; Goetz *et al.*, 2005). A Dual Clutch Transmission can be considered as an evolution of the AMT. An AMT is similar to a manual transmission, but the clutch actuation and the gear selection are performed by electro-hydraulic valves controlled by a TCU (Transmission Control Unit). The peculiarity of a DCT system is the removal of torque interruption during gear shift typical of AMT, through the use of two clutches: each clutch is connected on one side to the engine, and on the other side to its own primary shaft, carrying odd and even gears, respectively.

The role of engine and transmission electronic control units is steeply increasing, and new instruments are needed to support their development. Hardware In the Loop (HIL) systems are nowadays largely used in the automotive industry, in which the crucial role of electronic control systems requires new techniques for software testing and validation. HIL systems are designed for testing control units in a simulation environment, allowing to perform functional and failure tests on the control unit, by connecting it to a device able to simulate the behavior of the controlled system (engine, transmission system, vehicle, etc.) in real-time.

The aim of this work is the development of a control-oriented model of a DCT system designed to support model-based development of the DCT controller. The main original contribution of this work is a complete, real-time model of the hydraulic actuation circuit and of the DCT kinematics and dynamics, designed with a control-oriented approach. With respect to existing publications (Lucente *et al.*, 2007; Montanari *et al.*, 2004), the presented model (Cavina *et al.*, 2012) is in fact an innovative attempt to reproduce the fast dynamics of the actuation system while maintaining physical equations with a simulation step size large enough for real-time applications.

The model has been developed in Simulink environment; the reason for using such tool is twofold: Simulink allows the user to fully define the differential equations that represent the dynamic behavior of each component, allowing the introduction of mathematical simplifications when needed, without losing the physical validity of the simulation. Also, Simulink models are easily implemented in a fixed-step simulation model (which is necessary due to real-time constraints), and may be seamlessly integrated in the dSPACE HIL environment.

The developed model has been integrated in a Hardware In the Loop application for real-time simulation and the test of different software releases implemented inside the TCU is being carried out (Cavina *et al.*, 2013). Test automation permits executing tests at the simulator without the interaction of the user; the complete repeatability of every test is fundamental for non-regression tests on new software releases; the possibility to plan in advance the sequence of actions that have to take place during the test permits to execute tests that would not be possible with the manual interaction of the user.

The paper introduces the main aspects of the project, and provides further insight both into modeling and application areas. All the elements of the model are presented, including detailed descriptions of synchronizers, clutch and vehicle sub-models, and a new (and more accurate) clutch torque model, based on separators temperature estimation. Also, complex phenomena (such as for example clutch hysteresis) and the resulting modeling issues have been discussed and clarified. Finally, all relevant HIL tests are discussed in the second part of the paper, demonstrating the potential of model-based real-time simulation.

1 SYSTEM DESCRIPTION

In the system considered in this paper, shown in Figure 1, the two secondary shafts carry four synchronizers and eight different gears (1st-7th and reverse). Thanks to the coordinated use of the two clutches, at the moment of gear shift the future gear is already preselected by the synchronizer on the shaft that is not transmitting torque; the only action performed during the gear shift is the opening of the currently closed clutch and the closing of the other one. Thanks to the control of clutch slips, the shifting characteristic is similar to the clutch-to-clutch shift commonly seen in conventional automatic transmissions (Kulkarni *et al.*, 2007); while in a conventional AT the gear shift smoothness is achieved through the action of the torque converter, which provides a dampening effect during shift transients, in a DCT transmission the shift comfort depends only on the control of clutch actuation (Liu *et al.*, 2009). The gearbox is actuated by a hydraulic circuit controlled by the TCU and divided in two different parts, one carrying the two clutches and the relative actuation and lubrication circuit, and the other actuating the four gear selectors.

1.1 Hydraulic Circuit Model

The hydraulic circuit scheme is shown in Figure 2. A rotative pump, directly connected to the engine *via* a fixed gear ratio, provides the necessary pressure level to the circuit, in which two different parts can be distinguished. The high pressure circuit provides oil to all servo valves that need a fast and well-calibrated actuation: clutch pressure control valves, which connect the high pressure circuit to the clutch pressure one, and gear selector valves, which feed the gear actuation circuit. The parking lock circuit and electronic differential one are also connected to the high pressure circuit.

The low pressure circuit, instead, controls the clutch lubrication; the lubrication oil, heated up by the thermal power generated in the clutches during their slip, is cooled down by a cooling system. The low pressure circuit is connected to the high pressure one through an orifice controlled by a servo valve: in this way the desired high pressure value can be controlled. If there is no need for oil in the circuit, the flow coming from the pump can be discharged to the sump thanks to the bypass circuit.

A generic pressure dynamics model was developed by considering the continuity equation of an incompressible fluid (Merritt, 1967), taking into account all input flows Q_{in} and output flows Q_{out} in the circuit, the total volume change from the initial value V_0 caused by the motion of mechanical parts, and the total bulk modulus β_{tot} of the oil circulating in the circuit, as shown in:

$$Q_{in} - Q_{out} - \frac{dV_0}{dt} = \frac{V_0}{\beta_{tot}}\frac{dp}{dt} \qquad (1)$$

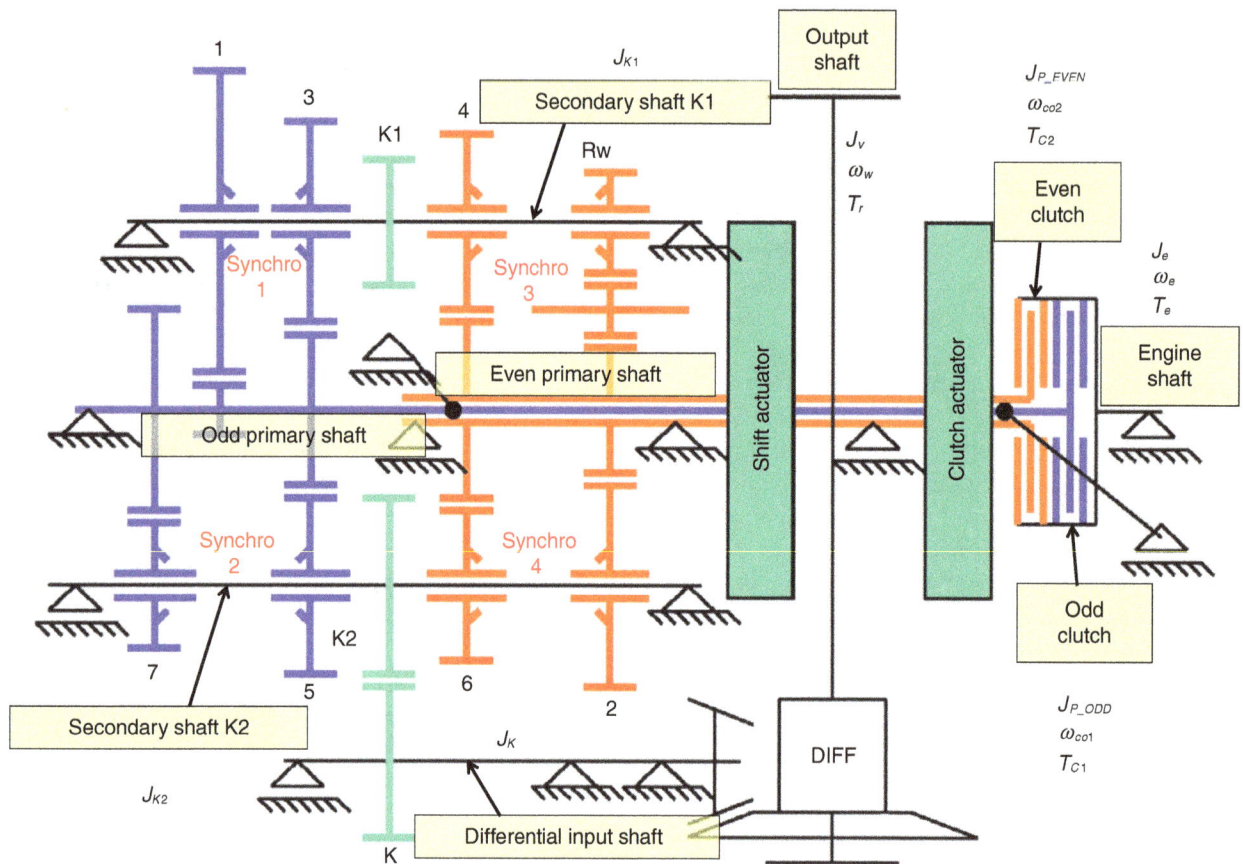

Figure 1

Dual Clutch Transmission (DCT) layout.

Equation (1) can be applied for every part of the circuit: in the high pressure circuit, the term dV_0/dt can be considered negligible, because only the valve spools are moving and the volume change is very low. Instead, this term is particularly important in the clutch pressure circuit, because of the clutch motion while closing or opening, and in the synchronizers' actuation circuit, due to the rod motion from a position to another (depending on the selected gear).

1.2 System Pressure Model

The system pressure value (*i.e.* the pressure level in the high pressure circuit) is calculated according to Equation (1). The control of the system pressure is performed by regulating the actuation current on a servo valve; its output flow actuates a hydraulic valve, which opens the orifice that connects the high pressure

circuit to the low pressure one. This part of the hydraulic circuit comprehends several flows and small orifices; consequently, to develop a model suitable for real-time applications, the calculation of the flow from the high pressure circuit to the low pressure one is provided by an experimental map rather than by a dynamic model. Inputs to such map are the current applied to the servo valve and the pressure level in the high pressure circuit. Furthermore, the input flow Q_{in} provided by the pump is calculated *via* an experimental map depending on engine speed, and considering also leakage flows, which depend on the pressure level in the high pressure circuit. All other flows needed for the calculation of the system pressure value, *i.e.* flows from the high pressure circuit to the clutch pressure circuit and to the gear actuation circuit, are dynamically calculated considering the servo valves model, as explained in the next paragraphs (*Eq. 4*).

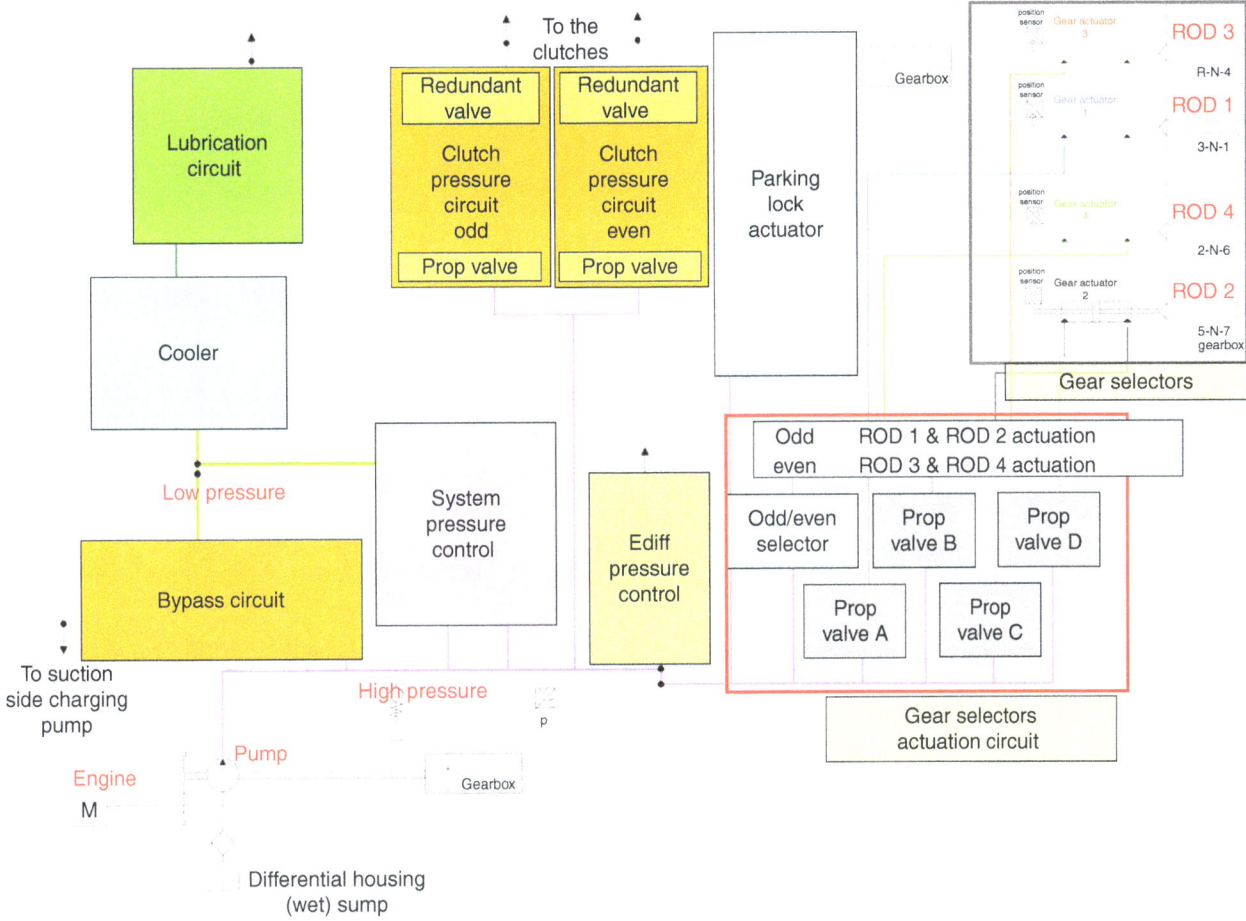

Figure 2
Hydraulic circuit layout.

1.3 Pressure Control Valve Model

The pressure level acting on clutches and gear actuators is controlled by proportional pressure control valves (*Fig. 3*), which are designed to act as closed loop systems (Lucente *et al.*, 2007): the proportionality between actuation current on the solenoid and pressure in the actuation chamber p_A is given by the feedback chamber, which generates a feedback force F_{fb} against the valve opening, whose proportionality factor is the feedback chamber area A_{fb} of the spool, as shown in:

$$F_{fb} = p_A A_{fb} \qquad (2)$$

The spool dynamics may be described by the mass-spring-damper analogy, as in:

$$m_{sp}\ddot{x}_{sp} + b_{sp}\dot{x}_{sp} + k_{sp}x_{sp} = F_{sol} - F_{fb} - F_{fl} - F_{pr\_sp} \qquad (3)$$

The electromagnetic force on the spool F_{sol} has been experimentally characterized, depending on the input current and on the spool position x_{sp}. The flow forces F_{fl} are calculated using experimental data, depending on flow, valve position and pressure difference between input and output ports. The spring preload F_{pr_sp} modifies the pressure range in which there is proportionality between current and pressure.

When there is no current acting on the solenoid, the actuator port A is connected to the return port T and the oil from the actuation circuit flows to the oil sump (*Fig. 3, 4a*). By supplying current, the solenoid force rises and the spool is moved forward; the spool reaches a position in which the port A is connected neither to the port T nor to the port P; this position is called "Dead Zone Start". Moving further, the spool connects the actuator port A to the inlet port P (*Fig. 4b*); the position of first connection between A and P is called "Dead Zone End".

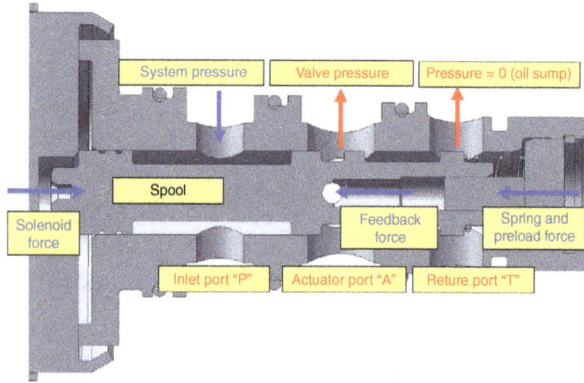

Figure 3

Proportional 3-way pressure regulation valve.

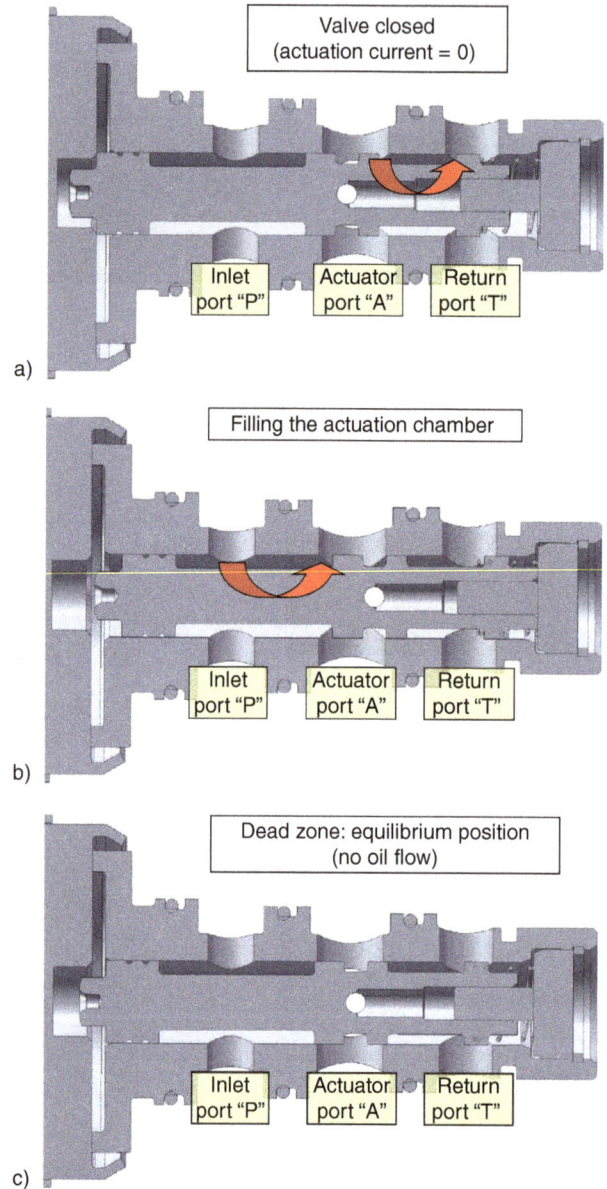

Figure 4

Proportional 3-way valve motion.

The oil flow raises the pressure in the actuation chamber, and the feedback force rises as well, forcing the spool to move back to the "Dead Zone", reaching the equilibrium position of the system (*Fig. 4c*).

The flow through the valve is given by (4), according to Bernoulli formulation for incompressible fluids (Lucente *et al.*, 2007; Montanari *et al.*, 2004):

$$
Q(x_{sp}, \Delta p) = \begin{cases} -sign(p_A - p_T)n_A C_d A_A(x_{sp})\sqrt{\frac{2|p_A - p_T|}{\rho_{oil}}} \\ \quad \text{if } x_{sp} < x_{DeadZoneStart} \\ 0 \text{ if } x_{DeadZonStart} \leq x_{sp} \leq x_{DeadZoneEnd} \\ sign(p_P - p_A)n_A C_d A_A(x_{sp})\sqrt{\frac{2|p_P - p_A|}{\rho_{oil}}} \\ \quad \text{if } x_{sp} > x_{DeadZoneEnd} \end{cases}
$$

(4)

The effective area connecting the actuation port depends on the geometrical area $A_A(x_{sp})$, the number of orifices n_A and the discharge coefficient C_d. To avoid stability problems during the simulation, when the pressure difference is very low the Bernoulli Equation (4) is replaced by the parabolic Equation (5) that provides smaller flow values when $|\Delta p| \leq \Delta p^*$ (a similar problem is solved with the same approach in Guzzella and Onder, 2004), as shown in Figure 5:

if $|\Delta p| \leq \Delta p^*$:

$$
Q(x_{sp}, \Delta p) = \pm sign(\Delta p)n_A C_d A_A(x_{sp})\sqrt{\frac{2\Delta p^*}{\rho_{oil}}}\frac{|\Delta p|^2}{\Delta p^{*2}} \quad (5)
$$

1.4 Clutch Model

The clutches of the considered DCT are wet clutches; the lubrication oil removes the heat generated while slipping, and torque is transmitted by the contact between the clutch discs, covered with high-friction materials, and the separator plates. The clutch motion is opposed by Bellville springs between every friction disc, giving also a preload force. For safety reasons, when no actuation is required, the clutch is open and engine shaft and transmission shaft are separated. The clutch closure is performed by pumping current in the

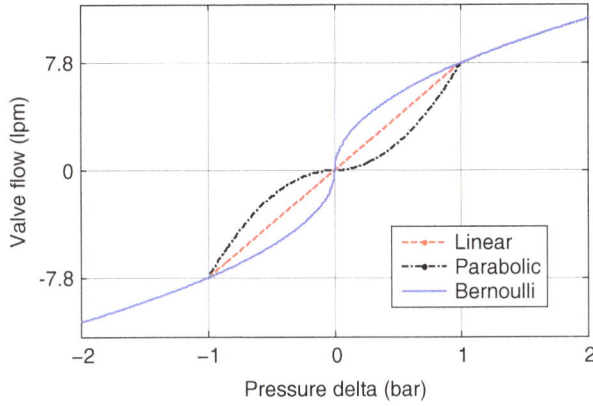

Figure 5

Parabolic interpolation for Bernoulli equation.

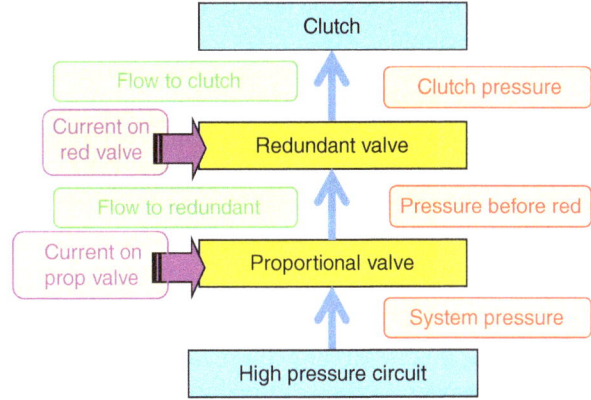

Figure 6

Clutch actuation scheme.

corresponding proportional valve, in order to raise the pressure level acting on the clutch actuator.

The clutches are not directly connected to their proportional actuation valve; for safety reasons, between the clutches and the proportional valves that control the clutch pressure, two on/off valves (called "redundant valves", Fig. 6), one for each clutch, permit the fast discharge of the oil from the clutch to the sump (and the consequent opening of the clutch) bypassing the other components of the circuit in case of fault, or when there is the need of opening the clutches as fast as possible during the gear shift. When no current is applied on the valves, they discharge the oil from the clutch actuation chamber to the sump; on the contrary, when current is supplied they allow the oil flow between the proportional valves and the clutches.

The redundant valve is constructively similar to the proportional one, but it presents an on/off behavior because of the absence of the feedback port; consequently, the position of the spool depends only on the current supplied to it; if the current level is high enough, it remains fully open, whatever pressure level is present in the actuation chamber. The dynamic model of the redundant valve spool motion is similar to the one of the proportional valve, but without the term which considers the feedback force:

$$m_{red}\ddot{x}_{red} + b_{red}\dot{x}_{red} + k_{red}x_{red} = F_{sol} - F_{fl} - F_{pr\_red} \quad (6)$$

The model of clutch actuation circuit between the proportional valve and the clutch is divided in two different parts, in order to consider the pressure dynamics through the redundant valve that, even when its spool is not moving but steady in the position of full opening,

is not negligible (for example during the clutch filling transient). Calculation of the pressure p_{b_red} in the chamber between the proportional valve and the redundant valve may be performed by applying the mass conservation principle of Equation (1), as shown in:

$$Q_{prop} - Q_{red} = \frac{V_{0b\_red}}{\beta} \frac{dp_{b\_red}}{dt} \quad (7)$$

Calculation of pressure p_c acting on the clutch may be performed in a similar way:

$$Q_{red} - Q_c - \frac{dV_{0c}}{dt} = \frac{V_{0c}}{\beta} \frac{dp_c}{dt} \quad (8)$$

The flow Q_c exiting the clutch is a leakage flow of oil going from the clutch actuation chamber back to the sump. This leakage flow is not negligible and it was designed to lubricate the clutch actuation moving elements. Its value is calculated according to an experimental map that takes into account the pressure in the actuation circuit and the oil temperature.

The volume change in the clutch, calculated in Equation (9), is not negligible while the clutch is being closed or opened; in this case, it can be evaluated by considering the longitudinal speed of the clutch $\dot{x}_c$ calculated in Equation (10):

$$\frac{dV_{0c}}{dt} = \frac{d(A_c x_c)}{dt} = A_c \dot{x}_c \quad (9)$$

The clutch closure procedure can be divided into different phases. At first, the clutch is completely open; when a clutch closure is required, current is pumped to the relative proportional valve, maintaining the relative redundant valve open; as soon as the oil reaches the clutch actuation chamber, the oil pressure inside the chamber rises, and consequently the force acting on the actuation piston, which is connected to the clutch discs. The clutch piston and discs longitudinal dynamics follows the mass-spring-damper Equation (10):

$$m_c \ddot{x}_c + b_c \dot{x}_c + k_c x_c = p_c A_c - F_{pr\_c} \qquad (10)$$

The clutch motion is contrasted by the resistant force $k_c x_c$ coming from the Bellville springs. Experimental tests showed that it is not a fully linear relation; consequently an experimental map was obtained.

As soon as the oil in the actuation chamber reaches a level higher than the preload force F_{pr_c}, the clutch is forced to move forward. The resistant force consequently rises and the clutch stops until more oil (and pressure) is provided. Therefore, in this phase the pressure in the chamber is not proportional to the current on the proportional valve, but to the piston longitudinal position ("filling phase"). The lowest pressure level for which the clutch is completely closed ("kiss point pressure") depends on preload pressure and spring stiffness. When the kiss point position is reached, the friction discs connected to the clutch output shaft and the separator plates connected to the clutch input shaft come in contact. Usually these two shafts have different speeds, and some friction torque is transmitted between them thanks to the friction material with which the discs are covered. From this moment, the pressure in the actuation chamber is directly proportional to the current on the valve. By exchanging torque, the two shafts synchronize their speeds, becoming a rigid system with only one degree of freedom; its final speed depends on the inertia of the two parts and on the applied torque.

1.5 Clutch Hysteresis

Analyzing Figure 7, which shows experimental data measured while actuating a current ramp on the valve, first rising and then falling, it can be noticed that the values of preload and kiss point pressure of the clutch are significantly different between the closing and the opening phase. This is due to the Coulomb friction between the clutch actuation seal and its seat: during the rising ramp, some force is needed to overtake the static force and move the clutch; consequently the pressure at the preload point is higher than during the falling ramp. When the clutch is completely closed some force is needed to move the clutch back through the action of the Belleville springs, and consequently a lower oil pressure level must be reached to start the clutch opening.

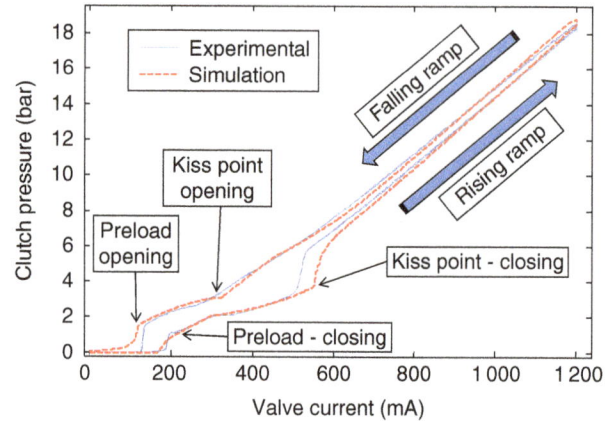

Figure 7

Rising and falling current ramp on clutch valve.

This consideration clarifies why the kiss point pressure is lower during the falling ramp.

Clutch hysteresis has been simulated in the model using two different levels of preload pressure for the two phases: the kiss point pressure level is a consequence of such choice, since once the preload level is defined, the kiss point pressure only depends on clutch spring stiffness. It is a rough simplification of a complex hysteretic behavior (Bertotti and Mayergoyz, 2006; Eyabi and Washington, 2006), but it is probably the most suitable, because of the real-time constraint and the lack of more experimental details about this phenomenon.

It can be also noticed that the relationship between actuation current and clutch pressure is not unique; it happens not only during the closing and opening phases, in which the process is ruled by the spring stiffness, but also at higher current values, for which the clutch is completely closed. This behavior is due to the fact that during the ramps the pressure variation is achieved by modifying the volume of oil inside the circuit. During a rising ramp, some oil flow has to enter the circuit, and the valve spool must be in a position that allows the connection between P and A ports (Fig. 4b). During a falling ramp, oil must be discharged to the sump, and the spool must be in a position that allows oil flow from port A to port T (Fig. 4a). Consequently, at the same pressure level, when the valve is filling the circuit the position of the spool is after the "Dead Zone", while when the valve is discharging the circuit the spool is before the "Dead Zone". In conclusion, during a rising phase more current is needed on the valve than during a falling phase, at the same pressure level, and it should be observed that in a static characterization of the valve this effect would not be present. The developed model, being based on physical laws, is capable to correctly reproduce this phenomenon, as shown in Figure 7.

1.6 Clutch Torque

The evaluation of the torque transmitted by a wet friction clutch is a very delicate topic to deal with. Even the use of simplified equations based on the assumption of constant fluid thickness and constant temperature over the clutch area (Deur *et al.*, 2005; Davis *et al.*, 2000; Greenwood and Williamson, 1966) would be too complicated for a real-time control-oriented application.

Therefore, the clutch torque model used in this work is based on experimental maps (*Fig. 8*) depending on the clutch slip (*i.e.* the difference between the engine speed ω_e and the clutch output speed ω_{co}) and on the "net" pressure acting on the clutch (the overall pressure level p_c decreased by the kiss point pressure p_{c_kp}). Different maps were obtained for different oil temperatures, and during the simulation such maps are interpolated by considering the actual oil temperature value. This torque value T_{cBasic}, called "Basic Torque", is further modified by taking into account the current operating conditions, since the temperature of the clutch separators strongly influences the friction coefficient of the friction material.

The temperature inside the clutch is calculated through a linear mean value model, based on energy conservation principle (Lai, 2007; Velardocchia *et al.*, 2000). The separators temperature T_{sep} is calculated by considering the net heat flow inside the clutches and the heat capacity $m_{sep} c_{p\_sep}$ of the separators:

$$P_c - P_{oil} = m_{sep}c_{p\_sep}\frac{dT_{sep}}{dt}$$
$$T_{sep} = \int \frac{P_c - P_{oil}}{m_{sep}c_{p\_sep}}\,dt \qquad (11)$$

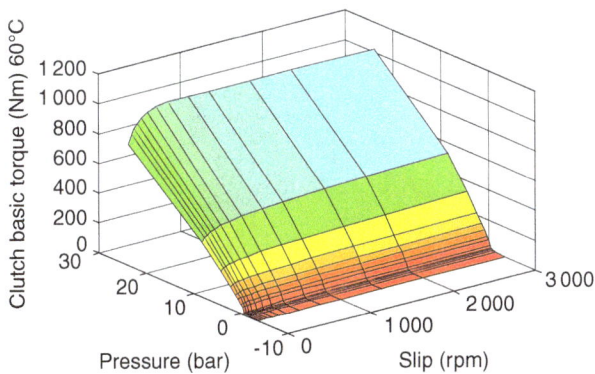

Figure 8

Clutch basic torque (oil temperature 60°C).

The heat power generated inside the clutches P_c is a consequence of the clutch discs slip and of the transmitted torque T_c; the heat power removed from the clutches by the lubrication oil P_{oil} depends on the amount of lubrication flow Q_{lb} and on the specific heat of the oil c_{p_oil}:

$$P_c = (\omega_e - \omega_{co})T_c \qquad (12)$$

$$P_{oil} = Q_{lb}c_{p\_oil}(T_{oil} - T_{cooler}) \qquad (13)$$

where T_{cooler} is the temperature of the oil out of the cooling system, that is assumed to be constant. An experimentally-derived function $f(Q_{lb})$ permits determining the temperature T_{oil} of the oil inside the clutches, if the temperature of the separators is known:

$$T_{oil} = T_{sep} - (T_{sep} - T_{cooler})f(Q_{lb}) \qquad (14)$$

The variation of the clutch friction coefficient $\Delta\mu_c(T_{sep}, \omega_e - \omega_{co})$, according to the temperature of the clutch separators and to the amount of slipping, was experimentally determined and used to calculate its effect ΔT_{cTsep} on transmitted torque:

$$\Delta T_{cTsep} = p_c A_c r_{mc} \Delta\mu_c \qquad (15)$$

where r_{mc} is the clutch mean radius.

The total friction torque T_{cFrict}, generated by the friction between the clutch discs, can therefore be expressed as:

$$T_{cFrict} = T_{cBasic} + \Delta T_{cTsep} \qquad (16)$$

When the pressure on the clutch is lower than the kiss point pressure, the clutch is not completely closed, and the drag torque T_{cDrag}, due to the viscosity of the oil inside the clutches, must be taken into account (it has been experimentally determined, depending on engine speed, clutch output speed, lube flow and oil temperature). Finally, the two clutches are not completely independent, being influenced by their mutual movement: an experimentally derived crosstalk torque ΔT_{cCross} is added to the total torque, depending on the clutch pressure. All these terms contribute to the calculation of the total transmitted torque T_c.

1.7 Synchronizers Model

The synchronizer mechanism consists of a gearshift sleeve with internal dog gearing, connected to the synchronizer body and to the transmission shaft, cone synchronizer rings with locking toothing, and a synchronizer hub with selector teeth and friction cone, connected to the gear which is idle on the transmission shaft thanks to needle roller bearings (Naunheimer *et al.*, 2011; Neto *et al.*, 2006).

The four rods used for the motion of the four synchronizers are hydraulically actuated by the high pressure circuit, as shown in Figure 2. Their position is controlled by four proportional pressure control valves, whose features have been described before, and four hydraulic double acting pistons, each controlled by two of those valves, one controlling the pressure level of the left chamber, and the other one of the right chamber. Actuating alternatively one valve or the other, the rod can be moved towards the desired position; every synchronizer controls the engagement of two gears, and the central position of the rod corresponds to the freewheel position.

If the synchronizer is not engaging any gear and no pressure is applied on its actuation piston, the gearshift sleeve is held in the middle position by a detent. Between the two chambers of the double acting piston with area A_{rod} a pressure difference p_{rod} is determined by providing different pressure levels p_l and p_r on the left and right chamber, respectively:

$$p_{rod} = p_l - p_r \qquad (17)$$

The selection between odd and even gears is executed by actuating an on/off valve, which moves a hydraulic selector, that remains in "odd" position if the valve is not actuated, while it moves to the "even" position if the valve is actuated. The developed model of this selector is a static model which calculates the position of the distributor according to the current on the valve and the system pressure level p_{sys}:

$$x_{sel} = \begin{cases} x_{0\_sel} \text{ if } current < threshold \\ x_{0\_sel} + \frac{p_{sys}A_{sel} - F_{pr\_sel}}{k_{sel}} \text{ if } current \geq threshold \end{cases} \qquad (18)$$

where A_{sel} is the actual area for pressure, F_{pr_sel} is the spring preload force and k_{sel} is the valve spring stiffness.

The pressure dynamics inside the chambers of the actuation piston can be calculated according to Equation (19), considering as input and output flows the ones coming from the valve model, while the leakage flow on the rod actuation chambers can be considered negligible:

$$Q_{in\_prop} - Q_{out\_prop} - \frac{dV_{0rod}}{dt} = \frac{V_{0rod}}{\beta}\frac{dp_{rod}}{dt} \qquad (19)$$

The motion of the hydraulic piston and of the sleeve, shown in Figure 9, is described as a spring-mass-damper

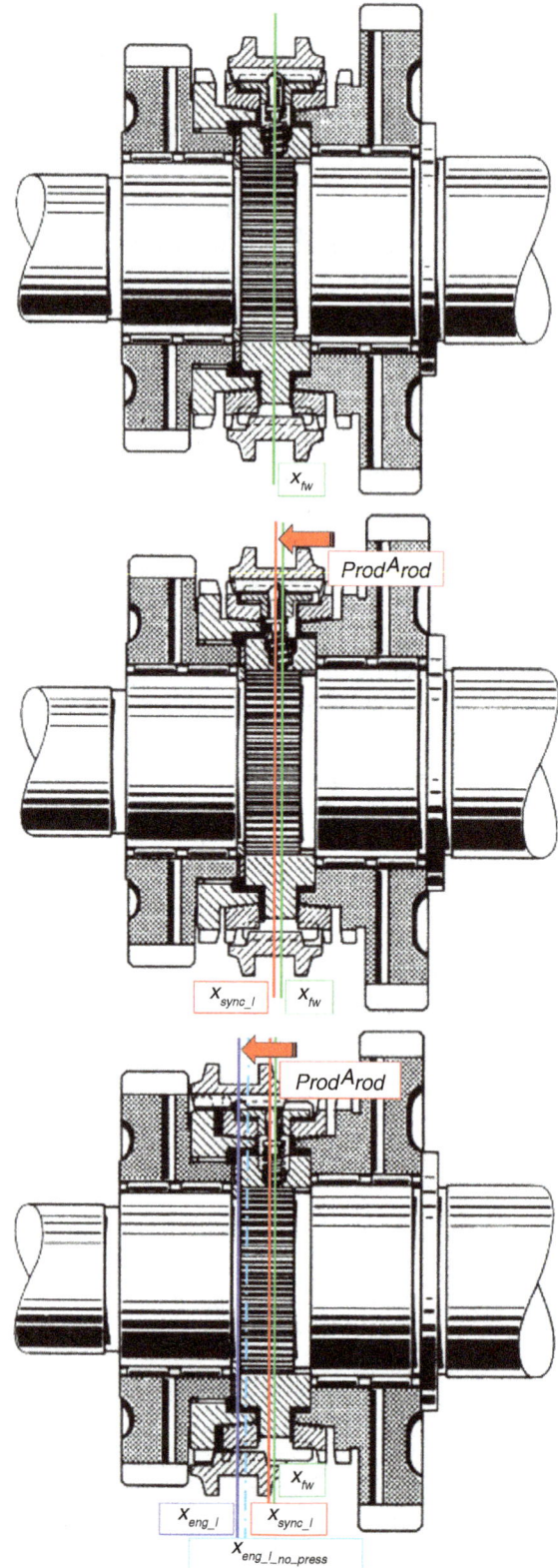

Figure 9

Synchronizing process: a) sleeve in freewheel position, no gear engaged; b) ring and hub synchronizing their speeds; c) sleeve on hub: left gear engaged.

dynamic equation. For simplicity, the coordinates taken into account are different depending on the position of the sleeve of the synchronizer; the position x_{rod} can be considered the "absolute" position, *i.e.* the position value as registered by the on-board sensor and that has to be reproduced by the model, always increasing from the left engaged position x_{l_eng} to the right engaged position x_{r_eng}; $x_{eng\_l\_no\_press}$ and $x_{eng\_r\_no\_press}$ are the engaged positions when no pressure is applied on the actuation piston anymore. The position x_{fw} is the free-wheel position of the sleeve, when no gear is selected. The different chosen intervals correspond to different phases:

$x_1 = x_{rod} - x_{eng\_l\_no\_press}$ when $x_{eng\_l} < x < x_{sync\_l}$; in this interval the left gear is selected:

$$m_{rod}\ddot{x}_1 + b_{rod}\dot{x}_1 + k_{eng}x_1 = p_{rod}A_{rod} \qquad (20)$$

$x_2 = x_{rod} - x_{fw}$ when $x_{sync\_l} < x < x_{sync\_r}$; in this interval no gear is selected:

$$m_{rod}\ddot{x}_2 + b_{rod}\dot{x}_2 + k_{fw}x_2 = p_{rod}A_{rod} \qquad (21)$$

$x_3 = x_{rod} - x_{eng\_r\_no\_press}$ when $x_{sync\_r} < x < x_{eng\_r}$; in this interval the right gear is selected:

$$m_{rod}\ddot{x}_3 + b_{rod}\dot{x}_3 + k_{eng}x_3 = p_{rod}A_{rod} \qquad (22)$$

where k_{eng} is the spring stiffness imposed when the gear is in engaged position, and k_{fw} the one when it is in free-wheel position.

During the synchronization phase, the longitudinal motion of the rod is temporarily stopped in the position x_{sync_l} or x_{sync_r}, waiting for the speed synchronization between the synchronizer rings and the hub; the synchronizer sleeve and ring rotate at the same speed of the secondary shaft, and consequently their speed is dependent on the vehicle speed; the gear, instead, rotates according to the speed of the primary shaft; in this position torque is transmitted between the two elements. The longitudinal motion of the synchronizer can restart when the synchronization phase is over.

It is supposed that during the synchronization process the correspondent clutch is open; thus, the only resistant torque is the viscous friction of the primary shaft bearings. The speed of the synchronizing gear ω_{gear} on the considered secondary shaft is given by Equation (23), depending on the relative primary inertia J_p referred to the secondary shaft, due to the gear ratio τ_{gear} between the two shafts:

$$T_{syn} - b_p\omega_{gear} = J_p\tau_{gear}^2\dot{\omega}_{gear} \qquad (23)$$

The torque T_{syn} exchanged between ring and hub can be calculated considering the torque transmitted by a cone friction clutch (Razzacki, 2004), according to the surface friction coefficient μ_{syn}, the cone mean radius R_{m_ring} and the cone angle :

$$T_{syn} = \frac{p_{rod}A_{rod}\mu_{syn}R_{m\_ring}}{\sin(\theta)} \qquad (24)$$

The viscous friction coefficient of primary shafts b_p was experimentally determined by analyzing the coast-down trend of the shaft, when the respective clutch is open, and the synchronizer is moved towards its free-wheel position, leaving the primary shaft and the gears free to slow down.

1.8 Vehicle Model

The vehicle model that has been designed and implemented for this project neglects elasticity effects of both shafts and tires, since the HIL simulator has been realized to support functionality analysis and software debugging of the DCT controller, rather than gearshift quality analysis (which is typically part of a drivability optimization activity carried out directly on the vehicle). For this reason, first order transmission system dynamics were not considered as an important feature to be included in the vehicle model, in an effort to keep it as simple (and fast) as possible; therefore, the implemented dynamic equations consider shafts with infinite stiffness.

The model consists of different sets of equations, depending on whether the clutch is completely closed or slipping, and further, whether a gear is currently selected, or not. The speeds of vehicle, gears and gearbox shafts are calculated by considering inertial effects and the torques acting on them (Kulkarni *et al.*, 2007; Liu *et al.*, 2009). When the clutch is completely closed and a gear is selected, engine, gearbox and vehicle are connected and only one differential equation provides the description of the entire system dynamics (under the hypothesis of infinite stiffness); the engine speed ω_e is calculated considering the net torque on engine shaft, *i.e.* the difference between the engine torque T_e and the resistant torque T_r, properly divided by the total gear ratio τ_{tot} of the currently selected gear:

$$T_e - \frac{T_r}{\tau_{tot}} = \left(J_e + J_{eq\_p} + \frac{J_{eq\_g}}{\tau_{tot}^2}\right)\dot{\omega}_e \qquad (25)$$

For simplicity, the inertia J_e of all the elements before the clutch and the equivalent inertia of primary shafts

J_{eq_p} are referred to the engine shaft, while the equivalent inertia J_{eq_g} of shafts $K1$, $K2$, K ($Fig.\ 1$) and of the whole vehicle are referred to the wheel shaft:

$$J_{eq\_p} = J_{eq\_p\_ODD} + J_{eq\_p\_EVEN} \quad (26)$$

$$J_{eq\_g} = J_{eq\_K1} + J_{eq\_K2} + J_{eq\_K} + J_{eq\_v} \quad (27)$$

The equivalent primary inertia is calculated from the inertia of primary shafts referred to their own axes; the equivalent value referred to the engine axes depends on the current gear, $i.e.$ the gear whose clutch is closed and therefore transmitting torque to the wheels. The equivalent inertia is the axes inertia if the clutch of the considered gear is closed, otherwise a multiple gear ratio must be considered. If the current gear is odd:

$$J_{eq\_p\_ODD} = J_{p\_ODD}$$

$$J_{eq\_p\_EVEN} = \left(\frac{\tau_{k\_EVEN}}{\tau_{k\_ODD}} \frac{\tau_{pres}}{\tau_{sel}}\right)^2 J_{p\_EVEN} \quad (28)$$

where τ_{k_ODD} is the gear ratio between the secondary shaft ($K1$ or $K2$, depending on which odd gear is selected) and K shaft, the same for τ_{k_EVEN}, while τ_{sel} is the total gear ratio of the currently selected gear and τ_{pres} the one of the preselected gear on the other primary shaft. If the current gear is instead an even gear:

$$J_{eq\_p\_ODD} = \left(\frac{\tau_{k\_ODD}}{\tau_{k\_EVEN}} \frac{\tau_{pres}}{\tau_{sel}}\right)^2 J_{p\_ODD}$$

$$J_{eq\_p\_EVEN} = J_{p\_EVEN} \quad (29)$$

The equivalent inertia of all the other shafts inside the gearbox (the secondary shafts and the differential input shaft) and the vehicle inertia are referred to the wheels, so the transmission ratio between the secondary shaft and the differential input shaft τ_{K1} and τ_{K2} and the differential ratio τ_{diff} must be taken into account:

$$J_{eq\_K1} = J_{K1}(\tau_{K1}\tau_{diff})^2 \quad (30)$$

$$J_{eq\_K2} = J_{K2}(\tau_{K2}\tau_{diff})^2 \quad (31)$$

$$J_{eq\_K} = J_K \tau_{diff}^2 \quad (32)$$

$$J_{eq\_v} = M_v R_w^2 \quad (33)$$

where M_v is the vehicle total mass and R_w the wheel radius. The total transmission ratio is defined as the ratio between engine speed ω_e and wheel speed ω_w:

$$\tau_{tot} = \frac{\omega_e}{\omega_w} \quad (34)$$

The vehicle speed v is calculated considering the total gear ratio τ_{tot} and the wheel radius R_w (assuming no wheel slip):

$$v = \frac{\omega_e}{\tau_{tot}} R_w \quad (35)$$

The resistant torque acting on the vehicle can be calculated following a simple standard approach (Guzzella and Sciarretta, 2008); it is due to the aerodynamic force (depending on air density ρ_a, frontal area of the vehicle A_v and vehicle drag coefficient C_x), the rolling friction resistance (depending on rolling friction coefficient f_v and vehicle total mass M_v) and the braking torque T_{br}:

$$T_r = \left(\frac{1}{2}\rho_a A_v C_x v^2 + f_v M_v g\right)R_w + T_{br} \quad (36)$$

Aerodynamic and rolling resistant torques can also be calculated considering the coast-down behavior of the vehicle, as shown in Equation (37), with the use of three curve fitting parameters f_0, f_1, f_2:

$$T_{aer} + T_{roll} = (f_0 + f_1 v + f_2 v^2)R_w \quad (37)$$

The braking torque is proportional to the pressure acting on the brake circuit, that is the pressure of the brake fluid p_{br} (proportional to the pressure on the brake pedal), the actual areas for pressure A_f, A_r of front and rear discs respectively, the friction coefficients of the linings μ_f, μ_r, and the mean radius of brake discs R_f, R_r:

$$T_{br} = 2p_{br}A_f\mu_f R_f + 2p_{br}A_r\mu_r R_r \quad (38)$$

When the clutch is slipping, the system has one more degree of freedom; it must be divided into two different parts, one from the engine to the clutch (39), and the other (40) from the clutch to the wheels:

$$T_e - (T_{c1} + T_{c2}) = J_e\dot{\omega}_e \quad (39)$$

$$T_{c1}\tau_1 + T_{c2}\tau_2 - T_r = (J_{eq\_p}\tau_{tot}^2 + J_{eq\_g})\dot{\omega}_w \quad (40)$$

where T_{c1} and T_{c2} are the torques respectively transmitted by clutch 1 and clutch 2. Equation (40) is referred to the wheel axes, thus the clutch torques must be multiplied by τ_1 and τ_2, which are the total gear ratios of the currently selected gears on odd and even shafts, respectively:

$$\tau_1 = \frac{\omega_{co1}}{\omega_w}; \tau_2 = \frac{\omega_{co2}}{\omega_w} \quad (41)$$

where ω_{co1} and ω_{co2} represent the clutch output speeds.

A particular case that must be taken into account is when no gear is selected: the primary and secondary shafts of the gearbox are not connected anymore, adding one more degree of freedom to the system. The vehicle, not influenced by the clutch or engine torque anymore, slows down because of the resistant torque experienced during coast down:

$$-T_r = J_{eq\_g}\dot{\omega}_w \tag{42}$$

The primary shafts are accelerated by the clutch torque; the dynamics are very fast because the only resistant torque is the viscous friction due to the bearings, and the primary shaft inertia is very low:

$$T_{c1} - b_p\omega_{co1} = J_{p\_ODD}\dot{\omega}_{co1}$$

$$T_{c2} - b_p\omega_{co2} = J_{p\_EVEN}\dot{\omega}_{co2} \tag{43}$$

where the clutch torque T_{ci} is equal to the engine torque if the clutch is closed (slip = 0), or it is calculated by the clutch torque model if the clutch is slipping.

In all these cases, the speed of every gear and shaft inside the gearbox can be calculated from the wheel speed (or from the clutch output speeds ω_{co1} and ω_{co2}, for this last case) by considering the corresponding gear ratios.

When one of the clutches is closed, the case with only one degree of freedom must be considered: this condition is called "no-slip" condition. When both clutches are slipping or open, they both are in "slip" condition and the cases with more degrees of freedom must be taken into account. At every gear shift and drive away event the clutch passes from one condition to the other; the modeling issue is to provide continuity in the calculation of the clutch speed while changing its condition from "no slip" to "slip" and *vice versa*. When during the "slip" phase the engine speed and one of the clutch output speeds become equal, the relative clutch goes to "no-slip" condition, and the entire system gets only one degree of freedom. On the contrary, the switch from "no-slip" to "slip" condition happens when during the "no-slip" phase the engine torque becomes greater than the transmissible clutch torque calculated by the clutch torque model, because in these conditions the clutch is not able to transmit all the required torque anymore. This passage happens also when the clutch pressure is lower than the kiss point pressure, because the friction discs are not in direct contact with the clutch separator plates anymore.

1.9 Engine Model

The model presented in the previous paragraphs may be considered complete once all the vehicle parts generating and transmitting torque are simulated, from the engine to the wheels. The engine model implemented for real-time application is a real-time zero-dimensional mean value model (He and Lin, 2007), and comprehends the control logic of the Engine Control Unit (ECU). It can therefore maintain idle speed set-point during idling, and it can correctly respond to torque and speed requests coming from the TCU, when the ECU becomes "slave" and the DCT controller (TCU) becomes "master" (for example during drive away and gear shift operations). All CAN (Controller Area Network) bus messages between the ECU and TCU have been reproduced in simulation.

2 SIMPLIFIED MODEL FOR HIL APPLICATION

The presented model needs, as inputs, only the electrical currents from the TCU, the driver inputs (accelerator pedal, brake pedal pressure and gear shift request) and environmental data. However, it is not directly suitable for real-time application, because both the pressure dynamics and the mass-spring-damper dynamics are very fast, while there is a strict lower limit in the simulation step size: as the simulation must be real-time, the step size for the considered model should be set to around 0.5 ms. The simulation of the whole system with this step size may cause instability and undermine the possibility of plausible results.

To ensure stability of the simulation, the model must be tested under the worst conditions, which in this case are represented by a step current input on the clutch actuation valve, that forces the fastest pressure and motion dynamics, related one to each other. Hence, before comparing the simulation results to experimental data, the model was tested by performing a simulation of the clutch dynamics, having as input on the actuation valve a current step of 700 mA and considering progressively smaller step sizes: as shown in Figure 10, to avoid instability even a step size as small as 1 μs may not be sufficient, and such value is absolutely not acceptable for a real-time application. Therefore, a simplification of the model dynamics is needed. The issue is to maintain the basic dynamic behavior, trying to recognize, isolate and modify only those parts that produce instability during real-time simulation.

Considering the mass-spring-damper dynamics of the valve spool, the instability is caused by the mass that is generally very small and determines a very large natural (and resonating) frequency of the system. The consequent oscillation is very fast and not reproducible in a

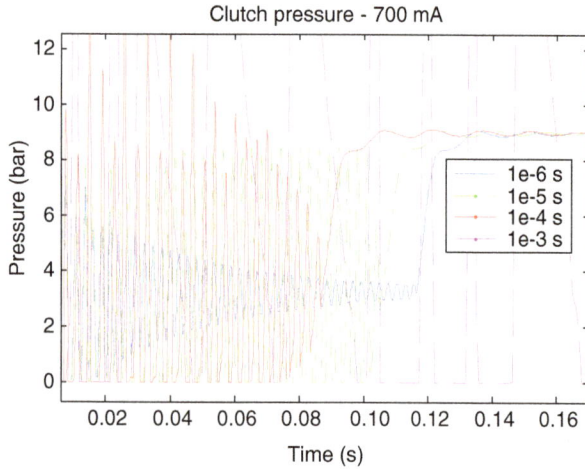

Figure 10

Analysis of simulation results while applying current steps with different simulation step sizes.

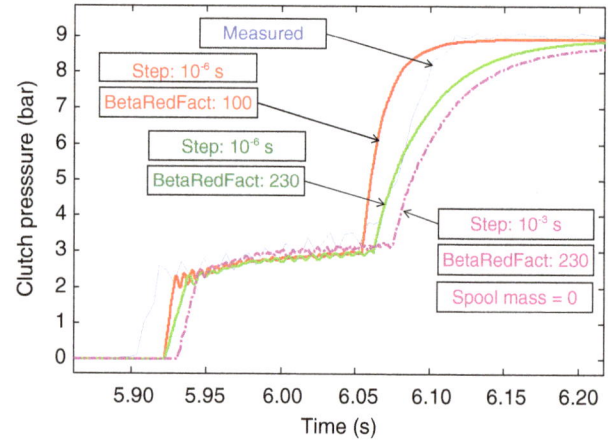

Figure 11

Simulation of clutch model with a fast current ramp from 0 to 700 mA.

real-time simulation. The solution that has been found is a "simplified dynamic equation" that does not consider masses anymore, and the viscous friction coefficient b_{sp} is replaced by a dummy b_{sp}^* much larger than the real one. Therefore, the second-order dynamic equation is simplified as a first-order dynamics ruled by the viscous coefficient and the spring stiffness. Equation (3) is replaced by Equation (44):

$$b_{sp}^* \dot{x}_{sp} + k_{sp} x_{sp} = F_{sol} - F_{fb} - F_{fl} - F_{pr\_sp} \qquad (44)$$

Instead, there is no need to simplify the dynamic equation of clutches and synchronizers, since their masses are large enough to prevent fast oscillations.

At the same time, the pressure dynamics in the different parts of the circuit are very fast, because the oil is theoretically incompressible and, even considering the theoretical value of the bulk modulus of the oil, a single drop of oil would be enough to raise the pressure to very large values in a single simulation step. Furthermore, in the real system other aspects must be taken into account: first of all, the effect of the percentage of air entrained inside the oil circuit (expressed in Eq. 45 as the ratio V_g/V_{tot}) is not negligible, and leads to a significant decrease of the bulk modulus of the whole system β_{tot}, because the air bulk modulus β_g is very small:

$$\frac{1}{\beta_{tot}} = \frac{1}{\beta_{oil}} + \frac{V_g}{V_{tot}} \frac{1}{\beta_g} \qquad (45)$$

The percentage of air inside the oil is not experimentally known, and the total bulk modulus should also

comprehend the stiffness of the pipes and the small leakages in the components (only the controlled leakage from the clutches to the oil sump is explicitly considered in the model). Therefore, the actual bulk modulus is much lower than the theoretical value, and the use of a lower value in the model, in order to allow a stable simulation of the system, can be considered physically realistic. The minimum equivalent value β_{tot} for a stable real-time simulation is 230 times lower than the theoretical value; it approximately corresponds to 1.4% of air volume inside the oil volume, which can be considered realistic.

In Figure 11, the effects of these two variations are separately analyzed, by comparing the model results with experimental data, measured while exciting the clutch control valve with fast current ramps, and analyzing the corresponding pressure ramps. First, the effect of the bulk modulus value is analyzed by comparing the results of simulations performed both with a step size of 1 μs, that allows to freely modify the bulk modulus to match the results of an on board measure: with a bulk modulus reduction factor set to 100, the simulation with a step size of 1 μs matches the measured pressure ramp; setting this value to 230 (the smallest reduction factor for a stable real-time simulation with step size of 0.5 ms) the pressure ramp is slower and reaches the steady-state value with a certain delay with respect to the measured signal, but such delay can be considered negligible for real-time application. The time delay between measured and simulated signals at the beginning of the pressure rise trend ($t = 5.9$ s) is negligible for the proposed application, and it is mainly due to an overestimated circuit volume fraction initially empty (to be filled with oil before the pressure can rise).

Despite all the simplifications introduced to reach a stable real-time simulation of a system with very fast dynamics, the model described in these paragraphs can be considered reliable for HIL applications. Before being implemented in the real-time application, a first offline calibration and validation of the model is needed.

3 OFFLINE SIMULATION RESULTS

The first analysis of the results obtained with the developed model, which comprehends all the different parts previously described, is made through an offline simulation, without the connection to the TCU; the input values (*i.e.* currents and driver's requests) are taken from measurements performed on board the vehicle. The simulation step size is set to 0.5 ms.

3.1 Clutch Pressure

Clutch motion is a particularly delicate parameter to deal with (*Fig. 12*). When the clutch is open, the clutch pressure is maintained at a level greater than zero, to be ready for a quick closing actuation, but it is still completely open, thanks to the clutch spring preload. When the clutch is not moving, the clutch pressure is proportional to the input current on the clutch valve; it is the case of completely open clutch (until $t = 27.35$ s) and completely closed clutch (from $t = 27.60$ s). Answering a clutch closure request, the TCU provides a current peak for the clutch filling phase, the clutch valve's spool starts moving and the oil flows from the input port P to the actuation port A (*Fig. 3, 4*) and then to the clutch circuit: the pressure level in this circuit rises. The value reached before the complete clutch closure is proportional to the clutch spring stiffness. Once the clutch is completely closed, the TCU provides current to maintain the desired pressure level, calculated according to the torque that must be transmitted.

The validity of the simulation results can be verified by analyzing the clutch circuit pressure signal measured on board (this pressure level is called clutch pressure). As shown in Figure 12, the simulated pressure signal matches the measured one closely.

3.2 System Pressure

The TCU controls the actuation current on the system pressure control valve according to the target pressure value. By supplying current to the valve, a controlled amount of oil is discharged to the low pressure circuit

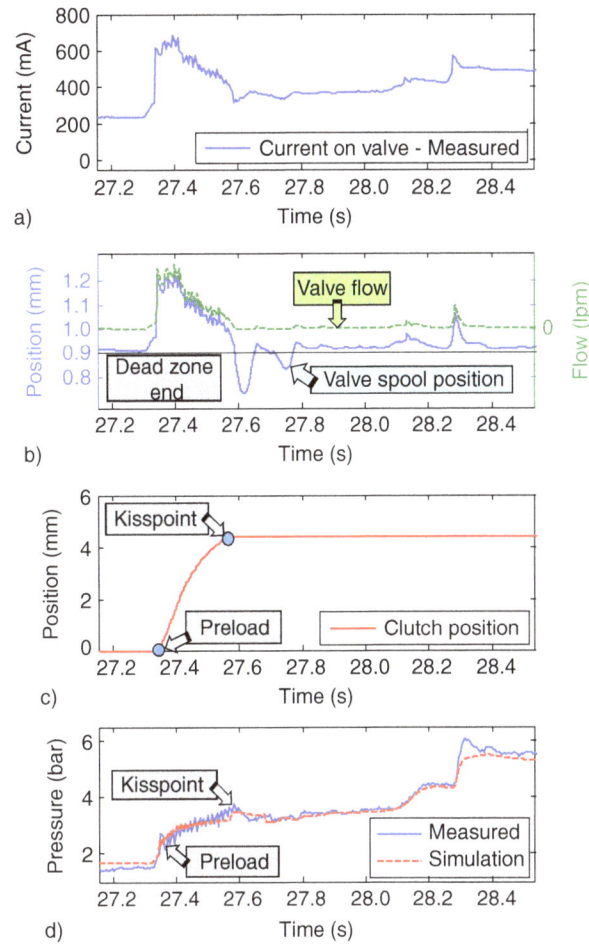

Figure 12

Offline simulation of clutch closing procedure.

from the high pressure circuit, causing a decrease of the system pressure level; all other flows (to the clutch pressure and to the rod pressure circuits) act as disturbances on the target system pressure, causing a quick fall of the pressure level; consequently, the actuation current is reduced by the TCU to maintain the desired pressure level. As in the clutch pressure simulation, the flows signals are not measured on board and the results of the simulation can be compared only with the pressure signal. The real-time simulation cannot reproduce the high frequency oscillations of the experimental measurements, but the obtained low-frequency content is correct (*Fig. 13*).

3.3 Synchronizers

In Figure 14, the gear selection on the odd shaft of the transmission is shown; during all this process the engine

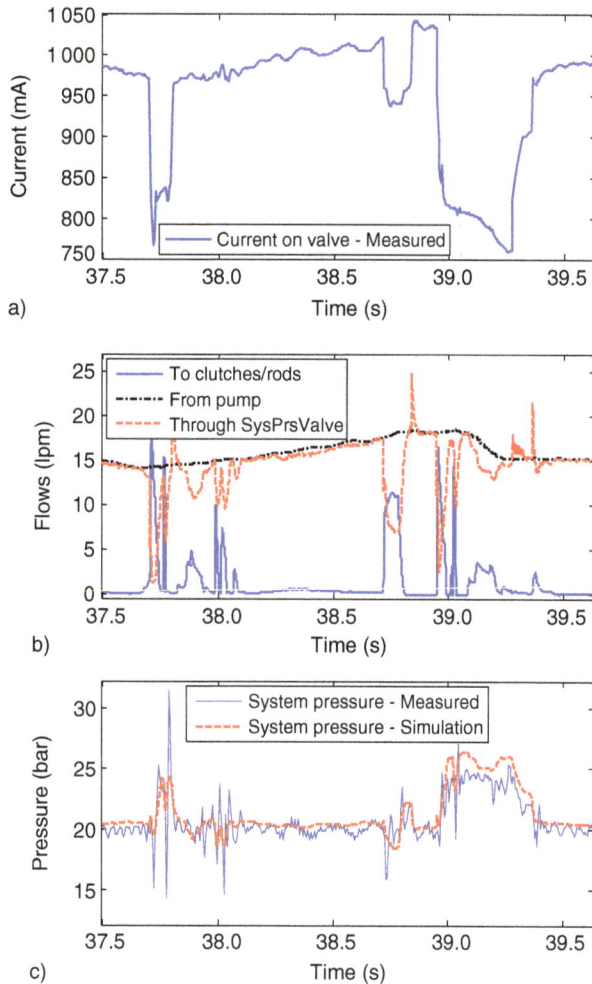

a)

b)

c)

Figure 13

Offline simulation of system pressure circuit.

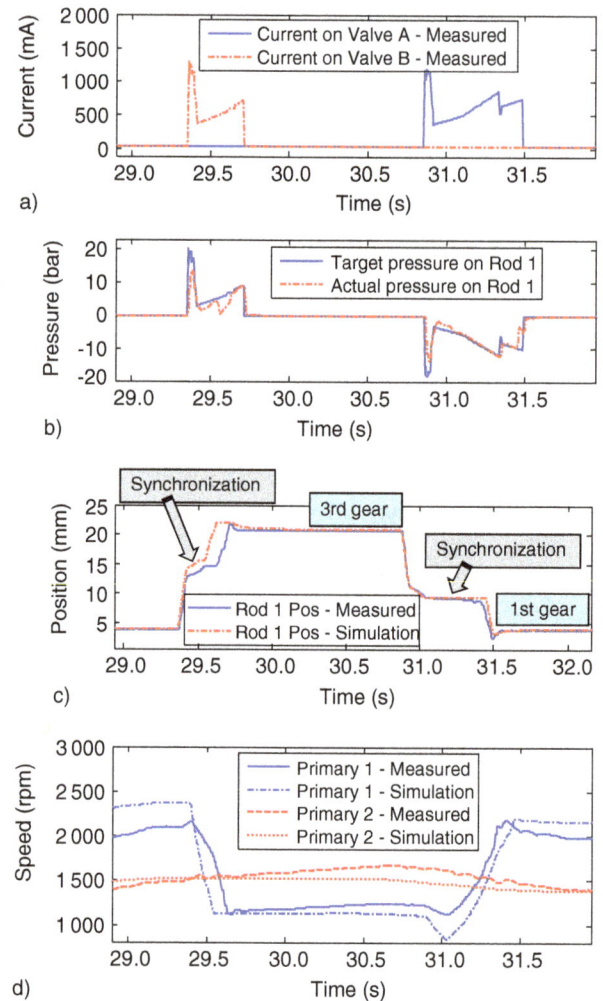

a)

b)

c)

d)

Figure 14

Offline simulation of gear selection procedure.

is transmitting torque through the even shaft in 2nd gear, while the odd clutch is completely open; therefore, the engine speed and the even shaft speed are coincident. The pre-selection of gears is controlled by the TCU, which predicts the "future" gear according mainly to the engine speed trend and to the torque request. The simulation results can be evaluated by analyzing the rod position signals measured on board; in this case the pressure value inside the gear actuators is not measurable.

At first, rod 1 is engaging the 1st gear, which corresponds to the low position measured by the sensor (*Fig. 14c*, until $t = 29.4$ s); when the TCU triggers the pre-selection of the 3rd gear, it provides on valve A (*Fig. 2*) an actuation current with a very impulsive shape (*Fig. 14a*, at $t = 29.4$ s), in order to move the

synchronizer from the engaged position; then the current changes into a ramp profile when the 3rd gear (which corresponds to the high position measured by the sensor) is synchronizing (*Fig. 14c, d*, from $t = 29.5$ s to $t = 29.7$ s), to prevent the risk of gear scratching. The net pressure inside the rod (*i.e.* the difference between the pressure on the left chamber and the one on the right chamber, *Fig. 14b*), follows the target pressure imposed by the TCU; the oil flow from the rod pressure circuit to the rod chamber in the phases with a fast movement of the rod acts as a disturbance on this target pressure, because the movement determines a volume increase that must be filled by oil, causing a temporary pressure level decrease (*Fig. 14b*, $t = 29.6$ s). When the speed synchronization is complete, the 3rd gear is correctly engaged and there is no more need for pressure in the rod chambers. At time

$t = 30.7$ s the engine starts slowing down and at time $t = 30.9$ s the TCU triggers the pre-selection of 1st gear, pumping current to valve A and following the same procedure.

The simulation output matches the experimental one closely; the most difficult part is the simulation of the synchronization phase, because a closed loop control would be necessary to recognize the moment in which the gear is engaged and consequently reduce the input current.

4 HARDWARE

The hardware chosen for the Hardware in The Loop application of the model developed during this project is based on a *dSPACE* "Mid-Size" Simulator, shown in Figure 15. It is equipped with a remote-controlled power supply unit (with an upper current limit of 50 A and a voltage regulation from 0 to 20 V controlled by the real-time system), a "ds1005" PPC board and two "ds2210" I/O boards, load cards, Failure Insertion Units (FIU) and ECU connectors.

The ds1005 PPC board comprehends a Real-Time Processor unit (RTP), RAM, flash and cache memory and timer interrupts, and is connected to the host PC. The PPC board is connected to the two I/O boards, containing sensors and actuators interfaces that provide a typical set of automotive I/O functions, including A/D conversion, digital I/O, and wheel speed sensor signal generation. In the application described in this paper, Analog/Digital Converter (ADC) is used for the actuation signals of proportional valves (pressure regulation of clutches, rods and electronic differential, and lubrication valves), Digital/Resistance converter (D/R) for temperature sensors, paddles and reverse button, Digital/Analog Converter (DAC) for pressure and position sensors, Pulse Width Modulation (PWM) signal generator for the actuation of on/off valves (clutch redundant valves, bypass, electric and hydraulic parking lock command); a CAN controller gives the possibility to set two

CAN lines per board. A load card permits the connection of the TCU to the loads it controls and actuates; the connected loads are the real valves and actuators of the DCT transmission, to obtain the best possible match with the real system. Each load channel is connected to a Failure Insertion Unit (FIU), to give the possibility to simulate failures in the TCU wiring. Three types of failure can be simulated: TCU output shorted to battery voltage, TCU output shorted to ground, and TCU output open circuit.

The developed Simulink model of the physical system has been integrated with a proper I/O conversion submodel, as shown in Figure 16, in order to connect the physical model to the I/O boards. It is capable to convert the physical signals in electric signals, considering the characteristic of every sensor, and send them to the TCU. The actuation currents controlled by the TCU are read and sent as inputs to the model.

A graphical interface was developed using the *dSPACE* Control Desk environment in the host PC, to allow full control of the simulator also by inexperienced users. Figure 17 shows the dashboard interface, developed to provide into one single window all the main controls and indicators needed to perform manual tests.

5 HIL APPLICATION RESULTS

The developed Simulink model is compiled with the use of "Real-Time Workshop" tool and then flashed inside the HIL real-time processor unit. The HIL application

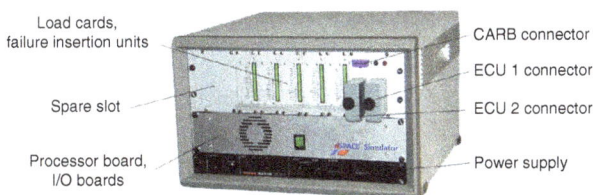

Figure 15

dSPACE "Mid-Size" Simulator.

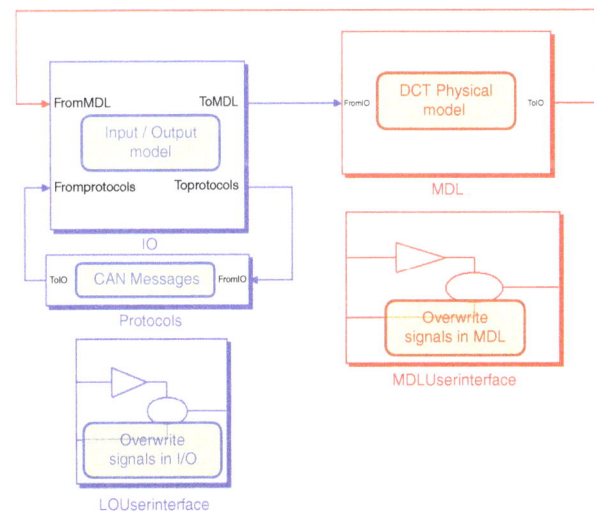

Figure 16

Complete Simulink model.

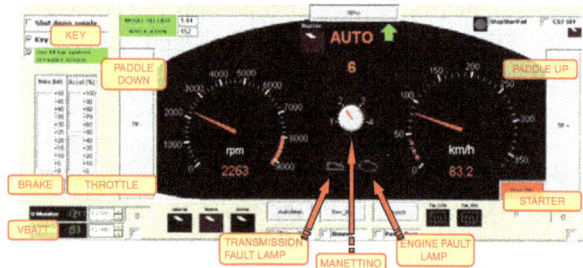

Figure 17

Dashboard interface.

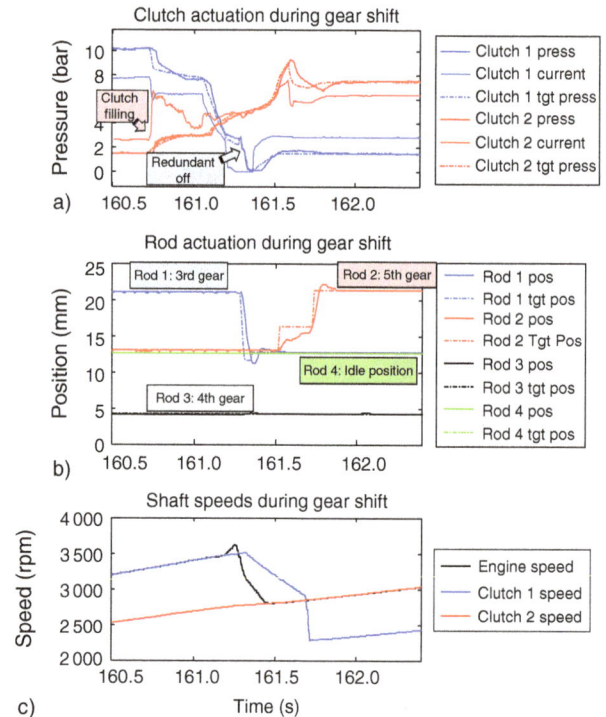

Figure 18

Gear shift from 3rd to 4th gear.

permits to execute a wide range of tests to check and validate the functionality of the DCT controller; in general, these tests can be of different kinds. Functional tests permit to check the main functionality of the TCU, as drive away, gear shift, performance launch and interaction with the driver through paddles and buttons. The possibility of introducing a mechanical, hydraulic or electrical fault in the system allows verifying the TCU capability to recognize the problem and recover to a safe state. The capability of the controller to adapt to changes in the controlled system during its lifetime can be checked by analyzing the result of adaption procedures.

5.1 Functional Tests

First of all, the basic functions implemented in the TCU have been tested. In a dual clutch transmission, the gear shift is performed only opening the offgoing clutch and closing the ongoing one, thanks to the possibility of preselecting gears on odd and even shaft before the gear shift. Figure 18 considers a gear shift from 3rd to 4th gear.

Figure 18a shows the pressure profile during the gear shift. The clutch torque transmitted by the offgoing clutch (clutch n.1 or "odd" clutch) is reduced at the beginning of the gear shift (time 160.7 s), by reducing the pressure in the clutch actuation; at the same time, the TCU starts pumping current on the ongoing clutch (clutch n.2 or "even" clutch); in this transient (time between 160.7 s and 161.1 s), the clutch pressure is not proportional to the current supplied to the proportional valve, because the oil is filling the clutch, which is moving forward towards the kiss point position, maintaining a low pressure value with high current on the valve. At time 161.1 s, the clutch kiss point is reached in the even clutch; the current on the even proportional valve is reduced and the pressure on the two clutches is regulated to achieve the best shift comfort; the two pressure signals are crossing themselves, in a way that some torque is

always transmitted during all the gear shift and the wheels are never left without torque: the torque interruption during gear shift typical of MT and AMT is eliminated. At the end of the process, the pressure in the odd clutch is discharged as fast as possible by connecting the relative redundant valve to the sump, to ensure a complete clutch opening; the valve is then actuated again, and the pressure level in the clutch actuation is maintained at around 1.5 bar, ready for the next gear shift.

The actual pressures calculated by the simulator are able to follow the target ones closely. A small delay can be noticed between the simulated signals and the target ones (set by the TCU) during the fastest transients; this delay is due to the simplifications needed for a real-time simulation, and it is small enough not to affect the TCU control.

In Figure 18b, the rod actuation is shown. During all the gear shift process, the rods of the even gears are not moved; the 4th gear (correspondent to a low position of rod 3) was already selected before the gear shift, while the even shaft was not transmitting torque because its clutch was open; the odd gear is closed and transmitting torque in 3rd gear (rod 1 in high position). As soon as the oil inside the odd clutch is discharged by the redundant valve, the 3rd gear is deselected, moving the correspondent rod from engaged to idle position, pumping current on the correspondent valve. Then, while the odd clutch is open, the rod 2 is moved to preselect the 5th gear (high position), ready for the next gear shift. During the gear shift the TCU is "master" and the engine ECU is "slave"; this means that the target engine speed and torque are set by the TCU. The engine speed follows the odd clutch speed as far as the odd clutch pressure is not reduced; then, after accelerating for a short while to improve the driver's feeling, it reaches the even clutch speed during the closure of the even clutch (*Fig. 18c*). Meanwhile, the odd clutch speed slows down, reaching the speed correspondent to the 5th gear as soon as the synchronization process is complete.

5.2 Mechanical Failures Simulation

Figure 19 shows the simulation of a problem with the synchronizer of 1st and 3rd gear: in the real system, there is the possibility that the synchronizer cones would not transmit the usual torque to the gear anymore; it can happen because of unusual wear (the durability of the cone friction material is not infinite), or when, due to some failure in the clutch actuation circuit, during the selection of the gear some torque is transmitted by the engine to the primary shaft, and consequently to

the gear; in this condition the cones covered with friction material heat up abnormally and lose their friction characteristic. In both cases, the effect is the loss of capacity in synchronizing the speed of synchronizer (and output shaft) and gear (rigidly connected to the input shaft). This behavior is simulated inside the model setting to zero the friction coefficient on the specific cones. Consequently, the synchronizer cannot select the gear anymore; the TCU detects a problem with the gear, tries the engagement other three times consecutively, pumping all the possible pressure inside the piston of the rod; if the procedure is not successful, the TCU validates the failure identification and sets the recovery mode, that consists in the impossibility to select that particular gear.

In Figure 19b, the 3rd gear cannot be selected, and after three more attempts the gearbox shifts directly from 2nd to 4th gear. This gear shift, in which offgoing and ongoing gears are both on the same shaft, is performed with torque interruption: the even clutch is opened, 2nd gear (rod 4 in high position) is deselected and 4th gear is selected (rod 3 in low position), then the even clutch is closed again (*Fig. 19a*, from time 27.5 s). During the three trials, the odd shaft is not correctly moving to the speed of 3rd gear (*Fig. 19c*); immediately after the TCU has set the error, the 5th gear is preselected on it.

5.3 Hydraulic Failure Simulation

Figure 20 shows a failure due to an anomalous pressure drop in the odd clutch actuation: it can be caused by an augmented leakage flow from the actuation chamber to the sump, due to a broken hydraulic seal in the clutch actuation piston, or because the pressure regulation valves, the proportional one or the redundant one, are stuck and their spools cannot move properly anymore when actuated with a certain current. All these kinds of failures can be simulated in the model.

Before the failure injection, a gear shift from 2nd to 3rd gear is performed, and the odd clutch is being closed, reaching the engine speed. When the failure is inserted in the system, the pressure level drops and the clutch cannot transmit to the wheels all the torque provided by the engine anymore; consequently, the engine speed cannot follow the odd clutch speed and the engine shaft starts revving up. The TCU recognizes that the actual odd clutch pressure is not following the target pressure; the consequent action is the closure of both redundant and proportional valves which control the odd clutch; the odd gear selection is then excluded and the respective synchronizers are put in idle position, to disconnect the odd primary shaft from the wheels.

From this moment, only the even clutch can be used; not to leave the car without torque on the wheels,

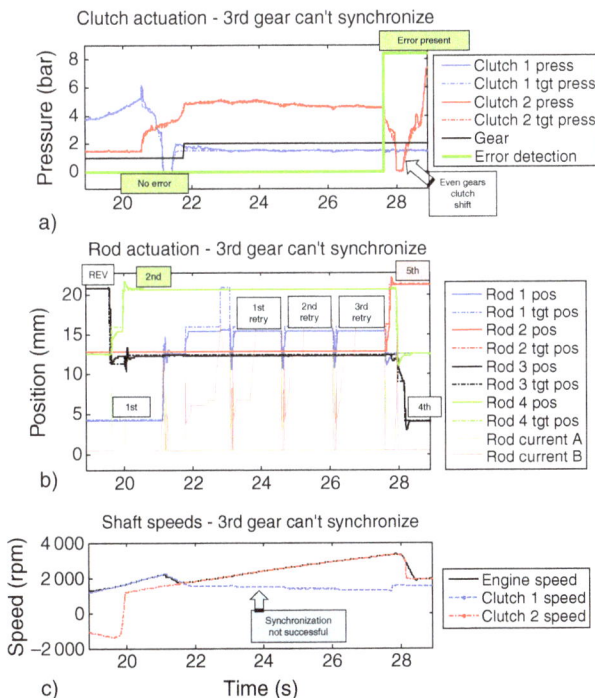

Figure 19

Failure while synchronizing 3rd gear.

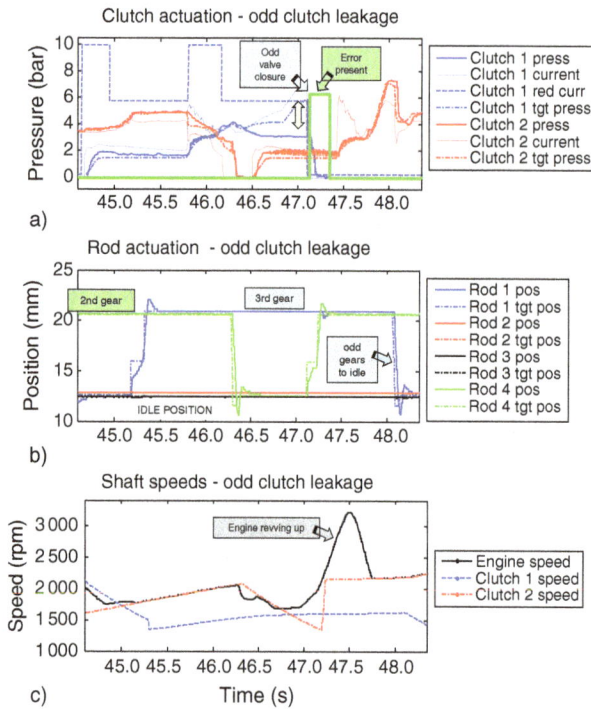

Figure 20

Failure in the odd clutch hydraulic actuation.

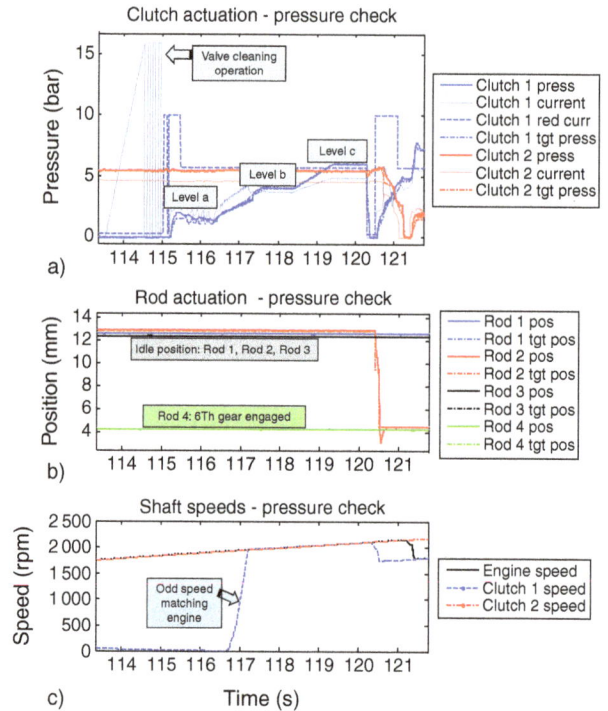

Figure 21

Recovery after failure in the hydraulic actuation of odd clutch.

the even clutch is closed immediately after recognizing the failure, performing an automatic gear shift not requested by the driver. The engine speed slows down, matching the even shaft speed.

The TCU regularly tries to restore the full functionality of the gearbox (*Fig. 21*), checking if the clutch actuation is capable again to give the desired pressure; a procedure of valve cleaning is performed on both the proportional and redundant odd valves, shaking them with an impulsive actuation (*Fig. 21a*, from time 114.5 s to time 115.1 s); then, the TCU checks if the proportional valve can provide the desired pressure level again, setting three different levels of target pressure to follow (A, B, C). During this test both the synchronizers of odd gears are maintained in idle position (*Fig. 21b*); the odd clutch speed joins the engine speed because the clutch is being closed (*Fig. 21c*). If the actual pressure follows the target pressure correctly, the complete functionality of the gearbox is restored.

5.4 Electrical Failure Simulation

The reproduction of electrical failures is a core aim of HIL applications; tests that would be complicated,

time consuming and cost relevant on the real system can be rapidly carried out on the simulated system. Thanks to the Failure Injection Unit (FIU) described in the previous paragraphs, it is possible to simulate electrical faults – short to battery, short to ground, open circuit – on all the actuation valves and on all the sensor signals.

Figure 22 shows a failure imposed on the system pressure actuation valve (*i.e.* the valve that regulates the pressure level in the high pressure circuit): at time 9.9 s the actuation wiring is shorted to ground. Before the failure injection, the system pressure is regulated at 15 bar by the TCU; when the electrical failure occurs, the valve cannot be kept open anymore, because no current can be supplied to it, being connected to the ground. Consequently, the oil cannot flow from the high pressure circuit to the low pressure one anymore, but at the same time the input flow from the pump cannot be stopped because the pump is rigidly connected to the engine. The amount of oil flowing inside the circuit cannot be discharged, and the pressure level rises, reaching a value of around 40 bar, level at which the safety hydraulic valve present inside the circuit opens, permitting to discharge some oil to the sump. In the meanwhile,

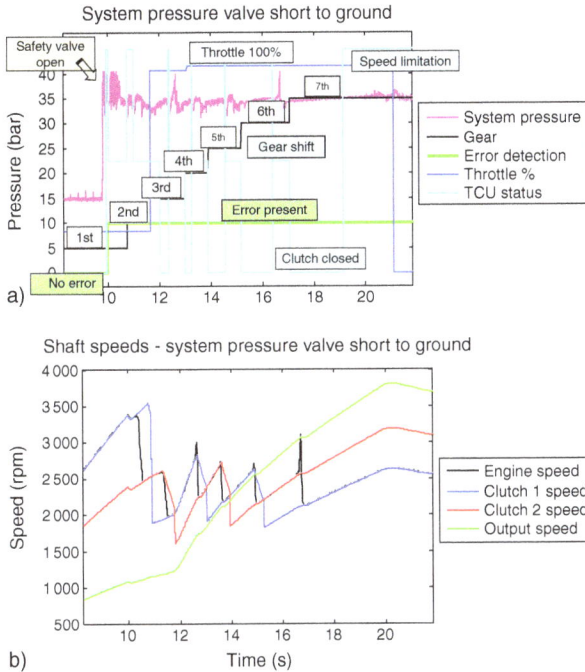

Figure 22

Simulation of electrical failure on system pressure regulation valve.

the TCU recognizes the failure in the electrical circuit and sets its recovery operation: in order to reduce the engine speed, and consequently the flow from the pump, consecutive gear shifts are performed. When the 7th gear is reached, the engine speed is limited by the TCU in order to maintain an acceptable pressure level inside the circuit.

6 ADAPTION PROCEDURES

To perform a precise and comfortable control of the transmission, every TCU must be adapted to the gearbox it is connected to; this is the reason why some "end of line" calibrations of the TCU are needed after coupling a TCU to a specific gearbox. The adaption procedures comprehend rod calibration, clutch preload and kiss point detection, clutch filling procedure, and detection of the solenoid characteristic of proportional clutch pressure regulation valves. These procedures are regularly performed by the TCU during its lifetime: the adaptive values can change from the ones identified with a new gearbox, because of many factors, such as wear, flexion of mechanical parts, augmented leakage in the hydraulic circuit, and change in oil properties. In the

HIL application, the characteristic of the components inside the model can be changed as desired; it is then possible to check the capability of the TCU to correctly adapt to the new settings, updating its calibrations to new values.

6.1 Rod Calibration

Figure 23a shows the adaption procedure on rod number 2. The calibration of a rod consists in moving the synchronizer from its idle position, forcing it to engage both of the gears it controls, and checking the 5 characteristic steady-state positions of the rod: idle position (N), gear engaged with pressure acting on the synchronizer (A2, B2), engaged gear without pressure on the synchronizer (A1, B1), on each of the two gears. It is performed three times per gear and the mean values of the 3 trials are memorized in the TCU as the adapted positions. During the gearbox lifetime, the TCU recognizes a failure in the synchronizer actuation when the position measured from the sensor does not match the expected one registered during the online self-calibration. The simulation results show a good capability of the TCU to adapt to the values set in the model; Figure 23b shows the original values memorized inside the TCU (on the left), and the new values registered by the TCU during the adaption procedure (on the right).

6.2 Detection of Clutch Valve Solenoid Characteristic

To perform an accurate pressure control in the clutch actuation circuit, the TCU needs to know with a high level of accuracy the current–pressure characteristic of the proportional valves that regulate the clutch actuation. For this reason, a default map is not sufficient, and the characteristic must be adapted to the specific hydraulic valve and plate. The solenoid characteristic adaption procedure shown in Figure 24a is performed by the TCU by setting different steady-state target pressure values to be reached inside the clutch actuation chamber, and measuring the input current needed to reach them. This procedure is performed for both clutches, and with all the rods in idle position. A correction map is then calculated to adapt the default map; the input value to the correction map is the default current calculated by the TCU, the output value is the correction to impose on the current itself to perform a precise actuation. The simulation results confirm the capability of the TCU to react to the characteristic being simulated in the model. Figure 24b shows the difference between the previous correction map and the last one, memorized after the adaption procedure.

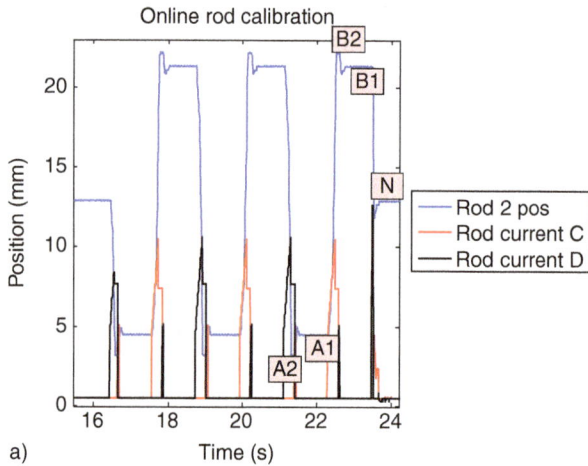

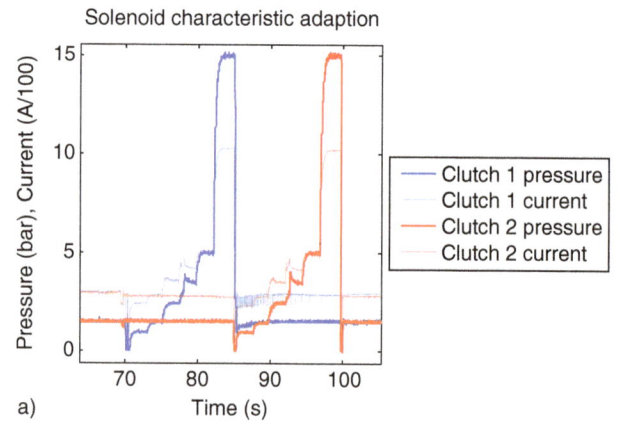

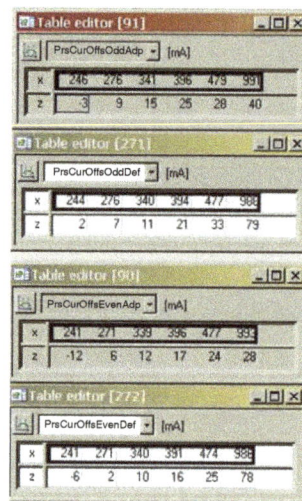

Figure 23

a) Calibration procedure of rod 2; b) comparison between default and adapted values.

Figure 24

a) Adaption of clutch actuation circuit; b) comparison between default values (white background) and adapted values (gray background).

7 AUTOMATION OF TESTS

All the tests on the Hardware In the Loop application have been initially performed manually with the use of the real-time interface. A list of tests has then been defined and every test in the list has been automated. A script developed in Python software environment controls the communication with the simulator, giving the proper commands and inserting the needed failures. During the test, all the relevant signals are automatically recorded in a .dat file, while a .txt log registers the error codes and the environmental conditions thanks to a diagnostic interface. During the post-processing stage, a Matlab code checks automatically the .dat measures one by one, according to a .xls file where all the condi-

tions to be checked had previously been defined. Only if all the conditions are fulfilled, the test is passed. This automation chain allows performing a large number of tests without human interaction. When the pattern of automatic tests is large enough, a complete validation of new software releases can be performed automatically.

CONCLUSIONS

A control oriented model of a Dual Clutch Transmission (DCT) has been developed, by considering the dynamic physical equations of hydraulic circuit, clutches and synchronizers, of all the shafts inside the gearbox (considered with infinite stiffness), and a vehicle model that takes into account the resistant forces acting on the vehicle and inertia effects. The different parts of the model were at first tested and validated separately; a simplification of some dynamics was carried out to adapt the model to real-time applications, in order to reproduce the fast dynamics of the hydraulic circuit while maintaining a sufficiently large simulation step size.

To validate the DCT model, the results of offline simulation of clutch pressure, system pressure, rod motion and shafts speeds have been compared to on-board measurements. An input/output signal scaling sub-model was then added, and the complete model was finally implemented in a HIL simulator, composed of a real-time processor, input/output boards, a TCU plate and a load plate.

Several tests have been performed on the HIL simulator: functional tests, as well as mechanical, hydraulic and electrical failure tests have been executed, analyzing the behavior of the model and the transmission control unit reaction. Adaption procedures were performed to verify the TCU ability to correctly adapt to specific gearbox characteristics (simulated by the HIL application). The results demonstrate the capability of the HIL model to correctly simulate the behavior of the real system, and to respond to the requests coming from the TCU both during functional tests and during failure recovery procedures.

A test automation procedure has been developed to meet the requirement of performing a pattern of tests without the direct interaction of the user. Tests automation allows saving time, reproducing identical tests on different software releases, and designing tests with a tight actions timing – sometimes with just a few milliseconds between them – that would not be possible to be executed manually by the user. Once a pattern of automatic tests can be considered complete, (*i.e.* once it stimulates all the relevant TCU functionalities), a fully automatic non-regression validation of the TCU software is possible in a safe, time-saving and cost-effective simulation environment.

REFERENCES

Bertotti G., Mayergoyz I.D. (2006) *The Science of Hysteresis: Mathematical modeling and applications*, Academic Press, Germany, ISBN: 978-0-080-54078-8.

Cavina N., Olivi D., Corti E., Poggio L., Marcigliano F. (2012) Development of a Dual Clutch Transmission Model for Real-Time Applications, *2012 IFAC Workshop on Engine and Powertrain Control, Simulation and Modeling* **Vol. 3**, Part 1, France, pp. 440-447, ISBN: 978-3-902823-16-8. DOI: 10.3182/20121023-3-FR-4025.00006.

Cavina N., Olivi D., Corti E., Mancini G., Poggio L., Marcigliano F. (2013) Development and implementation of Hardware In the Loop simulation for Dual Clutch Transmission Control Units, *SAE Int. J. Passeng. Cars – Electron. Electr. Syst.* **6** 2, 458–466, DOI: 10.4271/2013-01-0816.

Davis C.L., Sadeghi F., Krousgrill C.M. (2000) A Simplified Approach to Modeling Thermal Effects in Wet Clutch Engagement: Analytical and Experimental Comparison, Congrès STLE/ASME Tribology Conference, Orlando, Fl, 10 Nov. 1999, *J. Tribol.* **122**, 1, 110-118, DOI:10.1115/1.555370.

Deur J., Petric J., Asgari J., Hrovat D. (2005) Modeling of Wet Clutch Engagement Including a Thorough Experimental Validation, *SAE Technical Paper* 2005-01-0877, DOI: 10.4271/2005-01-0877.

Eyabi P., Washington G. (2006) Nonlinear Modeling of an Electromagnetic Valve Actuator, *SAE Technical Paper* 2006-01-0043, DOI: 10.4271/2006-01-0043

Goetz M., Levesley M.C., Crolla D.A. (2005) Dynamics and control of gearshifts on twin-clutch transmissions, *Proc. of the IMECHE, Part D: J. Automobile Engineering* **219**, 951-962.

Greenwood J.A., Williamson J.B.P. (1966) Contact of nominally flat surfaces, *Proceedings of the Royal Society of London Series A, Mathematical and Physical Sciences* **295**, 1442, 300-319, DOI: 10.1098/rspa.1966.0242.

Guzzella L., Onder C.H. (2004) *Introduction to Modeling and Control of Internal Combustion Engine Systems*, Springer, ISBN: 978-3-642-10775-7.

Guzzella L., Sciarretta A. (2008) *Vehicle Propulsion Systems*, Springer, ISBN: 978-3-642-35913-2.

He Y., Lin C. (2007) Development and Validation of a Mean Value Engine Model for Integrated Engine and Control System Simulation, *SAE Technical Paper* 2007-01-1304, DOI: 10.4271/2007-01-1304.

Kulkarni M., Shim T., Zhang Y. (2007) Shift dynamics and control of dual-clutch transmissions, *Mechanism and Machine Theory* **42**, 2, 168-182, DOI: 10.1016/j.mechmachtheory.2006.03.002.

Lai G.Y. (2007) Simulation of heat-transfer characteristics of wet clutch engagement processes, *Numerical Heat Transfer, Part A: Applications* **33**, 6, 583-597, DOI: 10.1080/10407789808913956.

Liu Y., Qin D., Jiang H., Zhang Y. (2009) A Systematic Model for Dynamics and Control of Dual Clutch Transmissions, *Journal of Mechanical Design* **131**, 6, 061012, DOI: 10.1115/1.3125883.

Lucente G., Montanari M., Rossi C. (2007) Modelling of an automated manual transmission system, *Mechatronics* **17**, 2-3, 73-91, DOI: 10.1016/j.mechatronics.2006.11.002.

Matthes B. (2005) Dual Clutch Transmissions – Lessons learned and future potential, *SAE Technical Paper* 2005-01-1021, DOI:10.4271/2005-01-1021.

Merritt H.E. (1967) *Hydraulic Control Systems*, John Wiley and Sons Inc, ISBN: 978-0-471-59617-2.

Montanari M., Ronchi F., Rossi C., Tilli A., Tonielli A. (2004) Control and performance evaluation of a clutch servo system with hydraulic actuation, *Control Engineering Practice* **12**, 11, 1369-1379, DOI: 10.1016/j.conengprac.2003.09.004.

Naunheimer H., Bertsche B., Ryborz J., Novak W. (2011) *Automotive Transmissions: Fundamentals, Selection, Design and Application*, Springer, ISBN 978-3-642-16214-5

Neto D.V., Florencio D.G., Fernandez J., Rodriguez P. (2006) Manual Transmission: Synchronization Main Aspects, *SAE Technical Paper* 2006-01-2519, DOI:10.4271/2006-01-2519.

Razzacki S.T. (2004) Synchronizer Design: A Mathematical and Dimensional Treatise, *SAE Technical Paper* 2004-01-1230, DOI: 10.4271/2004-01-1230.

Velardocchia M., Amisano F., Flora R. (2000) A Linear Thermal Model for an Automotive Clutch, *SAE Technical Paper* 2000-01-0834, DOI:10.4271/2000-01-0834.

Zhang Y., Chen X., Zhang X., Jiang H., Tobler W. (2005) Dynamic Modeling and Simulation of a Dual Clutch Automated Lay-Shaft Transmission, *ASME Journal of Mechanical Design* **127,** 2, 302-307, DOI: 10.1115/1.1829069.

3

Energy Management of Hybrid Electric Vehicles: 15 Years of Development at the Ohio State University

Giorgio Rizzoni[1] and Simona Onori[2]*

[1] Center for Automotive Research and Department of Mechanical and Aerospace Engineering,
The Ohio State University Columbus, OH 43212 - USA
[2] Automotive Engineering Department, Clemson University, Greenville, SC 29607 - USA
e-mail: rizzoni.1@osu.edu - sonori@clemson.edu

* Corresponding author

Abstract —   The aim of this paper is to document 15 years of hybrid electric vehicle energy management research at The Ohio State University Center for Automotive Research (OSU-CAR). Hybrid Electric Vehicle (HEV) technology encompasses many diverse aspects. In this paper, we focus exclusively on the evolution of supervisory control strategies for on-board energy management in HEV. We present a series of control algorithms that have been developed in simulation and implemented in prototype vehicles for charge-sustaining HEV at OSU-CAR. These solutions span from fuzzy-logic control algorithms to more sophisticated model-based optimal control methods. Finally, methods developed for plug-in HEV energy management are also discussed.

Résumé — Gestion énergétique des véhicules hybrides électriques : 15 ans de développement à l'université d'État de l'Ohio — Le but de cet article est de documenter 15 ans de recherche sur la gestion énergétique des véhicules hybrides électriques, effectuée au centre de recherche automobile de l'Ohio State University (OSU-CAR). La technologie VHE (Véhicules Hybrides Électriques) englobe divers aspects. Dans cet article, nous nous concentrons exclusivement sur l'évolution des stratégies de contrôle de surveillance pour la gestion énergétique embarquée dans les VHE. Nous présentons une série d'algorithmes de contrôle qui, à l'OSU-CAR, ont été développés en simulation et mis en œuvre dans des véhicules prototypes pour les VHE avec maintien de la charge. Ces solutions couvrent tant les algorithmes de contrôle par logique floue que les méthodes sophistiquées de contrôle optimal basé sur un modèle. Enfin, les méthodes développées pour la gestion énergétique des VHR (Véhicules Hybrides Rechargeables) sont également abordées.

INTRODUCTION

The aim of this paper is to document 15 years of Hybrid Electric Vehicle (HEV) energy management research at the Ohio State University Center for Automotive Research (OSU-CAR). The activities described in this review paper began in the second half of the 1990s, and have taken place in parallel with the commercial introduction of hybrid vehicles, dating back with the first offering of the *Toyota* Prius in Japan in 1998, and of the *Honda* Insight in the USA in 1999. In this paper, we focus exclusively on the evolution of energy management strategies for HEV, and not on the underlying hardware or architectures. It should be noted that the evolution of technology has brought forth new possibilities (for example, Plug-in Hybrid Electric Vehicles (PHEV)), and that these changes have provided new ideas and motivation. Many other researchers, in industrial and academic research organizations, have addressed similar and diverse aspects of HEV research. While today many institutions worldwide conduct research on HEV, at the beginning six main research groups have been extremely active in the development of energy strategies for HEV. The most significant contributions in this area are from researchers at: OSU-CAR (this paper deals with 15 years worth of research results produced by researchers at this institution); ETH Zurich (Sciarretta *et al.*, 2003, 2004; Rodatz *et al.*, 2005; Sundström *et al.*, 2008, 2010; Sundström and Guzzella, 2009; Ambuhl and Guzzella, 2009), University of l'Aquila, (Anatone *et al.*, 2005; Cipollone and Sciarretta, 2006); IFP Energies nouvelles (IFPEN) in France (Chasse *et al.*, 2009); IFPEN and ETH (Guzzella and Sciarretta, 2007); University of Michigan Ann Arbor (Lin *et al.*, 2001, 2003, 2006; Wu *et al.*, 2004; Tate *et al.*, 2007; Liu and Peng, 2008); Eindhoven University of Technology (Kessels *et al.*, 2006, 2008; Kessels, 2007).

This paper is not intended to be a comprehensive review of all of the developments in this field. We hope that the bibliographical references provide a reasonably broad overview of what (many) others have done.

Historical Notes

In 1993, eight agencies of the US government formed a partnership with the three major North-American automotive OEM to advance vehicle technology, with the goal of producing highly fuel-efficient vehicles. The Partnership for a New Generation of Vehicles, PNGV, involved *DaimlerChrysler*, *Ford*, and *General Motors*, through the United States Council for Automotive Research (USCAR); its most widely publicized (but not only) goal was to put in production vehicles capable of achieving 80 miles per gallon (approximately 3 liters per 100 km) by 2003. The program ended in 2001, due to the transition between the Clinton-Gore and the Bush administrations, with the automakers having demonstrated (but not launched production of) the *GM* Precept, the *Ford* Prodigy and the *Chrysler* ESX. All of these vehicles were characterized by the use of lightweight materials, hybrid powertrains, and other technological innovations. PNGV provided the opportunity for a number of US universities to collaborate with USCAR and with federal agencies towards the development of fuel-efficient vehicles.

HEV Research and Development at Ohio State

The Ohio State University was engaged in programs focused on the development of vehicle prototypes and on the development of energy management strategies and algorithms, as early as 1996. In particular, during the PNGV years the US Department of Energy collaborated with USCAR in creating a series of competitions that were part of the Advanced Vehicle Technology Competitions (AVTC) program and which focused on the development of high-fuel-economy vehicle prototypes that were in practice almost invariably hybrids. Through these competitions, which have continued without interruption since 1996, OSU faculty and students have developed 7 hybrid vehicle prototypes based on mid-size sedans (FutureCar 1996-97 and 1998-99, and EcoCAR 2 2012-14), full-size SUV (FutureTruck 2000-01 and 2002-04), and crossover SUV (ChallengeX 2005-08 and EcoCAR 2009-11). Further, the OSU-CAR has been continuously engaged in research programs related to hybrid vehicle development with a number of industry and government research sponsors, and focusing on military, commercial and passenger vehicles. Supervisory energy management of the hybrid powertrain is a critical element in each of these projects.

1 BACKGROUND

1.1 Hybrid Vehicles

Hybrid electric vehicles encompass two (or more) energy storage sources and associated energy converters. Most typically, the architecture of these vehicles includes an internal combustion engine with an associated fuel tank and one or more electric machine(s), requiring a battery system to store electrical energy. HEV are generally classified according to their powertrain architectures. A series hybrid employs a large electric motor to propel the vehicle while using the Internal Combustible Engine (ICE) and a second electric machine to generate electricity for the battery. A parallel hybrid can combine power from the ICE and the electric motor(s) to deliver mechanical power to the road or to recharge the battery as necessary. A third configuration is the power-split or multi-mode hybrid with the properties of both a series and parallel hybrids. Today, this last configuration is the most commonly employed in commercially produced hybrids. A further distinction among HEV is their electrical power system autonomy: an HEV is considered charge sustaining if the electric energy storage system is recharged only by power supplied by the ICE or by regenerative braking. If, on the other hand, the vehicle is designed to deplete stored energy in the battery during the course of a trip, ending the trip with a lower state of charge than at the start and requiring re-charging, the vehicle is called charge depleting, or plug-in hybrid (PHEV).

1.2 The Energy Management Problem

Control strategies for hybrid electric vehicles are aimed at meeting several simultaneous objectives. The primary one is usually the minimization of the vehicle fuel consumption, but minimizing engine emissions and maintaining or enhancing driveability are also important objectives. Regardless of the topology of these components, the essence of the HEV control problem is the instantaneous management of the power flows from two or more energy converters to achieve the overall control objectives. One important characteristic of this general problem is that the control objectives are mostly integral in nature (fuel consumption and emission per mile of travel), or semi-local in time, such as driveability, while the control actions are local in time. Furthermore, the control objectives are often subject to integral constraints, such as maintaining the battery State of Charge (SOC) within a prescribed range in charge-sustaining hybrids. The global nature of both the objectives and the constraints does not lend itself to traditional global optimization technique, as the future is unknown in actual driving circumstances. Much can be learned from global optimization exercises over a priori known driving cycles. However, these solutions do not directly lend themselves to practical implementations. In this paper, we review the evolution of practical and theoretical HEV optimal energy management strategies that optimize the power split between energy converters, while accounting for global constraints. The optimal energy management problem in a hybrid electric vehicle consists in finding the

sequence of controls $u(t)$ that leads to the minimization of a performance index J, defined as:

$$J(x(t), u(t)) = \int_{t_0}^{t_f} L(x(t), u(t), t)dt \tag{1}$$

where t represents the time, $u(t)$ is the control action, $x(t)$ is the state variable, $[t_0, t_f]$ is the optimization horizon, $L(\cdot)$ is the instantaneous cost function. If all fast dynamics in the powertrain are neglected, as well as the thermal phenomena, the vehicle can be described as a system in which the battery state of charge, SOC$=x$, is the only state variable and whose dynamics are given by:

$$\dot{x}(t) = f(x, u, t) = -\frac{1}{Q_{batt}} I_{batt}(x, u, t) \tag{2}$$

where I_{batt} is the battery current (positive in discharging and negative in charging) and Q_{batt} the battery charge capacity.

The control variable $u(t)$ is the vector of control variables. The number of control variables in the energy management problem depends on the number of degree of freedom in the powertrain architecture. If the powertrain only has one degree of freedom, then $u(t) = P_{batt}(t)$.

If fuel consumption minimization is the only optimization objective, the instantaneous cost is the fuel flow rate, or the power equivalent to it:

$$L(u, t) = P_{fuel}(u, t) = Q_{lhv}\dot{m}_f(u, t) \tag{3}$$

where Q_{lhv} is the lower heating value of the fuel and $\dot{m}_f$ the fuel flow rate. The energy management problem can be cast into a constrained optimization problem where the objective is to minimize (1) with L given as in (3) subject to dynamic constraints (2) with the inclusion of the following additional constraints:

State and input instantaneous constraints:

$$u_{min}(t) \leq u(t) \leq u_{max}(t) \ \forall t \in [0, t_f] \tag{4}$$

$$x_{min} \leq x(t) \leq x_{max} \ \forall t \in [0, t_f] \tag{5}$$

Global constraints on the state of charge:

$$x(t_0) = x_{ref}, \ x(t_f) = x_{ref} \tag{6}$$

This latter constraint could be expressed in other ways, for example by forcing $x(t)$ to stay within prescribed bounds. The formulation above has some advantages in formally defining the problem, as it will become clear in a later section.

The optimal control law is denoted as $u^*(t)$ and the corresponding state trajectory as $x^*(t)$.

2 ENERGY MANAGEMENT STRATEGIES

2.1 Rule-Based Approaches

The first energy management formulation we considered was motivated by the FutureCar 1996 competition, and is documented in Baumann et al. (2000). In this paper, the authors present a rule-based control method for energy management of hybrid electric vehicles that forces vehicle to act at or near either peak point of efficiency or its lowest fuel use (Brake-Specific Fuel Consumption, BSFC) at all times. It turns out that the second option provides better fuel economy. The paper introduces the concept of Degree Of Hybridization (DOH), which provides a quantitative measure of the relative importance of electrical vs mechanical power in the hybrid powertrain. The DOH is defined by:

$$DOH = 1 - \frac{|P_{max,EM} - P_{max,ICE}|}{P_{max,EM} + P_{max,ICE}} \qquad (7)$$

Graphically, Equation (7) is represented in Figure 1. The specific application described in the paper consists of an (ICE)-dominant hybrid vehicle with a DOH of 0.465.

In this study, an empirical methodology described as load-leveling is chosen to be the vehicle operation strategy. The idea behind "load-leveling an ICE dominated hybrid" is to move the actual ICE operating points as close as possible to some predetermined value for every instant in time during the vehicle operation. For instance, if best fuel economy is sought, then vehicle operating points will be forced to take place at the lowest BSFC points that are compatible with the driver power request, and the resulting power difference will be provided by the electric machine. The possible power contribution of the electric machine to the overall drivetrain is limited by the state of charge of the battery pack, and by its torque and power limitations. Maintaining the battery SOC within a prescribed range is a second objective of the control policy, which results in the engine operating at higher power when appropriate to re-charge the battery and maintain the desired SOC range.

In order to implement the operation strategy, the authors propose Fuzzy Logic Control (FLC) as a means of implementing a set of heuristic rules in a systematic manner. Three steps are used to design the FLC:
- fuzzification, in which rules are expressed in the form of fuzzy logic statements;
- inference process;
- defuzzification, to arrive at a crisp output value.

The controller inputs are the driver torque request and the state of charge, and the outputs are the power split between ICE and the electric machine. The formal process generates a total of 847 rules that are implemented in the FLC. The controller so designed has a sufficiently general form that it can work with multiple control strategies. For example, the authors illustrate its use in a "peak efficiency" strategy and in a "minimum BSFC" strategy. The latter results in superior results.

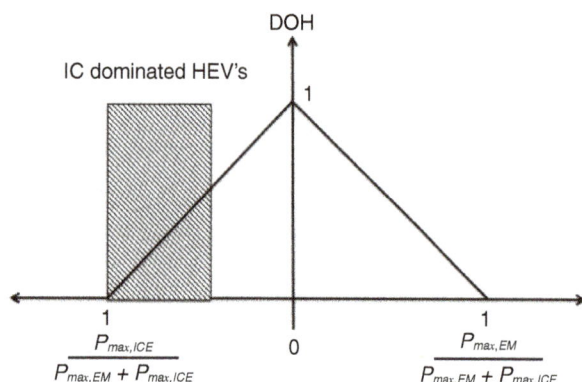

Figure 1

Degree of hybridization for HEV (Baumann et al., 2000).

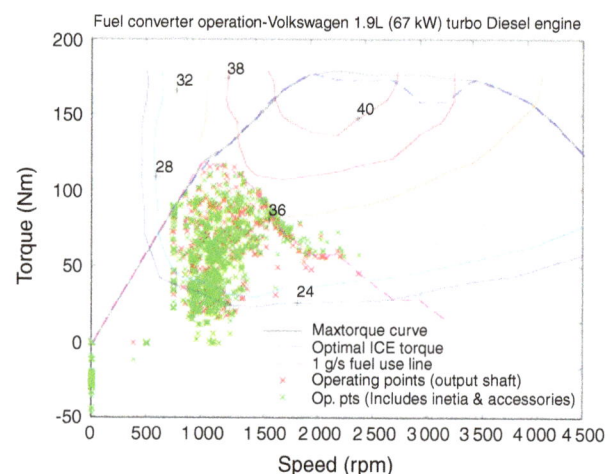

Figure 2

Efficiency map for HEV using the fuel use strategy (Baumann et al., 2000).

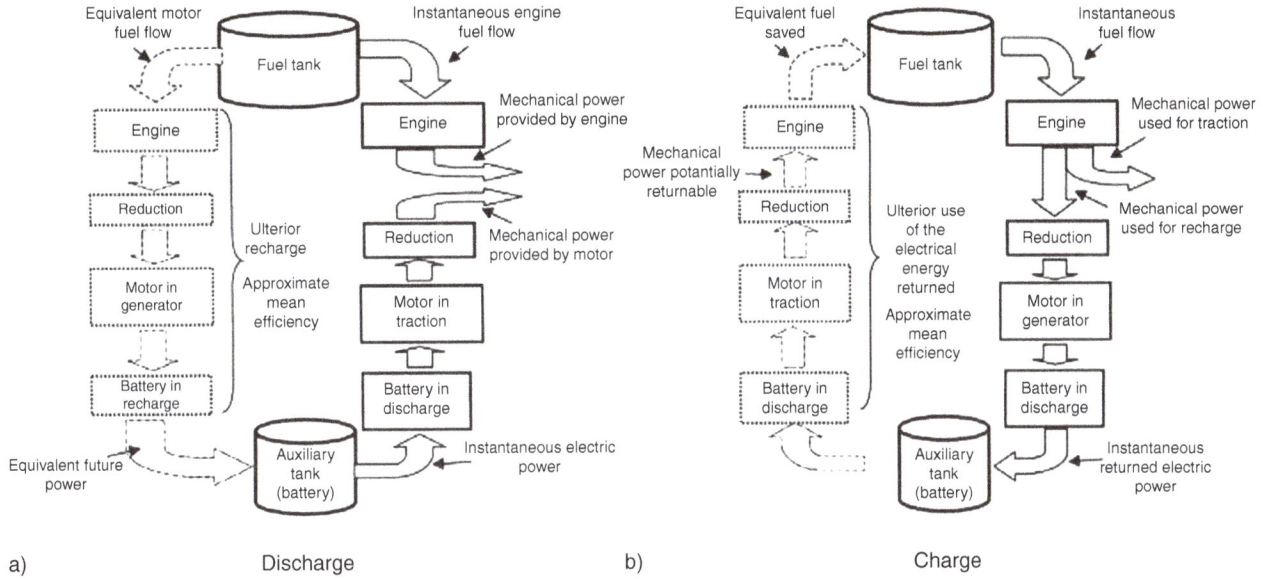

Figure 3

Energy path during charge and discharge in a parallel HEV Paganelli et al. (2000).

As one can see in Figure 2, the minimum BSFC control scheme forces the majority of operating points to be in the vicinity of the points of the best fuel economy. This early approach to the HEV energy management problem showed some initial success, and was implemented in simulation and in a prototype vehicle. The calibration complexity (847 rules!) and the heuristic nature of the strategy, with no explicit consideration of optimality, left much to be desired. In 1999, Gino Paganelli joined our group as a post-doctoral scholar, having completed his doctorate at Université de Valenciennes in France. His arrival revolutionized our thinking with regard to the HEV energy management problem.

2.2 Equivalent Consumption Minimization Strategy

The Equivalent Consumption Minimization Strategy (ECMS) was first introduced by Paganelli in 1999 (Paganelli, 1999; Paganelli et al., 2001a,b, 2002) as a method to reduce the global optimization problem to an instantaneous minimization problem to be solved at each instant, without use of information regarding the future. This strategy is based on the concept that, in charge-sustaining HEV, the battery is used only as an energy buffer, and all the energy ultimately comes from fuel (Fig. 3). Thus, the battery can be seen as an auxiliary, reversible fuel tank which is never refilled using energy from outside the vehicle. In order to keep the

vehicle charge-sustaining, the electricity used during the battery discharge phase must be replenished later using the fuel from the engine (either directly or indirectly through a regenerative path). In both charge and discharge phase, a virtual fuel consumption can be associated with the use of electrical energy, and summed to the actual fuel consumption to obtain the instantaneous equivalent fuel consumption:

$$\dot{m}_{eqv}(t) = \dot{m}_f(t) + \dot{m}_{batt}(t) \tag{8}$$

$$= \dot{m}_f(t) + \frac{s}{Q_{lhv}} P_{batt}(t) \cdot (1 - p(x)) \tag{9}$$

where $\dot{m}_f(t)$ is the engine instantaneous fuel consumption, $\dot{m}_{batt}(t)$ is the virtual fuel consumption associated with the use of the battery, $P_{batt}(t)$ the battery power, and $p(x)$ is a correction function that takes into account the deviation of the current SOC, x, from the reference SOC, x_{ref}. The correction term $p(x)$ is shown in Figure 4. The factor s is called equivalence factor and is used to convert electrical power into equivalent fuel consumption; it plays an important role in the ECMS, as will be shown later.

Depending on the sign of $P_{batt}(t)$ (i.e., on whether the battery is charged or discharged), the virtual fuel flow rate can be either positive or negative, therefore the equivalent fuel consumption can be higher or lower than the actual fuel consumption.

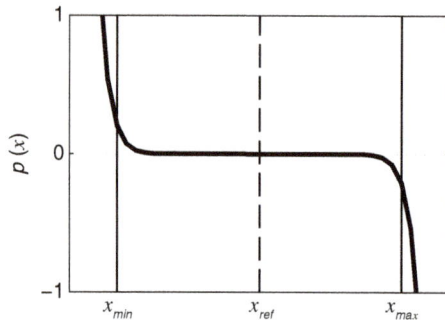

Figure 4

SOC correction term for ECMS, $p(x)$.

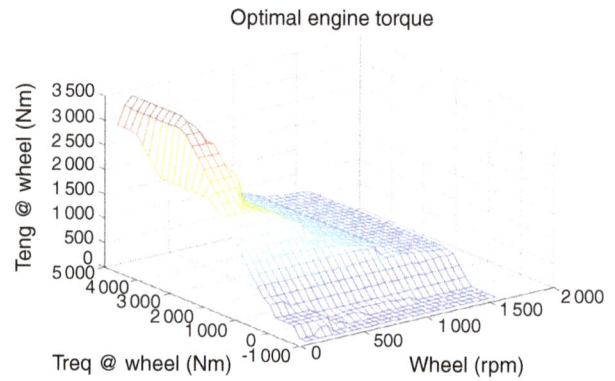

Figure 5

Optimal ICE torque resulting from calibration process (Paganelli *et al.*, 2001c).

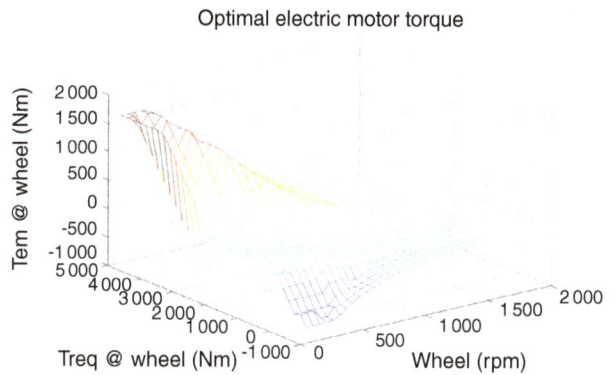

Figure 6

Optimal EM torque resulting from calibration process (Paganelli *et al.*, 2001c).

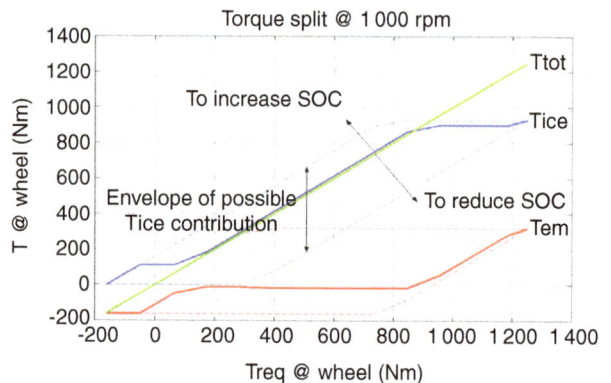

Figure 7

Effect of ICE and EM torques on SOC (Paganelli *et al.*, 2001c).

In the simplest implementation of ECMS, the equivalence factor is a constant, or rather a set of constants that represent the chain of efficiencies through which fuel is transformed into electrical power and vice-versa. In particular, there are at least two equivalence factors, one to apply during battery charge, and another during battery discharge. In each mode, the equivalence factor represents the average overall efficiency of the electric path, during a specific driving cycle. The equivalence factor can be interpreted intuitively as the future cost of the fuel that will be required to replenish the battery charge used at the present time. Thus, its value depends on the driving cycle.

The values of the equivalence factors affect the vehicle fuel consumption and the trend of the battery state of charge, which tends to be discharged if the equivalence factor is too low (charge-depleting behavior) or to be charged if it is too high (charge-increasing behavior). In order to obtain a charge-sustaining solution and minimize the total fuel consumption during a driving cycle, it is necessary to tune all the equivalence factors for the specific driving cycle. For example, it is possible to define a charge and a discharge equivalence factors (s_{chg} and s_{dis}), corresponding respectively to negative and positive values of battery power P_{batt}.

In a practical implementation of ECMS, the calibration process consists of pre-computing the optimal value of the instantaneous (equivalent) fuel consumption of each energy converter (ICE and Electric Machine, EM), so that, given a specific torque request the combination of ICE and EM torques selected is the one that results in the minimum instantaneous fuel consumption. Figures 5 and 6 show such pre-computed maps for the case of the FutureTruck 2000 hybrid-electric SUV (Paganelli *et al.*, 2001c). Figure 7 shows the possible ranges of ICE and EM torque contributions to the total requested torque at one particular engine speed, illustrating how the contributions would affect the state of charge of the battery. Figure 8 shows experimental

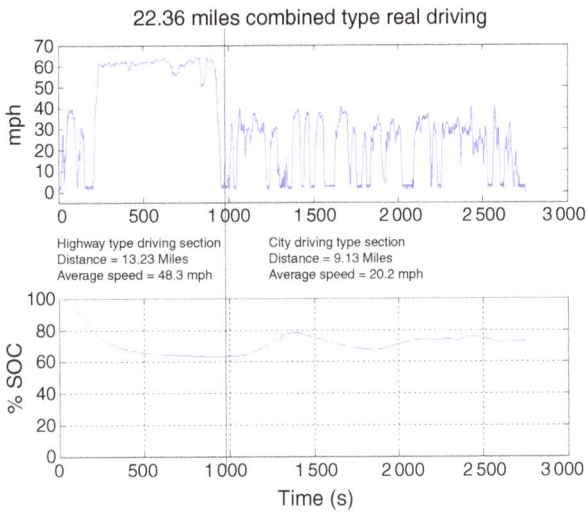

Figure 8

Actual driving profile and measured battery SOC from on-vehicle implementation of ECMS (Paganelli et al.. 2001c).

results for a hybrid SUV (*GM* Suburban), the first vehicle that saw the implementation of the ECMS strategy at the Ohio State University, in the summer of 2000.

2.3 Dynamic Programming and Optimal Control Methods

While ECMS provided a viable and practical solution to the energy management problem, the desire for a more formal optimization approach started surfacing. In Brahma et al. (2000) it was first proposed that the optimization problem of the instantaneous mechanical/electrical power split in parallel hybrid electric vehicles could be solved using Dynamic Programming (DP). ECMS is based on the premise that a local optimal solution can approach global optimality, which is addressed by means of DP. The price, though, one pays for the global optimum is that physical realizability is not possible, as it depends on knowing the precise vehicle driving cycle a priori. In Brahma et al. (2000), the authors proposed an approach that applies DP algorithm to the the optimization process for the power split between both sources of energy, with realistic cost calculation for all considered paths for the IC engine, EM machines and battery efficiencies, and a penalty function formulation for the deviation of the ideal SOC to be sustained over the length of time considered. In this work, charge sustaining was a continuous modulation of the battery SOC within certain operational bounds. The overall integral charge sustaining constraint causes the instantaneous

power flows in the system to become sequentially coupled to each other. Thus, power flow values at one instance of operation have an effect on the allowable power flow values at a later time. This transforms the nature of the optimization problem by rendering it non-local in time which implies that a method such as dynamic programming is extremely appropriate for the solution of a problem of this nature. This approach was formalized by others, as explained in a later section, and has become a standard method for establishing the benchmark performance of energy management algorithms.

Formal approaches based on optimal control were also proposed in Wei (2004) and in Wei et al. (2007), which included applying Pontryagin's Minimum Principle (PMP) and Variable Structure Control (VSC). This work was expanded in Serrao et al. (2009), Serrao (2009), and is explained in Section 2.5.

2.4 Adaptive ECMS Methods – Early Results

As mentioned earlier, ECMS can generate the optimal energy management solution for a given cycle, provided that the strategy is properly tuned by choosing the appropriate value of equivalence factor. The equivalence factor plays a crucial role in the charge sustaining ECMS; it trades off chemical against electric power. If the equivalence factor is very large, then the ECMS tends to recharge the battery in almost all operating points. If the equivalence factor is very small, then the ECMS favors pure electric driving. Since perfect tuning is possible only with *a-priori* knowledge of the cycle, research efforts have been directed towards online adaptation of ECMS, in order to achieve quasi-optimal results even without *a-priori* tuning of the strategy.

Two categories of methods have been initially proposed at CAR to design A-ECMS. They are:
- adaptation based on driving cycle prediction (Musardo et al., 2004a, 2005a, b);
- adaptation based on driving pattern recognition (Gu and Rizzoni, 2006; Gu, 2006).

Adaptation Based on Driving Cycle Prediction

The driving principle behind this class of methods is that when no information on future driving conditions is available, optimal fuel economy cannot be guaranteed. Thus, this family of algorithms aims at using any sort of future information to feed the ECMS control module with the more suitable value of equivalence factor.

Historically, this was the first adaptation approach. In fact, A-ECMS was first proposed in Musardo et al. (2004b) (and Musardo et al., 2004a, 2005a,b), by the same group of authors. In this series of papers, the term A-ECMS was coined and conceived as a real-time energy management strategy obtained adding to the ECMS framework an on-the-fly algorithm for the estimation of the equivalence factor according to the driving conditions. The main idea being a periodical refresh of the control parameter according to the current road load, based on prediction of driving conditions. The identification of the driving mission combined with past and predicted data are used to determine the optimal equivalence factor over the optimization segment, according to the scheme shown in Figure 9. The ECMS module is effectively augmented with a device able to relate the control parameters to the current velocity profile. The reference SOC is kept constant in this A-ECMS prediction scheme. In the same papers, the authors formally use Dynamic Programming as a benchmark to assess the performance of the adaptive ECMS. A comparison of the performance of ECMS and A-ECMS *versus* the DP

solution is summarized in Table 1. In this table, the "optimal ECMS solution" is one in which the optimal equivalence factor has been determined for each driving cycle using prior knowledge, whereas A-ECMS implements a real-time solution.

Much more recently, in Fu *et al.* (2011), the authors use a Model Predictive Control (MPC) based strategy and utilize the information attainable from Intelligent Transportation Systems (ITS) to establish a prediction based real-time controller structure. A constant reference SOC is considered and A-ECMS implemented as in Musardo *et al.* (2004a) is compared with a MPC type controller based on the prediction of future torque demand. The performances of the two controllers are very similar, indicating that A-ECMS with driving mission prediction is somehow equivalent to MPC. What emerges from the paper is also the importance of information provided by ITS and the impact of the accuracy of ITS information on HEV energy consumption.

Adaptation Based on Driving Pattern Recognition

In Gu and Rizzoni (2006), an approach for A-ECMS based on driving pattern recognition is presented. In this research, a driving pattern recognition method is used to obtain better estimation of the equivalence factor in different driving conditions. While the vehicle is running, a time window of past driving conditions is analyzed periodically and recognized as one of the representative driving patterns, according to the scheme of Figure 10.

A finite number of possible driving patterns is recognized, each corresponding to a pre-defined value of the equivalence factor (pre-computed from offline optimization). The battery SOC management is also maintained using a PI controller to keep the SOC around a nominal value (thus using feedback from SOC). Differently from

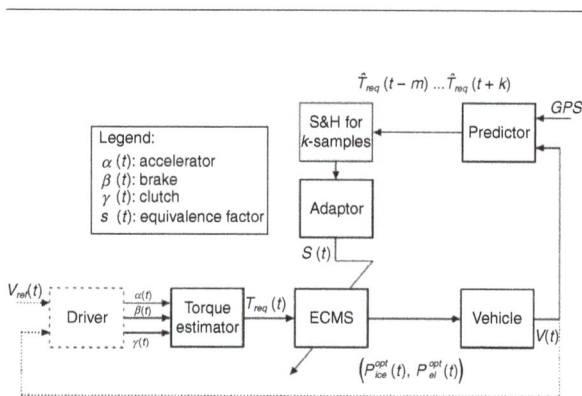

Figure 9

Control diagram of adaptive ECMS with online optimization (Musardo *et al.*, 2004a).

TABLE 1

Comparison of three energy management strategies over various driving cycles (Musardo *et al.*, 2005a)

Driving Cycle	Pure thermal	DP		ECMS optimal		A-ECMS	
	mpg	mpg	Improv.	mpg	Improv.	mpg	Improv.
FUDS	22.1	25.7	16.4%	25.7	16.3%	25.5	15.5%
FHDS	24.8	26.0	4.9%	25.8	4.1%	25.8	3.9%
ECE	20.8	24.5	18.2%	24.5	18.0%	24.5	17.9%
EUDC	23.3	24.8	6.3%	24.7	6.2%	24.7	6.1%
NEDC	22.2	24.5	10.7%	24.5	10.7%	24.4	10.1%
JP1015	21.0	25.2	20.1%	25.1	19.8%	24.8	18.2%

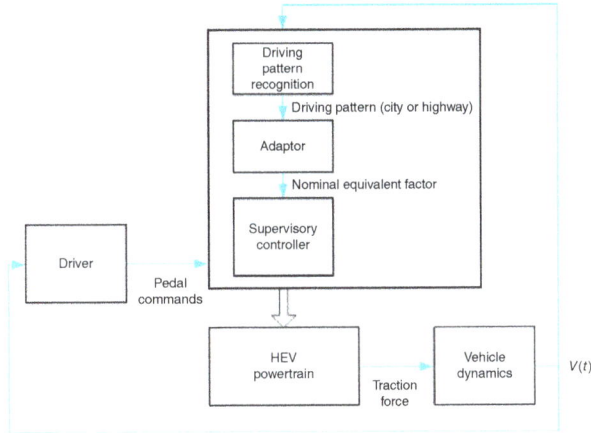

Figure 10

A-ECMS scheme based on pattern recognition (Gu and Rizzoni. 2006).

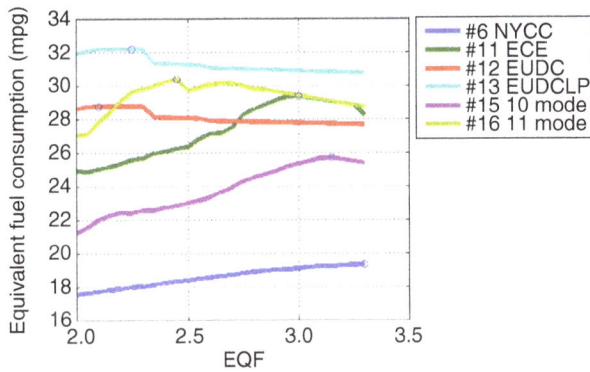

Figure 11

Fuel economy *versus* equivalence factor for different driving cycles (Gu and Rizzoni. 2006).

the proposed algorithm readily distinguishes among three types of driving conditions, representing two variants of urban driving and highway driving.

2.5 Formalizing ECMS: Pontryagin's Minimum Principle

When the HEV energy management problem is cast into an optimal control problem, tools from the optimal control theory, such as analytical optimization methods, can be used. These methods use an analytical problem formulation to find the solution in closed, analytical form. Among these methods are PMP and the Hamilton-Jacobi-Bellmann equation Kirk (2004). It has been initially shown in Sciarretta and Guzzella (2007a) and further developed in Serrao *et al.* (2009) that methods based on PMP are equivalent, in that they generate the same solution, to the ECMS based approach.

The PMP provides necessary conditions for optimality (Geering, 2007). Every solution that satisfies the necessary conditions is called an extremal solution. If the optimal solution exists, then it is also extremal. The opposite, however, is not true: a solution may be extremal without being optimal. However, if the problem has a unique optimal solution, and the application of the minimum principle gives only one extremal solution, then this is the optimal solution.

In practical applications, the minimum principle can be used to find solution candidates by computing and minimizing the Hamiltonian function at each instant, which generates, by construction, extremal controls. If the Hamiltonian is a convex function of the control, then there is only one extremal solution, which is therefore optimal.

In the HEV energy management problem, the Hamiltonian is defined as:

$$
\begin{aligned}
&H(x(t), u(t), \lambda(t), t) = \\
&\dot{m}_f\big(u(t), P_{req}(t)\big) - \lambda(t) \cdot f(x(t), u(t), t)
\end{aligned}
\tag{10}
$$

where $f(x(t), u(t), t)$ is given by Equation (2), and the control $u(t) = P_{batt}(t)$ is obtained at each instant as the value that minimizes Equation (10):

$$
P_{batt}^*(t) = u^*(t) = \arg\min_{P_{batt}} H(x(t), u(t), \lambda(t), t)
\tag{11}
$$

The co-state variable $\lambda(t)$ appearing in Equation (10) is obtained as the solution of:

$$
\dot{\lambda}(t) = -\lambda(t)\frac{\partial f(x(t), u(t), t)}{\partial x}
\tag{12}
$$

the methods seen before, such control algorithm does not require the knowledge of future driving cycles and has a low computational burden but higher memory requirement. Results obtained in this research show that the driving conditions can be successfully recognized and good performance can be achieved in various driving conditions while sustaining battery SOC within desired limits. Figure 11 shows the fuel consumption changes as a function of the equivalent factor (EQF) for different driving cycles. The optimal equivalence factors, which come out with the best fuel economy when ECMS is applied, are marked as a circle. It is obviously that, the optimal equivalence factors are widely spread and they significantly influence the fuel economy. In the paper,

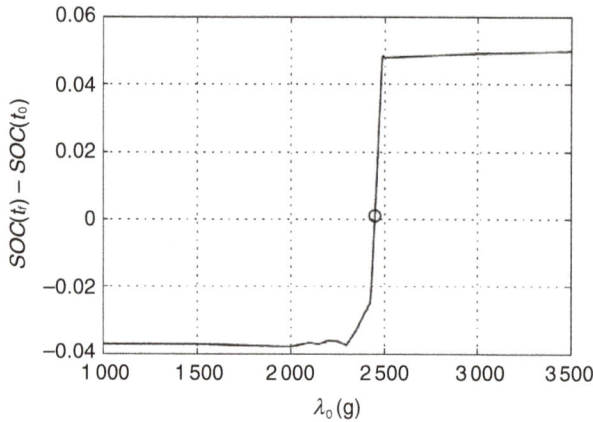

Figure 12

Effect of initial co-state value on SOC variation (Serrao *et al.*, 2011).

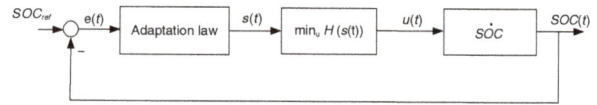

Figure 13

A-ECMS feedback scheme.

Equation (12), together with Equation (2), represents a system of two differential equations with two variables, $x(t)$ and $\lambda(t)$. The solution requires two boundary conditions. These are the initial and final value of the state: $x(t_0) = x_{ref}$ and $x(t_f) = x_{ref}$.

Despite being completely defined, this two-point boundary value problem can be solved numerically only using an iterative procedure, because one of the boundary conditions is defined at the final time. The procedure is known as shooting method and consists in replacing the two-point boundary value problem with a conventional initial-condition problem, starting from an initial guess for $\lambda(t_0)$. The solution of the problem is then obtained by integration in time of Equation (12) and Equation (2), replacing at each time the value of P_{batt} resulting from the minimization of Equation 11. If the final value of the state does not match the desired terminal condition $x^*(t_f) = x_{ref}$, the value of $\lambda(t_0)$ is adjusted iteratively until the terminal condition on the state is met. A bisection procedure can be used to obtain convergence in few iterations, making the minimum principle sensibly faster than DP. The solution is very sensitive to the initial co-state value, as shown in Figure 12.

The existence and uniqueness of the solution cannot be proved formally in the general case, but it is reasonable to assume that at least one optimal solution exists for the energy management problem, in the sense that there must necessarily be at least one sequence of controls giving the lowest possible fuel consumption. If the minimum principle generates only one extremal solution, that can be considered the optimal solution; if there is more than one extremal solution, they are all

compared (*i.e.*, the total cost resulting from the application of each is evaluated) and the one yielding the lowest total cost is chosen.

2.6 Adaptive ECMS: Recent Results

The most recent and interesting approach developed at CAR to design A-ECMS is based on the feedback of the current SOC (Onori *et al.*, 2010; Onori and Serrao, 2011). This method tries to change dynamically the value of the equivalence factor in order to contrast the SOC variation (and thus maintain its value around the reference level). The following discrete time adaptation law was proposed:

$$s(x_k) = \frac{s(x_{k-1}) + s(x_{k-2})}{2} + k_P^d(x_{ref} - x_k) \quad (13)$$

where k is an integer number indicating the k-th fixed time interval of length T seconds, and $s(x_k)$ is the value of the equivalence factor in the interval $[(k-1)T \; kT]$, x_k is the value of SOC at the beginning of said interval. Equation (13) is in the form of Auto-Regressive Moving-Average (ARMA) model, with two autoregressive terms and one moving average term. The key feature of (13) is that the adaptation takes place at regular intervals of duration T.

The correction of the equivalence factor s is achieved with a feedback on the system state, according to the scheme of Figure 13.

2.7 Rule-Based Approaches Based on Optimization Methods

While the DP is an invaluable tool to find the global optimal solution to the energy management problem, it cannot be implemented in a real time setting as it requires complete knowledge of the driving cycle in advance in addition to a high computational load. Nonetheless, the behavior obtained by the DP solution could in principle be mimicked and reproduced by means of a set of rules which are of easier implementation. Thus, inspired by Lin *et al.* (2003), Bianchi *et al.* (2010, 2011) and Biasini *et al.* (2012) we went through a re-thinking of a

Operative mode - cycle: man, WVU, random, APTA, man-WVI-man, 3WVI

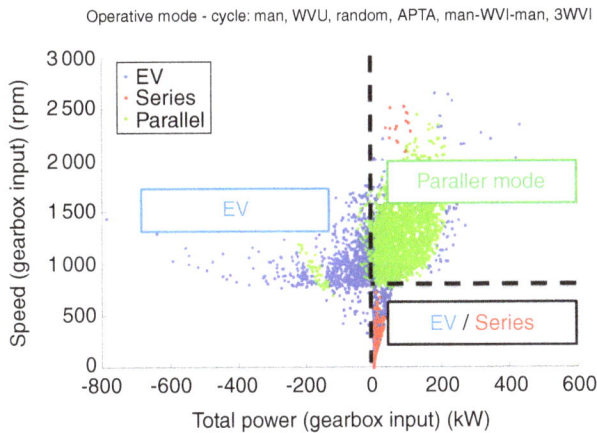

Figure 14

Hybrid powertrain mode of operation selection as a function of gearbox input torque and speed (Bianchi *et al.*, 2011).

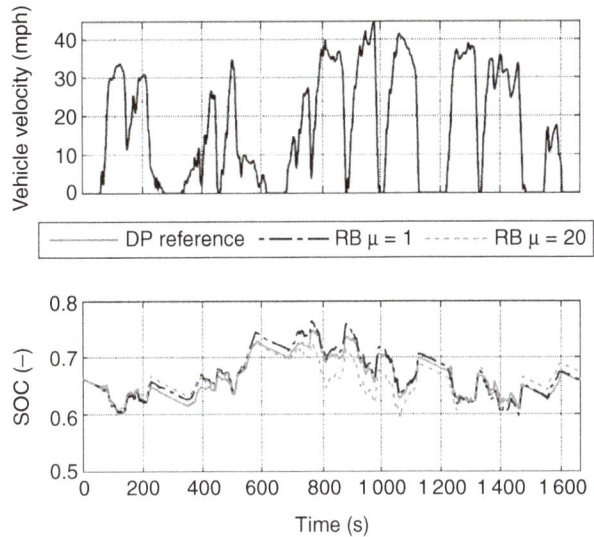

Figure 15

Comparison between RB and DP and effect of calibration parameter on the RB strategy over the WVU-suburban cycle.

rule-based approach in light of optimal control methods. The knowledge acquired from DP simulations over extensive driving conditions is captured and synthesized in the form of if-then-else rules which require very low computational load and relatively low calibration effort. These aspects have made the rule-based strategy the most appealing solution among all the others presented in this paper for product development process. The derivation of the rules from DP is a relatively fast process allowing to obtain results close to the optimal solution, and to reduce the number of calibration parameters. An example of data processing of DP solution is displayed in Figure 14, where, to understand the possible rule-based behavior of the supervisory control, the operating modes chosen by DP over all the analyzed driving cycles were plotted as function of the gearbox input torque and speed (a series-parallel hybrid medium duty truck was considered in this study). The plots shows that:

– at low speed and low torque, the powertrain works either in series or in pure electric (EV) mode;

– in the area limited by engine idle speed and positive gearbox torque only the parallel configuration is selected by the DP. The few points of series operation are only used to limit engine speed to its maximum;

– the third area includes all the points with a negative torque. In this case, the supervisory controller switches the engine off in order to save fuel since vehicle is decelerating.

In Figure 15, the comparison between the DP solution and RB strategy over the West Virginia Urban (WVU) suburban cycle shows how close the SOC profile from

RB is to the DP. The RB is shown for different values of the calibration parameter.

3 PHEV ENERGY MANAGEMENT

The problem of on-board energy management in PHEV offers different challenges from the HEV case. Compared to charge-sustaining HEV, PHEV are characterized by a larger energy storage system, as well as the ability of recharging the battery through direct connection to the energy grid. This provides the opportunity to deplete the battery during driving operations, further improving the vehicle fuel economy. Nevertheless, utilizing the battery with variable initial and final values of SOC and in a broader range poses complex challenges for the definition of a suitable energy management strategy that maximizes the overall energy efficiency, while complying with the limitations of vehicle components. In Tulpule *et al.* (2009a,b, 2010), the energy management problem in PHEV was initially investigated. In particular, two modes of operating a PHEV were considered: 1) EV mode control – the battery energy is used as quickly as possible followed by charge sustaining operation; and 2) Blended Mode control (BM) – the battery is discharged gradually throughout the trip. With respect to these two powertrain modes of operation different strategies (DP and ECMS) were compared and analyzed

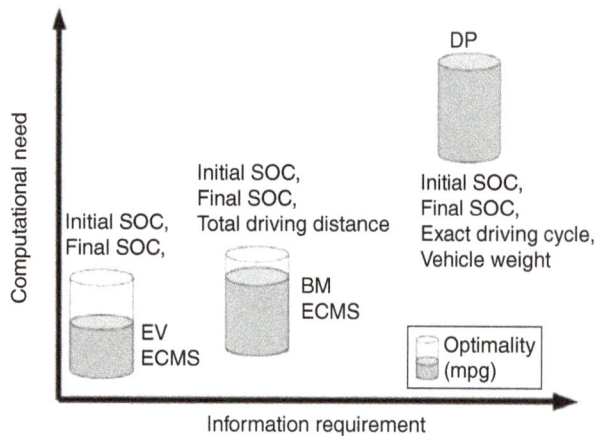

Figure 16

Comparison of the control strategies. DP: Dynamic Programming; BM ECMS: Blended Mode control using ECMS; EV ECMS: EV mode using ECMS and the ECMS and DP are compared in simulation (Tulpule *et al.*, 2009b).

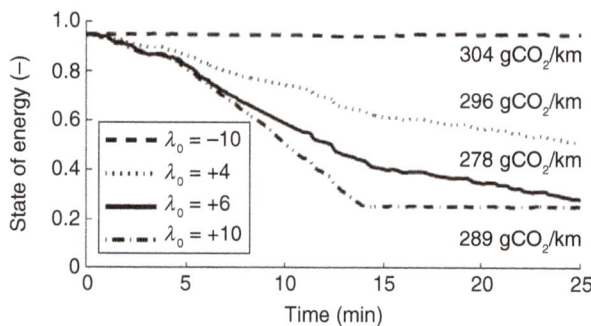

Figure 17

Evolution of the battery SOE as a function of the PMP control parameter.

in terms of amount of information needed to achieve optimality, as shown in the drawing in Figure 16.

In Stockar *et al.* (2011), a model-based control approach for PHEV energy management was presented based on the formulation of PMP implemented on a forward-based vehicle model simulator. In the optimization problem, a cost functional defined based on the cumulative CO_2 produced by the vehicle to account for both the fuel energy consumption and the use of the electrical energy from the grid was considered. In this study, it is shown that a near-optimal fuel economy and CO_2 emissions can be achieved with a minimal calibration effort, without the need of driving duty information. In Figure 17, it is shown how the CO_2 emission varies as a function of the PMP co-state λ_0, which is one of the control calibration parameter in the PMP.

4 STRATEGIES COMPARISON

The energy management strategies reviewed in this paper can be classified into two main groups (Sampathnarayanan, 2013):

- non-causal or non-realizable strategies. They require complete *a priori* knowledge of the driving cycle and are not applicable in real conditions (*e.g.*, DP, PMP);
- causal or realizable strategies. They do not require a priori knowledge of the driving cycle and are developed with the primary objective of realizability and do not guarantee optimality (*e.g.*, Adaptive-PMP, rule-based, ECMS).

Although, the primary objective is to design and implement causal strategies that can be eventually tested on real vehicles, the importance of finding non-causal optimal solutions resides in that:

- they provide a benchmark solution (global optimum) any causal strategy can be compared against;
- properly modified they can be used to develop on-line strategies (Sciarretta and Guzzella, 2007; Serrao *et al.*, 2011).

Although rule-based energy management strategies are relatively easy to develop and implement in a real vehicle, a significant amount of calibration effort is required to guarantee performances within a satisfactory range for any driving cycle. Moreover, rules are not necessarily scalable to different powertrain architectures and different component sizes.

Local optimization methods, such as ECMS and PMP have gained popularity over DP as methods to find the global optimum. These methods are used to find the optimum by performing an offline optimization when the drive cycle is known (using a forward-looking vehicle simulator). Moreover, they can also be employed to design adaptive optimal strategies (AECMS) to achieve near optimal performances when the driving cycle is unknown.

The foundation laid in this paper presents many opportunities to extend this work to include various other objectives in the optimal control cost function. For example, considerations of engine exhaust emissions, of battery aging, and of drivability in the cost function give rise to interesting problems and are the subject of current research. A further extension of this work involves the use of geographical and traffic information and of navigation systems to provide a more accurate prediction of the vehicle trajectory, so as to be able to compute adaptive solutions using model predictive control and other methods. The use of cloud computing as a means of implementing more ambitious optimization algorithms that would otherwise not be implementable in an on-board processor is also a topic of current research. The work presented in this paper offers a solid foundation for the study of these problems.

ACKNOWLEDGMENTS

The authors gratefully acknowledge the support of the United States Department of Energy through its GATE program, The United States Army TARDEC, Cummins, Inc., the OSU CAR Industrial Consortium, the OSU SMART@CAR Consortium, and numerous industry partners in making this research possible.

REFERENCES

Ambuhl D., Guzzella L. (2009) Predictive reference signal generator for hybrid electric vehicles, *IEEE Transactions on Vehicular Technology* **58**, 9, 4730-4740.

Anatone M., Cipollone R., Sciarretta A. (2005) Control-oriented modeling and fuel optimal control of a series hybrid bus, *SAE paper* 2005-01-1163.

Baumann B., Rizzoni G., Washington G. (2000) Mechatronic design and control of hybrid electric vehicles, *IEEE/ASME Transactions on Mechatronics*.

Bianchi D., Rolando L., Serrao L., Onori S., Rizzoni G., Al-Khayat N., Hsieh T., Kang P. (2010) A rule-based strategy for a series/parallel hybrid electric vehicle: an approach based on dynamic programming, *ASME Dynamic Systems and Control Conference*, Cambridge, MA.

Bianchi D., Rolando L., Serrao L., Onori S., Rizzoni G., Al-Khayat N., Hsieh T., Kang P. (2011) Layered control strategies for hybrid electric vehicles based on optimal control, *International Journal of Electric and Hybrid Vehicles* **3**, 191-217.

Biasini R., Onori S., Rizzoni G. (2012) A rule based energy management strategy for hybrid medium duty truck, *International Journal of Powertrains*, forth coming.

Brahma A., Guezennec Y., Rizzoni G. (2000) Dynamic optimization of mechanical/electrical power ow in parallel hybrid electric vehicles, *AVEC 2000, 5th International Symposium on Advanced Vehicle Contro*, Ann Arbor, MI.

Chasse A., Hafidi G., Poignant-Gros P., Sciarretta A. (2009) Online optimal control of a parallel hybrid with costate adaptation, *Proceedings of the 2009 Conference on Decision and Control (CDC09)*.

Cipollone R., Sciarretta A. (2006) Analysis of the Potential Performance of a Combined Hybrid Vehicle with Optimal Supervisory Control, *Proceedings of the 2006 IEEE International Conference on Control Applications*, 2802-2807.

Fu L., Ozguner U., Tulpule P., Marano V. (2011) Real-time energy management and sensitivity study for hybrid electric vehicles, *American Control Conference*.

Geering H.P. (2007) *Optimal Control with Engineering Applications*, Springer, Berlin Heidelberg.

Gu B. (2006) Supervisory Control Strategy Development for a Hybrid Electric Vehicle, *Master Thesis*, The Ohio State University.

Gu B., Rizzoni G. (2006) An adaptive algorithm for hybrid electric vehicle energy management based on driving pattern recognition, *Proceedings of the ASME International Mechanical Engineering Congress and Exposition*.

Guzzella L., Sciarretta A. (2007) *Vehicle Propulsion Systems: Introduction to Modeling and Optimization*, Springer.

Kessels J. (2007) Energy Management for Automotive Power Nets, *PhD Thesis*, Technical University of Eindhoven.

Kessels J., Koot M., van den Bosch P. (2006) Optimal adaptive solution to powersplit problem in vehicles with integrated starter/generator, *Proceedings of the 2006 IEEE Vehicle Power and Propulsion Conference*.

Kessels J., Koot M., van den Bosch P., Kok D. (2008) Online energy management for hybrid electric vehicles, *IEEE Transactions on Vehicular Technology* **57**, 6, 3428-3440.

Kirk D.E. (2004) *Optimal Control Theory: An Introduction*, Dover Publications.

Lin C., Filipi Z., Wang Y., Louca L., Peng H., Assanis D., Stein J. (2001) Integrated, feed-forward hybrid electric vehicle simulation in simulink and its use for power management studies, *SAE Paper* 2001-01-1334.

Lin C., Kim M., Peng H., Grizzle J. (2006) System level model and stochastic optimal control for a pem fuel cell hybrid vehicle, *Journal of Dynamic Systems, Measurement, and Control* **128**, 878.

Lin C., Peng H., Grizzle J., Kang J. (2003) Power management strategy for a parallel hybrid electric truck, *IEEE Transactions on Control Systems Technology* **11**, 6, 839-849.

Liu J., Peng H. (2008) Automated modelling of power-split hybrid vehicles, *Proceedings of the 17th IFAC World Congress*.

Musardo C., Staccia B., Bittanti S., Guezennec Y., Guzzella L., Rizzoni G. (2004a) An adaptive algorithm for hybrid electric vehicles energy management, *FISITA 2004 World Automotive Congress*, Barcelona, Spain.

Musardo C., Staccia B., Bittanti S., Guezennec Y., Guzzella L., Rizzoni G. (2004b) Predictive control for hybrid electric vehicles energy management, *FISITA 2004 World Automotive Congress*, Barcelona, Spain.

Musardo C., Rizzoni G., Guezennec Y., Staccia B. (2005a) A-ECMS: An adaptive algorithm for hybrid electric vehicle energy management, *European Journal of Control* **11**, 4-5, 509-524.

Musardo C., Staccia B., Midlam-Mohler S., Guezennec Y., Rizzoni G. (2005b) Supervisory control for nox reduction of an HEV with a mixed-mode HCCI/CIDI engine, *Proceedings of the 2005 American Control Conference*.

Onori S., Serrao L. (2011) On Adaptive-ECMS strategies for hybrid electric vehicles, *Les Rencontres Scientifiques d'IFP Energies nouvelles - RHEVE 2011*.

Onori S., Serrao L., Rizzoni G. (2010) Adaptive Equivalent Consumption Minimization Strategy for HEVs, *IASME Dynamic Systems and Control Conference*.

Paganelli G., Guerra T., Delprat S., Santin J., Delhom M., Combes E. (2000) Simulation and assessment of power control strategies for a parallel hybrid car, *Proceedings of the Institution of Mechanical Engineers, Part D: Journal of Automobile Engineering* **214**, 7, 705-717.

Paganelli G., Brahma A., Rizzoni G., Guezennec Y.G. (2001a) Control development for a hybrid-electric sport-utility vehicle: Strategy, implementation and field test results, *American Control Conference*.

Paganelli G., Ercole G., Brahma A., Guezennec Y., Rizzoni G. (2001b) General supervisory control policy for the energy optimization of charge-sustaining hybrid electric vehicles, *JSAE* **22**, 511-518.

Paganelli G., Guezennec Y., Rizzoni G. (2002) Optimizing control strategy for hybrid fuel cell vehicle, *SAE Transactions, Journal of Engines*.

Paganelli G., Tateno M., Brahma A., Rizzoni G., Guezennec Y. (2001c) Control development for a hybrid-electric sport-utility vehicle: strategy, implementation and field test results. **2**, 5064-5064.

Paganelli G. (1999) Conception et commande d'une chaîne de traction pour véhicule hybride parallèl ether-mique et électrique, *PhD Thesis*, Université de Valenciennes, Valenciennes, France

Rodatz P., Paganelli G., Sciarretta A., Guzzella L. (2005) Optimal power management of an experimental fuel cell-supercapacitor powered hybrid vehicle, *Control Engineering Practice* **13**, 41-53.

Sampathnarayanan B. (2013) Analysis and Design of Stable and Optimal Energy Management Strategies for Hybrid Electric Vehicles, *PhD Thesis*, The Ohio State University.

Sciarretta A., Guzzella L. (2007) Control of hybrid electric vehicles, *IEEE Control Systems* **27**, 2, 60-70.

Sciarretta A., Guzzella L., Onder C. (2003) On the power split control of parallel hybrid vehicles: from global optimization towards real-time control, *Automatisierungstechnik* **51**, 5, 195-203.

Sciarretta A., Back M., Guzzella L. (2004) Optimal control of parallel hybrid electric vehicles, *IEEE Transactions on Control Systems Technology* **12**, 3, 352-363.

Serrao L. (2009) A comparative analysis of energy management strategies for hybrid electric vehicles, *PhD Thesis*, OSU.

Serrao L., Onori S., Rizzoni G. (2009) ECMS as a Realization of Pontryagin's Minimum Principle for HEV Control, *Proceedings of the 2009 American Control Conference*.

Serrao L., Onori S., Rizzoni G. (2011) A comparative analysis of energy management strategies for hybrid electric vehicles, *ASME Transactions, Journal of Dynamic Systems, Measurement and Control* **133**, 03-12.

Stockar S., Marano V., Canova M., Rizzoni G., Guzzella L. (2011) Energy-optimal control of pluging hybrid electric vehicles for real-world driving cycles, *IEEE Transactions on Vehicular Technology* **60**, 2940-2962.

Sundström O., Ambühl D., Guzzella L. (2010) On implementation of dynamic programming for optimal control problems with final state constraints, *Oil and Gas Science and Technology – Rev IFP* **65**, 91-102.

Sundström O., Guzzella L. (2009) A generic dynamic programming matlab function, *Proceedings of the 18th IEEE International Conference on Control Applications*, 1623-1630.

Sundström O., Guzzella L., Soltic P. (2008) Optimal hybridization in two parallel hybrid electric vehicles using dynamic programming, *Proceedings of the 17th IFAC World Congress*.

Tate E., Grizzle J., Peng H. (2007) SP-SDP for fuel consumption and tailpipe emissions minimization in an EVT hybrid, *IEEE Transactions on Control Systems Technology*.

Tulpule P., Marano V., Rizzoni G. (2009a) Energy management for plug-in hybrid electric vehicles using equivalent consumption minimisation strategy, *International Journal of Electric and Hybrid Vehicles* **2**, 329-350.

Tulpule P., Stockar S., Marano V., Onori S., Rizzoni G. (2009b) Comparative studies of different control strategies for plug-in hybrid electric vehicles, *International 9th Conference on Engines and Vehicles*, Capri, Naples, Italy.

Tulpule P., Marano V., Rizzoni G. (2010) Energy management for plug-in hybrid electric vehicles using equivalent consumption minimisation strategy, *Electric and Hybrid Vehicles Int. J.* **2**, 4, 329-350.

Wei X., Guzzella L., Utkin V., Rizzoni G. (2007) Model-based fuel optimal control of hybrid electric vehicle using variable structure control systems, *Journal of Dynamic Systems, Measurement, and Control* **129**, 13.

Wei X. (2004) Optimal control of hybrid electric vehicles. *PhD Thesis*, The Ohio State University.

Wu B., Lin C., Filipi Z., Peng H., Assanis D. (2004) Optimal power management for a hydraulic hybrid delivery truck, *Vehicle System Dynamics* **42**, 1, 23-40.

4

Evaluation of Long Term Behaviour of Polymers for Offshore Oil and Gas Applications

P.-Y. Le Gac, P. Davies* and D. Choqueuse

IFREMER Centre de Bretagne, Marine Structures Laboratory, 29280 Plouzané - France
e-mail: peter.davies@ifremer.fr

* Corresponding author

Abstract — *Polymers and composites are very attractive for underwater applications, but it is essential to evaluate their long term behaviour in sea water if structural integrity of offshore structures is to be guaranteed. Accelerated test procedures are frequently required, and this paper will present three examples showing how the durability of polymers, in the form of fibres, matrix resins in fibre reinforced composites for structural elements, and thermal insulation coatings of flow-lines, have been evaluated for offshore use. The influence of the ageing medium, temperature, and hydrostatic pressure will be discussed first, then an example of the application of ageing test results to predict long term behavior of the thermal insulation coating of a flowline will be presented.*

Résumé — **Durabilité des polymères pour application pétrolière offshore** — Les polymères et composites sont largement utilisés pour les applications en mer et particulièrement dans le domaine de l'industrie pétrolière. La fiabilisation des structures utilisées pour ces applications nécessite la prise en considération du comportement à long terme des matériaux. L'évaluation de la durabilité des polymères et composites en milieu marin exige la mise en place de vieillissements accélérés spécifiques. La nature de ces vieillissements ainsi que les paramètres influant tels que la nature de l'eau, la température ou la pression hydrostatique seront discutés dans cet article à travers trois exemples. Le premier sera dédié au comportement à long terme de fibres polymériques utilisées principalement pour l'amarrage, le second traitera de la durabilité des composites. Enfin le dernier traitera de la prédiction du comportement à long terme de polyuréthane utilisé comme isolant thermique sur des pipes.

INTRODUCTION

Offshore exploration and production are pushing deeper and deeper, with ultra-deep water applications targeting beyond 4 000 metres depth, and for production applications a service lifetime of 20 years or more is often required. Despite their attractive specific properties the deep sea use of polymers and composites offshore is still quite limited, so there is little in-service experience to validate design predictions. It is generally necessary therefore to employ accelerated tests, in order to check the long term durability of these materials. Ageing can be accelerated by increasing the severity of environmental conditions (temperature, humidity) or mechanical loading (applied stress, hydrostatic pressure) but in all cases it is essential to ensure that the acceleration does not introduce degradation mechanisms which will not be encountered under service conditions. This is not an easy task, and this paper illustrates this by providing some examples in which accelerated tests have been performed on polymers intended for a range of offshore applications.

There is a vast literature on the hygro-thermal ageing of polymers and the aim here is not to provide a review. The reader is referred to books by Verdu [1] and Weitsman [2], on polymers, and Martin [3] in the area of polymer composites.

Researchers at ENSAM [4-5] have been particularly active in this area and detailed phenomenological studies have been published on a number of polymers such as PA11, of direct interest to the offshore oil & gas industry for flow-line liners. Their approach involves acquiring a detailed understanding of the chemical degradation reactions and their kinetics, so that a predictive model can be set up. This allows subsequent changes in formulation or test conditions to be analysed without having to restart the ageing tests and can provide a very powerful tool for durability analysis. The only disadvantage of this approach is that it is very time-consuming so that for many applications a more pragmatic approach is required.

In this paper, three case studies will be described. First, the influence of ageing medium and temperature will be discussed, using the ageing of polymer fibres as an example; synthetic fibres are finding increasing offshore use, in the form of mooring and handling ropes and in umbilical line reinforcements, but there is still relatively little experience to allow lifetime predictions to be made with confidence. Examples from studies on polyamide and aramid fibres will be discussed. Then the influence of mechanical loading will be shown, for an epoxy resin and its glass fibre reinforced composite, aged under different hydrostatic pressures. Finally, a more detailed case study will be presented, concerning the prediction of the long term behaviour of a polyurethane flow-line coating under a combination of ageing conditions.

1 INFLUENCE OF AGEING MEDIUM AND TEMPERATURE

The first decision to be made when defining an accelerated test procedure is the choice of the ageing medium. Ideally natural sea water should be used, but this is rarely available so tap water, distilled water or artificial sea water are employed. These media do not have the same chemical activities so the choice may affect the results. Once the ageing medium has been defined the second parameter to be selected is the range of temperature to be applied, as this is generally the main accelerating factor available to the experimenter. Upper ageing temperature is often limited by polymer transitions so these must be clearly identified before embarking on a long testing programme.

In order to discuss these parameters a form of polymeric material will be studied which has increasing applications offshore; polymer fibres. Various types of fibre are employed today, from the hundreds of tons of nylon and polyester fibre used for mooring and station-keeping [6-7] to the large high performance fibre ropes used for deep sea installation [8-9]. For fibres such as aramids high ageing temperatures can be used, as these fibres are very stable at high temperature, but when we consider melt spun fibres such as nylon and polyester, which show poorer high temperature behaviour than the aramid fibres, the choice of ageing temperature becomes more complicated. For example, the nylon 6 fibres used in Single Point Moorings (SPM) show a dry glass transition temperature around 60°C. Water will tend to reduce this value further, so a maximum ageing temperature of 40°C has been applied.

One direct approach to studying the influence of water for this application is to immerse sections of large ropes, but these may take many months to reach an equilibrium state. Measuring the diffusion kinetics of water in fibres is more accessible, and can be very useful if modeling is to be performed, but is not straightforward, as water absorbed at the surface makes it difficult to obtain accurate weight changes. In order to get round this DVS (Dynamic Vapour Sorption), involving continuous accurate (µg) weight change measurements on a small length of fibre exposed to different relative humidity and temperature conditions appears attractive. Figure 1 shows an example of results for a nylon (PA6) sample. The sample was first dried *in-situ* at 40°C, until constant weight was achieved, (about 1% water by weight was removed), then temperature was maintained at 40°C and relative humidity increased in 20% steps up to 80%. At 80% RH the weight gain was around 4% by weight. The humidity was then reduced and it can be seen that weight gain was completely reversible.

Such measurements can provide the first data needed for modeling, and also help to indicate the amount of water to expect when larger structures are being aged. The aim of studies on fibres is generally to establish whether long term exposure affects fibre mechanical properties. For this PA6 fibre, used in single point mooring lines, lengths of yarn were fully immersed in natural sea water at the Ifremer Brest laboratory at 40°C, and

samples were removed periodically and tested on a tensile test machine. Stiffness values were measured first, by cycling around different mean load levels, then the samples were tested to failure. Figure 2 shows an example of results.

Even after 10.5 months in water at 40°C, the yarn stiffness and strength values are unchanged. This is an interesting result, as it is generally accepted that wet strength of nylon ropes is about 10% lower than that of dry ropes. One might conclude that water affects the interactions between the elements (yarns, rope yarns, strands) in ropes, affecting their relative movements, more than the properties of the elements themselves. It is also interesting to compare this result to the effect of water on bulk PA6 properties. For the latter, a significant reversible effect of water is noted, revealed as a drop in Young's modulus after saturation as water plasticizes amorphous regions [10]. The lower sensitivity of fibre properties to water is the result of their more highly oriented microstructure, as detailed by Richards [11].

Another case where the long term behaviour of synthetic fibres is particularly critical is in the many umbilical control lines employed underwater. Here aramids are often used as the strength member, and these must resist both mechanical loads and the marine environment. A recent study focussed on the ageing of these materials and tests on single fibres were performed after different ageing periods under different conditions. Full details of testing can be found in [12-15]. Figure 3 shows the influence of ageing test temperature on tensile properties of *Twaron* 1 000 aramid yarns.

Tests in different environments also revealed the influence of water activity and pH, Table 1 [13].

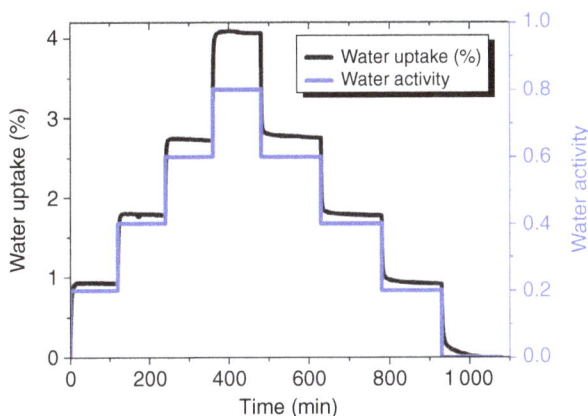

Figure 1

DVS example PA6 fibres, showing weight gain at increasing then decreasing humidity levels, 40°C.

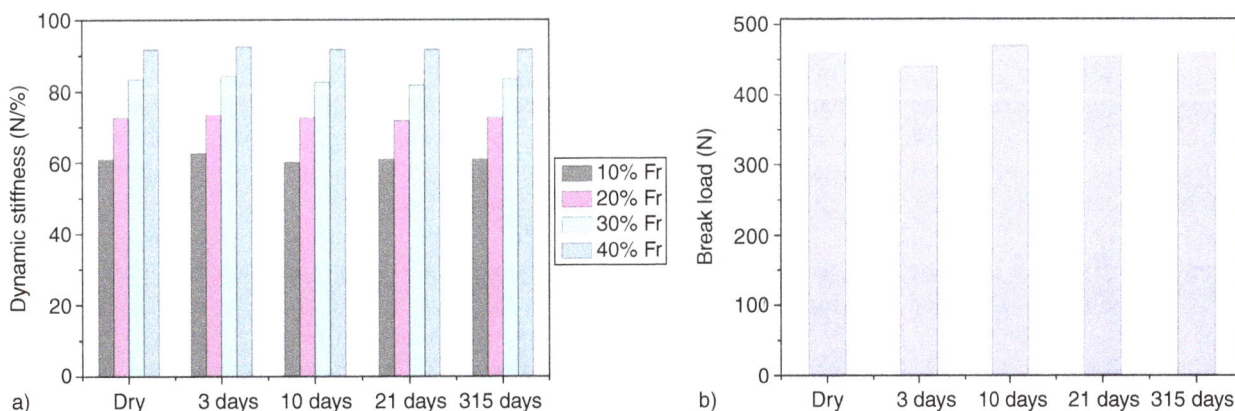

Figure 2

Influence of immersion on a) dynamic stiffness of PA6 yarns at different mean load levels, 10% amplitude, shown as % average dry break load (Fr), and b) break load, after ageing for different periods in sea water at 40°C.

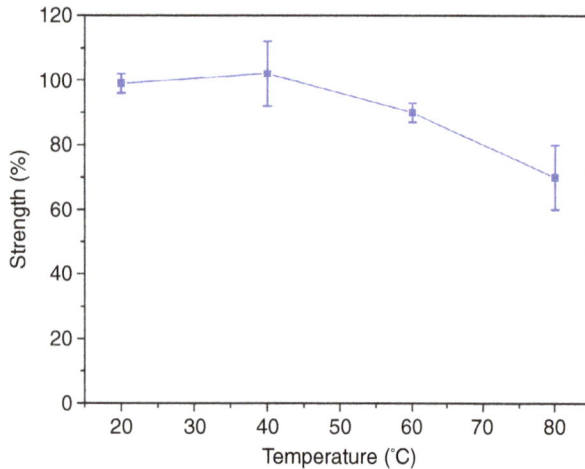

Figure 3

Tensile strength (tests performed at 20°C) of aramid fibres after 231 days seawater immersion at different water temperatures, as a percentage of unaged strength.

TABLE 1

Residual tensile strength of *Twaron* 1 000 fibres after one and a half years' ageing at 80°C in different environments, mean (± one standard deviation) values

	Residual tensile strength (%)
As-received	100
Air	92 ± 12
Sea water	72 ± 10
Deionised water	58 ± 7
pH9	62 ± 20
pH11	20 ± 7

After one and half years at 80°C in air, the strength loss is close to 8%: thermo-oxidation is thus not a serious concern for this fibre. However, the degradation is significant in the presence of water. After one and half years at pH9 and in deionised water the degradation in tensile strength is similar, and is slightly lower under sea water exposure. It appears that the degradation is significantly higher at pH11 than for the other conditions, and this may be due to an additional degradation mechanism, degradation of the crystallites, at the higher pH [14]. Tensile modulus remained unchanged for all these conditions.

2 INFLUENCE OF HYDROSTATIC PRESSURE

Another parameter which may have a strong influence on ageing behavior is mechanical loading, so under

Figure 4

Test facility to age resin and composite samples under temperature and pressure. Control specimens in water on left, pressure vessel on right.

certain conditions this may be used to accelerate ageing. For underwater applications the effect of hydrostatic pressure is particularly relevant, and several authors have examined this parameter. Extensive work in the 1970's on a wide range of polymers has shown that high pressures can significantly modify polymer properties [16-17].

Weight gain results are less abundant but some data exist. Pressure effects on composites were discussed by Fried [18] and he noted a difference between low and high void composites. In a study on flat composite panels, the effect of a 10 MPa pressure on water absorption was also examined and found to be very small [19]. Other authors have also noted rather small effects, with increases in moisture absorption [20-22], no effect [23] or decreases in absorption [24] being reported. However, some results from tests performed by the authors for specimens taken from filament wound cylinders [25] have shown strong pressure effects. The latter materials may be particularly susceptible to pressure effects as void contents tend to be quite high (several percent) in filament wound tubes. Voids influence directly the moisture pick-up [26].

It is clear that water absorption results are inconclusive, in some cases an influence of pressure is observed, in others there is no effect. A test program was therefore launched to examine this parameter in more detail. Pressure vessels were manufactured and placed in ovens in order to be able to immerse specimens at different pressures and control the temperature, Figure 4.

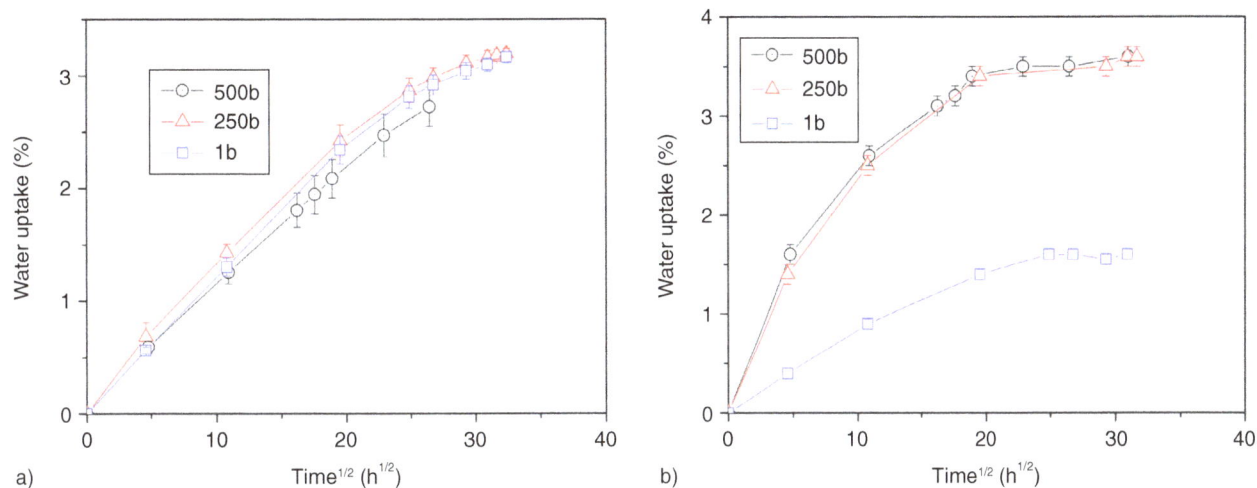

Figure 5

Weight gains, a) epoxy resin, b) glass fibre reinforced epoxy.

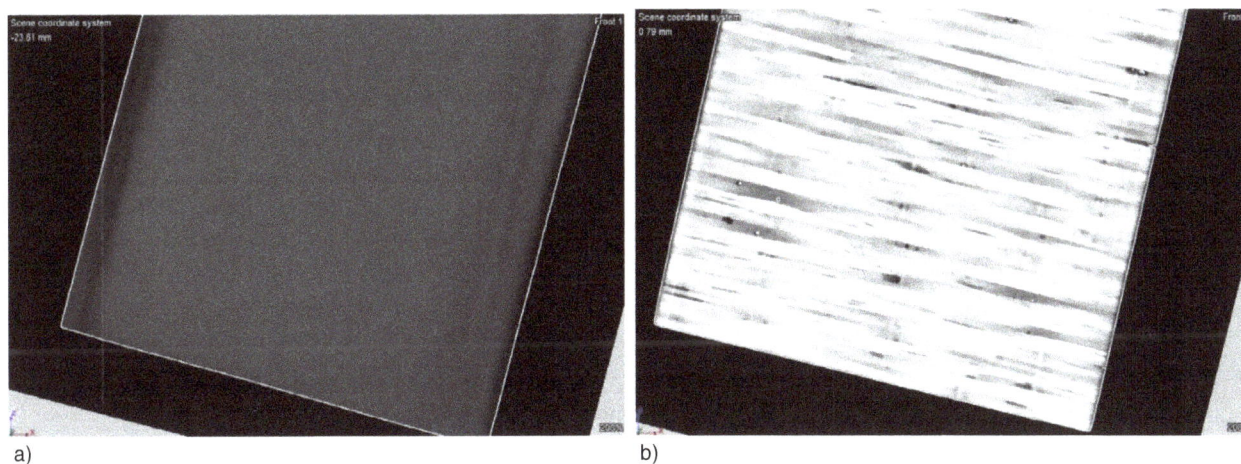

Figure 6

X-ray micro-tomography images, a) epoxy resin, b) glass reinforced epoxy composite.

First a series of 4 mm thick cast epoxy resin samples of dimensions 50×50 mm^2 was prepared. These were dried then 5 were immersed in tap water without pressure and 5 others were placed in the pressure vessels at 250 or 500 bar. All were held at 60°C in the oven. Figure 5a shows the weight gains measured periodically. For this resin, an amide cured epoxy with a Tg of 80°C, there is no influence of pressure even at 500 bar, suggesting that any effects of free volume changes and are not significant.

A second series of specimens was then prepared using the same epoxy resin to impregnate unidirectional glass fibres. The same procedure was followed and Figure 5b shows the result. At 60°C without pressure the weight gain is what would be expected based on the resin results

and the fibre fraction (around 50%). At higher pressure however, significantly higher weight gains are recorded. After drying however the weight returns to its initial value.

In order to investigate this, samples of both resin and composite were examined by X-ray micro-tomography. Figure 6 shows examples of images, which indicate significantly higher levels of porosity in the composite. The hydrostatic pressure may force additional water into these cavities, though the fibre/matrix interface may also accommodate some additional water.

These results show the difficulty in using hydrostatic pressure as an accelerating factor. For a polymer (with no porosity) there is no effect, while for a composite (with voids) a pressure of 250 or 500 bar will result in the same weight gain increase compared to a test without pressure. Thus while investigation of pressure effects remains important for deep sea applications of composites, this parameter will not be easy to employ to accelerate ageing tests.

3 PREDICTION OF DEGRADATION OF A STRUCTURE DURING AGEING: THERMAL INSULATION COATING

One of the key points for offshore oil and gas production is flow assurance. Under particular conditions of pressure and temperature, solid gas hydrates can appear in the oil situated inside the pipe leading to a total interruption of the production. In order to avoid gas hydrate formation, it is essential to limit the cooling of extracted hot oil during transport in the flowlines from the Christmas trees (at 2 000 metres depth for example) to the production structure (at the surface). Passive thermal insulation coatings are usually employed to limit thermal exchange between hot oil inside the steel pipe and the cold sea water outside (4°C), Figure 7 is a schematic representation of a flow line under service conditions. Due to their excellent thermal properties, polymers are widely used for this application, especially PolyPropylene (PP) and PolyUrethane (PU). Furthermore, to improve the thermal properties of the coating, glass bubbles can be included in these polymers to form so-called "syntactic foams". From a material point of view this application

is complex, the polymers are exposed to severe conditions (high temperature, sea water contact, high hydrostatic pressure) for long periods (up to 40 years). Thus it is important to study the polymer durability under these conditions using appropriate ageing tests. This section will discuss the influence of different parameters during accelerated ageing tests on a polyurethane coating polymer.

3.1 Water Diffusion

For this application, water diffusion in the thermal insulation coating is very important, as in the presence of water the thermal insulation properties of polymers decrease significantly [28]. The prediction of the water content in a coating as a function of time and field characteristics (such as temperature, depth) is therefore essential. Furthermore, if the polymer undergoes degradation in the presence of water, this reaction will be dependent on the water diffusion profile.

3.1.1 Water Diffusion Characterization

As discussed in the preceding sections, the easiest way to characterize water diffusion is by measuring weight changes. Using an appropriate protocol, it is possible to assess water diffusion characteristics by regularly measuring the weight gain of a polymer when immersed in water. In this section, we will discuss the effect of different experimental parameters on water diffusion.

Sample Thickness and Shape

In terms of characterization, the easiest way is to work in one-dimensional diffusion *i.e.* with a semi-infinite plate, in order to minimize water diffusion by the edges. This results in typical coupon dimensions of 50×50 mm^2 provided the thickness is less than 5 mm. It is also interesting to study mass evolution using samples with two different thicknesses, in order to check the validity of Fickian behaviour. Figure 8 shows an example for PU coating samples, which validates this approach as the two weight gain curves superpose.

Temperature

Water temperature has a strong effect on the diffusion kinetics of this material, Figure 9, whereas the water content at saturation does not change much (due to the small evolution of the solubility with temperature). On account of the temperature gradients in a flow-line coating during service it is necessary to assess water diffusion at different temperatures, to be able to predict water content in the coating as a function of time.

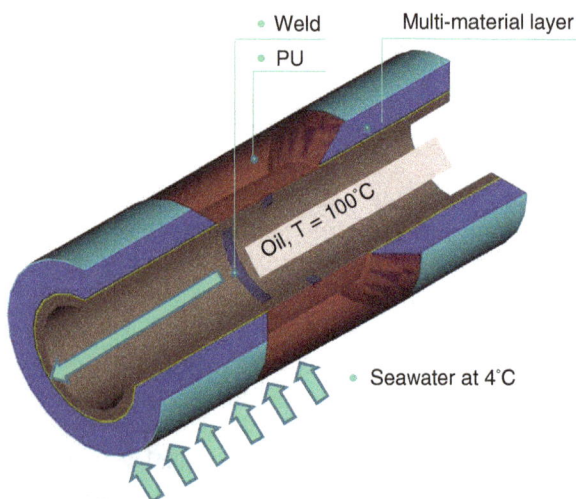

Figure 7

Schematic representation of a flow line under service conditions [27].

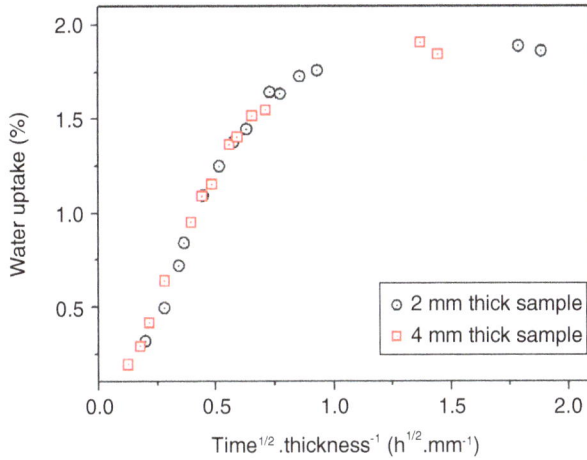

Figure 8

Weight evolution of solid PU samples of two thicknesses (2 mm and 4 mm) immersed in sea water at 80°C.

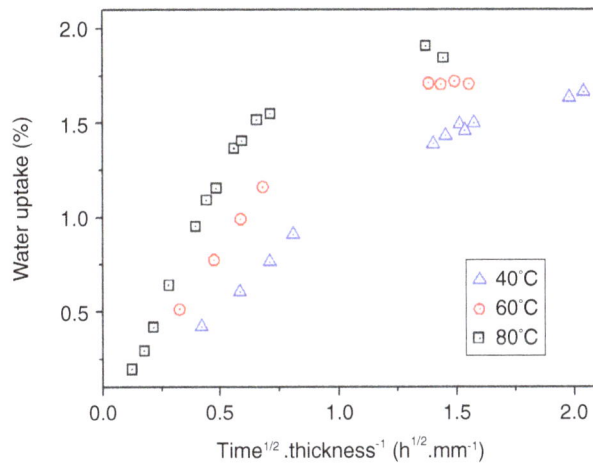

Figure 10

Influence of hydrostatic pressure on water uptake in PU with glass microspheres.

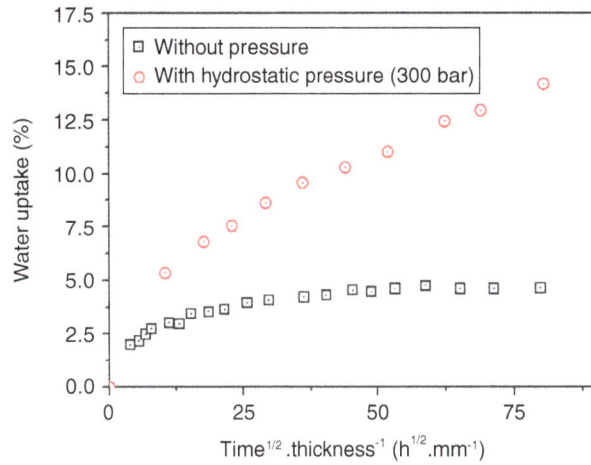

Figure 9

Influence of sea water temperature on water diffusion in PU coating polymer.

Pressure

The influence of pressure on water diffusion depends on the nature and composition of the polymer. For bulk polymers with no fillers, there may be a modification of free volume which leads to a slight modification of the water absorption. However for a polymer containing glass microspheres, hydrostatic pressure can significantly modify water absorption in the material due to collapse of these bubbles [29]. Figure 10 shows water uptake in a PU with glass bubbles in sea water at 40°C with and without hydrostatic pressure, for tests lasting over a year. Without pressure, water uptake is Fickian with a maximum weight gain of 4%, whereas with hydrostatic pressure (300 bar) water uptake increases continuously due to progressive collapse of the glass bubbles.

3.1.2 Water Diffusion Modeling

From Experimental Data

Because water absorption of a polymer depends on its nature and formulation there are many different behaviours, and hence many different models have been proposed [30-31]. Nevertheless, we have seen above that in the case of a PU coating Fickian modeling can be used, thus we will focus on the use of this model which can be presented as:

$$\frac{dC}{dt} = D.\frac{d^2C}{dx^2}$$

where C is water concentration, t is time (s), x is position in the sample (m) and D is the water diffusion coefficient (m^2/s).

From experimental data, it is possible to evaluate water diffusion coefficients and their evolution with temperature (Fig. 11). Water diffusion coefficients often show Arrhenius behaviour, and can then be written in the form:

$$D = D_o.exp\left(-\frac{E_a}{RT}\right)$$

where D is the water coefficient (m^2/s), D_o the pre-exponential factor (m^2/s), E_a the activation energy (J/mol),

R the perfect gas constant (m^2.Kg/(s^2.K.mol)) and T the temperature (K). These data will be useful to evaluate the water diffusion in the polymer coating in service.

Water Diffusion in a Polymer Coating During Service

Once sea water diffusion behaviour is known at different temperatures it is possible to predict the water content in the insulation coating as a function of time during service using numerical modeling. For example, we can consider a flow line with an internal temperature of 100°C immersed in sea water at 4°C and covered by an 80 mm thick PU coating. In this case, the temperature gradient during service can be modeled (*Fig. 12*) and thus water content and thermal properties can be predicted as a function of time and position (*Fig. 13*).

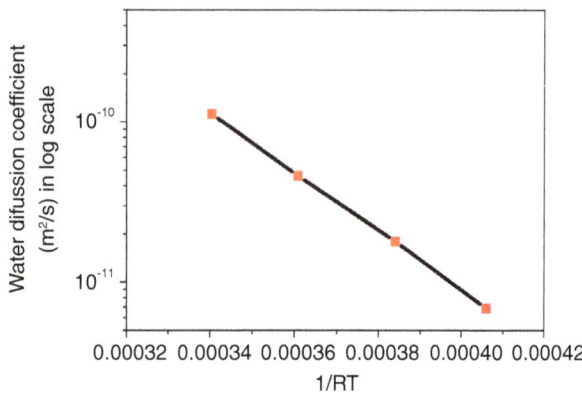

Figure 11

Evolution of the water diffusion coefficient with temperature.

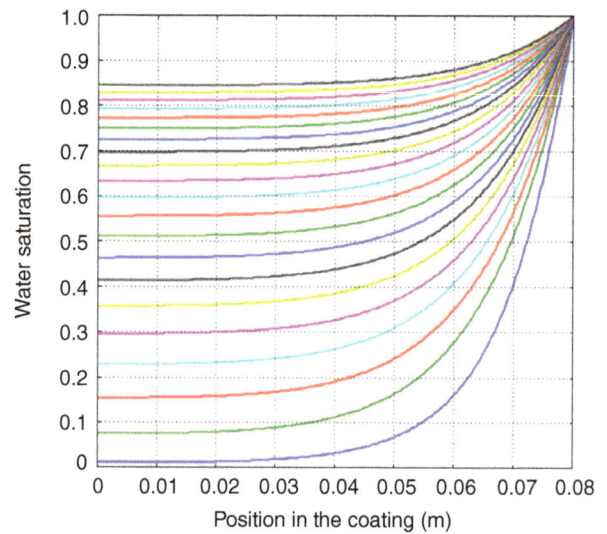

Figure 13

Evolution of the normalized water content as a function of time and position in the coating – the hot steel pipe is on the left side, at position 0 (one line for one year).

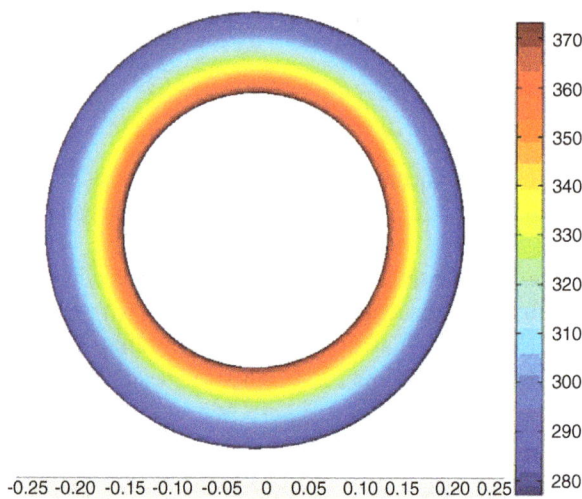

Figure 12

Temperature (K) profile in polymer coating in service.

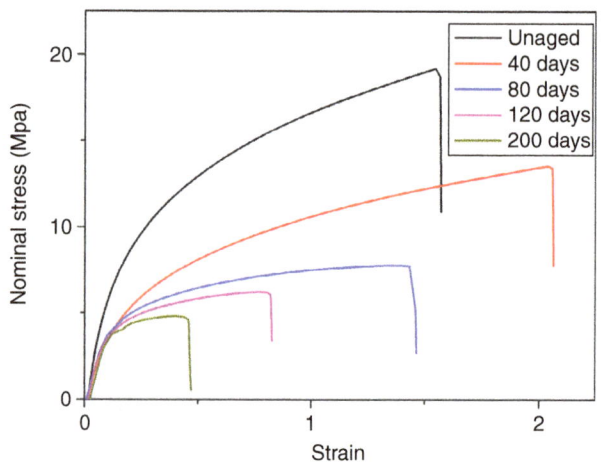

Figure 14

Evolution of tensile properties of PU during ageing.

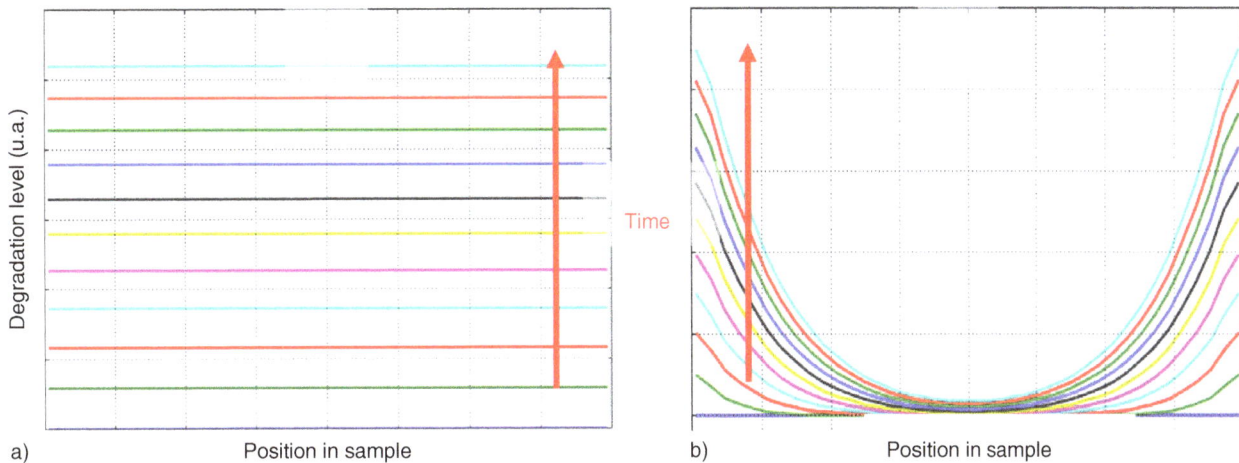

Figure 15

Degradation profiles through sample thickness showing competition between diffusion rate and reaction rate. a) Diffusion faster than reaction; b) reaction faster than diffusion (DLR).

3.2 Polymer Degradation in Sea Water

When a polymer is used in a marine environment, it absorbs water and may thus undergo reversible changes (plasticization) due to the presence of water within the polymer [32-33]. However, irreversible chemical reactions can also occur, due to reactions with water or oxygen [34-36]. Thus in addition to measuring the diffusion kinetics, it is necessary to establish how water affects the properties of the polymer and to characterize the degradation. Temperature is a useful accelerating factor and is widely used. Nevertheless it must be used carefully, especially for life time prediction.

3.2.1 Nature and Consequences of the Degradation

When temperature is used to accelerate polymer ageing, it is absolutely necessary to understand the chemical mechanisms involved in the degradation. At elevated temperature, the polymer degradation mechanisms can be changed, compared to what will occur at service temperature over a longer period. In the case of PU, the evolution of mechanical properties has been characterized at different temperatures in sea water from 110°C to 25°C for 18 months. Figure 14 shows the evolution of the tensile stress/strain curves during ageing for different periods at 100°C in sea water. In this case, loss in tensile properties is due to hydrolysis of the urethane bond [37-38].

Based on accelerated ageing and considering chemical reactions involved in the degradation, it is possible to propose a life time prediction. Nevertheless, to enhance

reliability of this prediction it is necessary to take care of diffusion/reactions aspects.

3.2.2 Diffusion-Reaction Phenomena

If the reagent is not present in the polymer but only in the external environment, then the reaction will occur only if it diffuses into the material. This is the case for most degradation in a marine environment (e.g. hydrolysis and oxidation). In this case, it is important to examine the heterogeneity of the degradation within the sample. In fact if the diffusion of the reagent is faster than its reaction by degradation then the degradation level is constant through the sample thickness (Fig. 15a). On the contrary, if the reaction occurs faster than diffusion, the reagent does not have time to diffuse deeply into the sample and thus the degradation will be heterogeneous in the sample (Fig. 15b). For oxidation this phenomenon is known as a Diffusion Limited Oxidation (DLO) [39-40].

Limitation of the degradation in the polymer thickness due to diffusion of reactant can complicate the life time prediction. In fact, the two phenomena (reaction and diffusion) do not have the same evolution with temperature, thus an Arrhenius extrapolation cannot be used. For example, it has been shown that the mechanical behaviour of a polychloroprene used for 23 years as a flow line can not be correctly predicted using accelerated ageing and Arrhenius extrapolation (Fig. 16). To overcome this limitation it is necessary to describe each phenomena involved in the overall degradation and characterize its evolution with temperature.

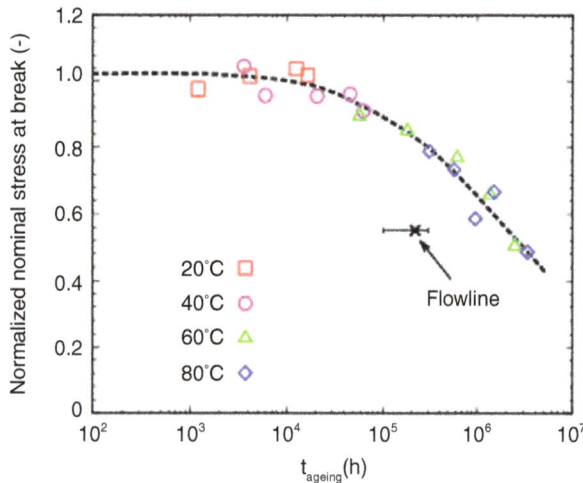

Figure 16

Comparison of stress at break predicted using Arrhenius extrapolation and experimental results on CR sample naturally aged during 23 years in service [41].

CONCLUSION

In this paper, a number of parameters affecting the ageing of polymers has been considered, in order to establish reliable and representative accelerated testing procedures for long term applications of these materials in offshore oil and gas applications. The influence of ageing temperature, medium and hydrostatic pressure has been examined using examples from previous studies, then the durability of a PU thermal insulation coating has been considered in detail. For the latter, a diffusion model has been determined which can be used to predict how the water profile within the insulation coating evolves with time under service conditions. This model can then be coupled with a mechanical model in order to establish the long term durability of such coatings. Further work is in progress to establish the kinetics of the reactions causing the changes in properties, in order to ensure that accelerated tests produce valid physical changes.

REFERENCES

1 Verdu J. (2012) *Oxidative ageing of polymers*, Wiley.

2 Weitsman Y. (2012) *Fluid Effects in Polymers and Polymeric Composites*, Springer.

3 Martin R. (2008) *Aging of Composites*, Woodhead Publishing.

4 Merdas I., Thominette F., Tcharkhtchi A., Verdu J. (2002) Factors governing water absorption by composite matrices, *Composites Science and Technology* **62**, 4, 487-492.

5 El Yagoubi J., Lubineau G., Roger F., Verdu J. (2012) A fully coupled diffusion-reaction scheme for moisture sorption–desorption in an anhydride-cured epoxy resin, *Polymer* **53**, 24, 5582-5595.

6 Hoppe L.F.E. (2003) Designing SPM hawsers to today's requirements, *OCEANS 2003 Proceedings,* San Diego, CA 22-26 Sept., **3**, 1353-1359.

7 Del Vecchio C.J.M. (1992) *Light weight materials for deep water moorings*, University of Reading.

8 Torben S. (2007) Fiber Rope Deployment System for Ultra Deep Water Installations, *2007 Offshore Technology Conference*, Houston, Texas, OTC, 30 April-3 May, OTC 18932.

9 Davies P., Reaud Y., Dussud L., Woerther P. (2011) Mechanical behaviour of HMPE and aramid fibre ropes for deep sea handling operations, *Ocean Engineering* **38**, 17-18, 2208-2214.

10 François M., Davis P., Grosjean F., Legerstee F. (2010) Modelling fiber rope load-elongation properties - Polyester and other fibers, *2010 Offshore Technology Conference*, Houston, Texas, 3-6 May, OTC 20468.

11 Richards A.F. (2005) Nylon fibres, in *Synthetic fibres*, McIntyre J. (ed), Woodhead Publishing.

12 Derombise G. (2009) *Comportement à long terme des fibres aramides en milieu neutres et alcalins*, École Nationale des Ponts et Chaussées, Paris.

13 Derombise G., Vouyovitch Van Schoors L., Davies P. (2009) Degradation of Technora aramid fibres in alkaline and neutral environments, *Polymer Degradation and Stability* **94**, 10, 1615-1620.

14 Derombise G., Chailleux E., Forest B., Riou L., Lacotte N., Van Schoors L.V., Davies P. (2011) Long-term mechanical behavior of aramid fibers in seawater, *Polymer Engineering and Science* **51**, 7, 1366-1375.

15 Derombise G., van Schoors L.V., Bourmaud A., Davies P. (2012) Morphological and physical evolutions of aramid fibers aged in a moderately alkaline environment, *Journal of Applied Polymer Science* **123**, 5, 3098-3105.

16 Radcliffe S. (1972) Effects of hydrostatic pressure on the deformation and fracture of polymers, in Deformation and Fracture of High polymers, in *Deformation and Fracture of High polymers*, Plenum Press.

17 Pae K.D., Bhateja S.K. (1975) The Effects of Hydrostatic Pressure on the Mechanical Behavior of Polymers, *Journal of Macromolecular Science, Part C* **13**, 1, 1-75.

18 Fried N. (1967) Degradation of composite materials, Proceedings of the *Fifth Symposium on Naval Structural Mechanics*, May, pp. 813-837

19 ASTM STP 1302 (1997) *Aging of composites in water*, pp. 73-96.

20 Tucker W.C., Lee S.-B., Rockett T. (1993) The Effects of Pressure on Water Transport in Polymers, *Journal of Composite Materials* **27**, 8, 756-763.

21 Tucker W.C., Brown R. (1989) Moisture Absorption of Graphite/Polymer Composites Under 2000 Feet of Seawater, *Journal of Composite Materials* **23**, 8, 787-797.

22 Pollard A., Baggott R., Wostenholm G.H., Yates B., George A.P. (1989) Influence of hydrostatic pressure on the moisture absorption of glass fibre-reinforced polyester, *Journal of Materials Science* **24**, 5, 1665-1669.

23 Bradley W.L., Chiou P.-L.B., Grant T.S. (1993) The effect of seawater on polymeric composite materials, *Proceedings of the First International Workshop*, Houston, Texas, 26-28 Oct, in *Composite Materials for Offshore Operations*, Wang S.S., Fitting D.W. (eds), NIST Special Publication, **887**, 193-202.

24 Avena A., Bunsell A.R. (1988) Effect of hydrostatic pressure on the water absorption of glass fibre-reinforced epoxy resin, *Composite* **19**, 5, 355-357.

25 Davies P., Choqueuse D., Mazéas F. (1997) Composites Underwater, *Proc Duracosys, 3rd International Conference Duracosys*, Blacksburg, Virginia, 14-17 Sept.

26 Harper B.D., Staab G.H., Chen R.S. (1987) A Note on the Effects of Voids Upon the Hygral and Mechanical Properties of AS4/3502 Graphite/Epoxy, *Journal of Composite Materials* **21**, 3, 280-289.

27 Le Gac P.-Y., Choqueuse D., Paris M., Recher G., Zimmer C., Melot D. (2013) Durability of polydicyclopentadiene under high temperature, high pressure and seawater (offshore oil production conditions), *Polymer Degradation and Stability* **98**, 3, 809-817.

28 Lefebvre X., Sauvant-Moynot V., Choqueuse D., Chauchot P. (2009) Durability of Syntactic Foams for Deep Offshore Insulation: Modelling of Water Uptake under Representative Ageing Conditions in Order to Predict the Evolution of Buoyancy and Thermal Conductivity, *Oil & Gas Science and Technology - Revue IFP* **64**, 2, 165-178.

29 Choqueuse D. (2012) Experimental study and analysis of the mechanical behaviour of syntactic foams used in deep sea, *PhD Thesis*, Université de Franche-Comté.

30 Frisch H.L. (1970) *Diffusion in polymers*, Crank J., Park G.S. (eds), Academic Press, London and New York, 1968; 452 pg, Journal of Applied Polymer Science **14**, 6, 1657-1657.

31 Park G.S. (1986) Polymer permeability, Comyn J. (ed.), Elsevier Applied Science Publishers, London, 1985, vii + 383, ISBN 0-85334-322-5, *British Polymer Journal*. **18**, 3, 209–210, DOI: 10.1002/pi.4980180315.

32 Xiao G.Z., Shanahan M.E.R. (1997) Water absorption and desorption in an epoxy resin with degradation, *Journal of Polymer Science Part B-Polymer Physics* **35**, 16, 2659-2670.

33 Gaudichet-Maurin E., Thominette F., Verdu J. (2008) Water sorption characteristics in moderately hydrophilic polymers, Part 1: Effect of polar groups concentration and temperature in water sorption in aromatic polysulfones, *Journal of Applied Polymer Science* **109**, 5, 3279-3285.

34 Xiao G.Z., Shanahan M.E.R. (1998) Irreversible effects of hygrothermal aging on DGEBA/DDA epoxy resin, *Journal of Applied Polymer Science* **69**, 2, 363-369.

35 Le Gac P.Y., Le Saux V., Paris M., Marco Y. (2012) Ageing mechanism and mechanical degradation behaviour of polychloroprene rubber in a marine environment: Comparison of accelerated ageing and long term exposure, *Polymer Degradation and Stability* **97**, 3, 288-296.

36 Jacques B., Werth M., Merdas I., Thominette F., Verdu J. (2002) Hydrolytic ageing of polyamide 11. 1. Hydrolysis kinetics in water, *Polymer* **43**, 24, 6439-6447.

37 Matuszak M.L., Frisch K.C., Reegen S.L. (1973) Hydrolysis of linear polyurethanes and model monocarbamates, *Journal of Polymer Science: Polymer Chemistry Edition* **11**, 7, 1683-1690.

38 Chapman T.M. (1989) Models for polyurethane hydrolysis under moderately acidic conditions: A comparative study of hydrolysis rates of urethanes, ureas, and amides, *Journal of Polymer Science Part A: Polymer Chemistry* **27**, 6, 1993-2005.

39 Celina M., Wise J., Ottesen D.K., Gillen K.T., Clough R.L. (1998) Oxidation profiles of thermally aged nitrile rubber, *Polymer Degradation and Stability* **60**, 2-3, 493-504.

40 Celina M., Wise J., Ottesen D.K., Gillen K.T., Clough R.L. (2000) Correlation of chemical and mechanical property changes during oxidative degradation of neoprene, *Polymer Degradation and Stability* **68**, 2, 171-184.

41 Le Saux V., Le Gac P.Y., Marco Y., Calloch S. (2014) Limits in the validity of Arrhenius predictions for field ageing of a silica filled polychloroprene in a marine environment, *Polymer Degradation and Stability*, **99**, 254-261.

5

Energy Management Strategies for Diesel Hybrid Electric Vehicle

Olivier Grondin*, Laurent Thibault and Carole Quérel

IFP Energies nouvelles, 1-4 avenue de Bois-Préau, 92852 Rueil-Malmaison Cedex - France
e-mail: olivier.grondin@ifpen.fr - laurent.thibault@ifpen.fr - carole.querel@ifpen.fr

* Corresponding author

Abstract — *This paper focuses on hybrid energy management for a Diesel Hybrid Electric Vehicle (HEV) with a parallel architecture. The proposed strategy focuses on the reduction of Nitric Oxides (NO_x) emissions that represents a key issue to meet Diesel emissions standards. The strategy is split in two separated functions aiming at limiting the NO_x in steady-state and transient operating conditions. The first functions, control the torque split between the engine and the electric motor. This energy management is based on the Equivalent Consumption Minimization Strategy (ECMS) where an additional degree of freedom is introduced to tune the optimization trade-offs from the pure fuel economy case to the pure NO_x limitation case. The second function adapts the torque split ratio between the motor and the engine, initially computed from the optimal control strategy during transient operations where NO_x are produced. The engine torque correction relies on mean value models for the EGR system dynamics and for the NO_x formation. This paper applies a methodology based on Software in the Loop (SiL) and Hardware in the Loop (HiL) simulations in order to understand the system performance according to the powertrain configurations and also to tune the proposed energy management strategy. The simulation results are confirmed by experiments performed on Hybrid-Hardware in the Loop (Hy-HiL) test bench. This work shows the potential of using the hybrid architecture to limit NO_x emissions by choosing the best operating point and by limiting the engine dynamics. The NO_x reduction has limited impact on fuel consumption.*

Résumé — **Lois de gestion de l'energie pour le véhicule hybride Diesel** — Cet article présente une stratégie de gestion de l'énergie pour un véhicule hybride parallèle muni d'une motorisation Diesel. Cette stratégie est consacrée à la réduction des émissions d'oxydes d'azote (NO_x) dont la dépollution constitue une problématique majeure pour l'homologation des véhicules Diesel. La stratégie comporte deux fonctions distinctes visant à limiter les émissions de NO_x en régime stabilisé et en régime transitoire. La première fonction agit sur la répartition de couple entre le moteur thermique et la machine électrique. L'approche ECMS (*Equivalent Consumption Minimization Strategy*, ou stratégie de minimisation de la consommation équivalente) est utilisée pour gérer la réparation de couple et dans notre cas elle intègre dans la fonction coût un compromis entre la consommation et les émissions de NO_x. La seconde fonction corrige la répartition de couple issue de l'ECMS mais uniquement en régime transitoire et afin de réduire les pics d'émissions de NO_x. L'adaptation de la répartition de couple repose sur des modèles moyens du système EGR et de la formation des NO_x. Cet article présente le principe de ces stratégies et leur comportement est analysé grâce à un simulateur représentatif des émissions polluantes. Ces résultats sont complétés par des essais expérimentaux sur un banc moteur HiL

(*Hardware in the Loop*). Ce travail montre qu'il est possible de limiter les émissions de NO$_x$ par le biais de l'hybridation en adaptant le choix du point de fonctionnement du moteur et en réduisant les sollicitations transitoires pendant lesquelles les NO$_x$ sont produits. La réduction des émissions de NO$_x$ n'est pas obtenue au détriment de la consommation.

NOMENCLATURE

PARAMETERS & VARIABLES

α_i	NO$_x$ model parameters (-)
F_1	Intake burned gas ratio (-)
β_d	EGR model time delay (s.rpm)
β_f	EGR model time constant (s.rpm)
k_{fc/NO_x}	Emissions weighting factor (-)
N_e	Engine speed (rpm)
R_1	Front axle ratio (-)
R_{gb}	Gear Box ratio (-)
SOC	Battery state of charge (%)
Θ	Temperature (K)
t	Time (s)
T	Torque (Nm)
u	Control input (Nm)
ξ	NO$_x$ reduction factor (%)

SUBSCRIPTS & SUPERSCRIPTS

b	Burned gas
cyl	Cylinder
eng	Engine
est	Estimated
f	Feasible
mot	Motor
pwt	Powertrain
ss	Steady-state
sp	Setpoint
t	Trajectory

ACRONYMS

BGR	Burned Gas Ratio
DPF	Diesel Particulate Filter
ECMS	Equivalent Consumption Minimization Strategy
EGR	Exhaust Gas Recirculation
EM	Electric Machine
EMS	Energy Management Strategy
FC	Fuel Consumption
FTP	Federal Transient Procedure
GB	Gear Box
HEV	Hybrid Electric Vehicle
HP	High Pressure (EGR)
LP	Low Pressure (EGR)
MVM	Mean Value Model
NEDC	New European Drive Cycle
NO$_x$	Nitrogen oxides
S&S	Stop & Start
SOC	State of Charge
SSG	Separated Starter-Generator
SSP	Steady-State Part
TP	Transient Part
VVA	Variable Valve Actuation
WLTP	Worldwide harmonized Light-duty Transient Procedure

INTRODUCTION

In a context of increasing demand for cleaner vehicles, hybrid electric vehicles are recognized as an effective way to reduce fuel consumption and emissions. The most common hybrid powertrains usually combine a gasoline engine with an electric motor with different architectures and several degrees of hybridization. The pollutant emissions of the gasoline engine are treated by the three-way catalytic converter. For the gasoline hybrid vehicle, the energy efficiency improvement is the main objective. Thus, the energy supervisor improves the fuel economy rather than the emissions in warm conditions while the thermal management is the main issue in cold conditions [1]. If we consider hybrid powertrains with a Diesel engine, the Nitrogen Oxide (NO$_x$) emissions must be considered because the operating regions of maximum efficiency and minimum NO$_x$ emissions are generally not the same. For this purpose, the energy management strategy must be adapted to maximize the use of the engine within its low NO$_x$ emissions operating points.

This control issue was studied in the literature mainly in simulation. Early work from Johnson *et al.* [2] deals with heuristic supervision strategies for a Diesel HEV. The latter takes into account fuel consumption weighted with the four main pollutants (NO$_x$, PM,

HC, CO). Other heuristic strategies are proposed in [3]. On the other hand, model-based techniques have already been proposed by Lin *et al.* [4] (ECMS) and Musardo *et al.* [5, 6] (dynamic programming). More recently, Dextreil and Kolmanovsky [7] propose an energy management controller based on the application of game theory. The simulation and experimental results from these papers indicate that a significant NO_x reduction is possible at the price of a slight drop in fuel consumption. These papers confirm the feasibility of integrating emission constraints into heuristic or model-based supervision strategies.

These solutions can be applied for Euro 6 engines but new constraints will arise from the transient nature of the upcoming Worldwide harmonized Light vehicles Test Procedures (WLTP). Indeed if we consider only the current European driving cycle, the transient part of the total NO_x emissions can clearly be neglected. But these transient emissions are actually significant with the new driving cycle which is more representative of real world conditions. In this context, the current strategies applying optimal control based on static maps are not sufficient and NO_x transient emissions should be taken into account.

The goal of this paper is to propose a global energy management strategy for Diesel HEV taking into account both static and transient NO_x emissions in addition to the fuel consumption objective. In our approach, the steady-state and transient parts are treated separately as described in Figure 1. The steady state optimal torque splits are found using the optimal control theory. The engine torque $T_{eng,ss}^{sp}$ and motor torque $T_{mot,ss}^{sp}$ setpoints are computed from an ECMS-based strategy including the steady-state NO_x emission maps into the cost function. The second function consists of a model-based strategy that adapts the static torque split during the transient phases where NO_x peaks can occur. This leads to a computation of two trajectories for the engine torque ($T_{eng,t}^{sp}$) and the motor torque ($T_{mot,t}^{sp}$).

The paper is organized as follows: the vehicle architecture is presented in the first part of the paper (*Sect. 1.1*)

with the modeling aspects (Sect. 1.2) and the presentation of the EMS based on the equivalent consumption minimization strategy (Sect. 1.3). The steady-state strategy performances are analysed in simulation and some experimental results are shown in Section 1.4. In the second part of the paper, we highlight the transient NO_x emissions problem in Diesel engines (Sect. 2.1). The principle of the engine torque setpoint control and the NO_x trajectory definition are detailed in Section 2.2. The strategy is validated in simulation and the results are discussed in Section 2.3. Finally, this paper ends up with conclusions.

1 ADAPTATION OF THE EMS FOR DIESEL ENGINE: MINIMIZATION OF NO$_X$ ALONG WITH FUEL CONSUMPTION

1.1 System Description

In this paper, we consider the vehicle architecture depicted in Figure 2. This is a parallel hybrid architecture that uses a Separated Starter Generator (SSG) in the pre-transmission side (only allowed to start the engine) and the post-transmission Electric Machine (EM) allows for power assist, full electric drive, regenerative braking and battery recharge. More details about the hybrid parallel vehicle can be found in [8]. The Diesel engine is a 1.6 liter, four-cylinder, direct-injection engine with a maximum power of 50 kW and a maximum torque of 150 Nm. The engine uses high EGR rate to operate under low temperature combustion mode such that the engine NO_x emissions are close to the Euro 6 emission standard without dedicated post-treatment. To supply the cylinder with EGR, the engine has a Low Pressure (LP) EGR loop. The LP EGR circuit takes

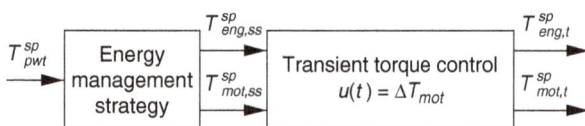

Figure 1

Static and dynamic torque split strategy based on a transient torque controller cascaded with a ECMS-based controller.

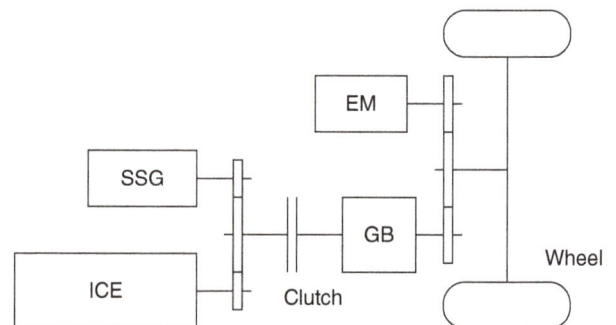

Figure 2

Parallel hybrid-electric propulsion system under consideration.

TABLE 1

Parameters of the hybrid powertrain considered in this paper: vehicle mass, electric motor power and battery energy

Vehicle mass (kg)	EM power (kW)	Battery energy (wh)
1 470	8	280
1 580	14	520
1 630	20	720

the exhaust gases downstream after the treatment system – composed of a Diesel Oxidation Catalyst (DOC) and a Diesel Particulate Filter (DPF) – to upstream the compressor. In this paper, we consider only the LP EGR operations. In this configuration, the system has a slower burned gas settling time compared to the HP EGR mode. The engine transients are then more drastic according to the NO_x emission peaks since they are mostly caused by EGR time lag. This is a fancy case study to develop and validate a transient NO_x limitation strategy using the power assist from the electric machine (Sect. 2). In this work, we perform a parametric study of the electric machines power. We adopt a basic rule to adapt the battery size in order to keep the battery energy to motor power ratio constant and close to 35 Wh/kW. This value is chosen according to the reference hybrid system where the electric motor has a rated power of 42 kW and the battery an energy of 1.5 kWh [9]. The vehicle mass is also adapted to account for the electric machine and battery size. This leads to the configuration detailed in Table 1.

1.2 System Modeling

The aim of this section is to introduce the mathematical model of the parallel hybrid powertrain shown in Figure 2. The model has a backward causality and it is of the quasi-static type. All the dynamics except for the vehicle longitudinal dynamics and the battery accumulator dynamics are neglected. These models equations are accepted to be relevant for optimization tasks. They are physically based, they capture the dominant phenomena and they can be implemented online. In backward modeling, the vehicle speed profile and the driver's torque demand are the main model inputs. The torque request at wheel, T_{pwt}^{sp}, depends on the driving conditions (speed and acceleration) and can be evaluated from drivers demand (throttle and brake pedal positions). Moreover:

$$T_{pwt}^{sp}(t) = R_1 T_{mot,ss}^{sp}(t) + R_{gb} T_{eng,ss}^{sp}(t) \qquad (1)$$

where R_{gb} is the gear ratio and R_1 is the front axle ratio. The torque setpoint can be positive or negative depending on the vehicle operating conditions (traction or braking). $T_{mot,ss}^{sp}$ is the requested electric motor torque and the engine torque is the control input:

$$u(t) = T_{eng,ss}^{sp}(t) \qquad (2)$$

The electric motor power, P_{elec}, is usually described by a map as a function of the motor torque and speed N_{mot}:

$$P_{elec}(t) = f_{elec}(T_{mot,ss}^{sp}, N_{mot}) \qquad (3)$$

The engine Fuel Consumption (FC) and the NO_x emissions are provided by steady state maps that depend on the engine torque and speed N_e:

$$\dot{m}_f(t) = f_{FC}(T_{eng,ss}^{sp}, N_e) \qquad (4)$$

$$\dot{m}_{NO_x}(t) = f_{NO_x}(T_{eng,ss}^{sp}, N_e) \qquad (5)$$

These two maps describe the quasi-static performances resulting from the engine calibration. They will be introduced in the energy management strategy.

The fuel mass injected $\dot{m}_f$ is transformed into a fuel power using the lower heating value of the fuel Q_{lhv} which is a constant for a given fuel:

$$P_{fuel}(t) = Q_{lhv} \dot{m}_f(t) \qquad (6)$$

The battery pack is modeled as a basic equivalent circuit comprising a voltage source U_0 placed in series with a resistance R_0. These two variables vary according to the battery State of Charge (SOC). The battery current and voltage are given by:

$$U_{bat} = U_0(SOC) - R_0(SOC)I_{bat} \qquad (7)$$

$$I_{bat} = \frac{U_0}{2R_0} - \sqrt{\frac{U_0^2}{4R_0^2} - \frac{P_{elec}}{R_0}} \qquad (8)$$

where the battery power is $P_{bat} = P_{elec} = U_{bat}I_{bat}$. The variation of battery SOC is computed from the battery current and power:

$$\widetilde{SOC}(t) = \begin{cases} -\dfrac{I_{bat}}{Q_0}, & \text{if } P_{elec} > 0 \\ -\eta_{bat}(SOC)\dfrac{I_{bat}}{Q_0}, & \text{else} \end{cases} \qquad (9)$$

where Q_0 is the battery capacity and η_{bat} represents its Faradic efficiency. In the next section, the battery SOC is the system state $(x(t) = SOC(t))$. The electrochemical battery power is:

$$P_{ec} = -Q_0 U_0(SOC)\dot{\widehat{SOC}}(t) \qquad (10)$$

1.3 Supervisory Control Strategy

For hybrid electric vehicle, the total power delivered to the wheels comes from two energy sources: the fuel and a battery. The energy management strategy computes the best power split between these two sources. The proposed strategy is based on the ECMS previously developed [10-12] and is already implemented on a gasoline hybrid electric vehicle [1, 13]. In previous studies considering SI engines, the main optimization criterion was the overall fuel consumption. Thus, the problem formulation was to determine the command $u(t)$ that minimizes the cost function expressed as:

$$J = \int_{t_0}^{t_f} P_{fuel}(u, t)dt \qquad (11)$$

For a Diesel HEV, this objective is modified and we chose to minimize a compromise between fuel consumption and NO_x emission. Then, the integral cost of the fuel is defined to be the integral of the weighted sum of the fuel consumption and the NO_x emission over the driving cycle. The idea is to merge the fuel map and the NO_x map in order to replace the fuel consumption-based cost by a weighted average of NO_x and fuel consumption. Thus, the fuel power defined by Equation (6) becomes:

$$P^*_{fuel} = Q_{lhv}\left[(1 - k_{fc/NO_x})\dot{m}_f + k_{fc/NO_x}\dot{m}_{NO_x}\right] \qquad (12)$$

where the parameter k_{fc/NO_x} is used to set the trade-off between fuel consumption and NO_x emission. In order to ease the tuning of this parameter, the two maps must be comparable. Thus, the NO_x map is normalized such that its mean value remains close to the mean value of the fuel maps. The fuel mass flow rate and the NO_x emission strongly depend on the engine operating condition. As a first approximation, we use two static maps depending on engine speed and torque (*Eq. 4, 5*) and the optimal control problem is then:

$$\min_{u \in \mathcal{U}} \int_{t_0}^{t_f} P^*_{fuel}(u, t)dt \qquad (13)$$

The state of the system is the State of Charge. Its dynamics writes:

$$\dot{x} = \alpha(x, u)\left(1 - \sqrt{1 - \beta(x)u}\right) \qquad (14)$$

where α and β depend on the battery equivalent circuit characteristics (resistance R_0, voltage U_0, efficiency η_{bat}) and the SOC as described in Equations (7-9). The optimization problem has a state constraint such that the final value of the SOC $x(t_f)$ should be equal to its initial value $x(t_0)$. The system input is the engine torque request that belongs to an admissible space such that $\mathcal{U} = [T_{eng_{min}} T_{eng_{max}}]$. This optimization problem can be solved using the Pontryagin's Minimum Principle (PMP) where the optimal command writes:

$$u^{opt} = arg \min_{u \in \mathcal{U}} \mathcal{H}(x(t), \lambda(t), u(t), t) \qquad (15)$$

where the Hamiltonian function is defined as:

$$\mathcal{H}(x(t), \lambda(t), u(t), t) = [(1 - k_{fc/NO_x})\dot{m}_f(u, t) \\ + k_{fc/NO_x}\dot{m}_{NO_x}(u, t)] \\ - \lambda(t)\alpha(x, u)\left(1 - \sqrt{1 - \beta(x)u}\right) \qquad (16)$$

In first approximation, if U_0 and R_0 are constant (the SOC does not vary so much), the Hamiltonian can be rewritten as:

$$\mathcal{H}(x(t), \lambda(t), u(t), t) = P^*_{fuel}(t) + sP_{ec}(t) \qquad (17)$$

The equivalence factor s weighs the electrochemical power. Its value modifies the balance between the electrochemical power use and the equivalent fuel power use. The equivalence factor is chosen such that the constraint on the final SOC is fulfilled $x(t_f) = x(t_0)$. In this paper, we assume the drive cycle is known and, under this assumption, the equivalence factor is a constant determined off-line. The sensitivity of this parameter and the online adaptation rule was studied by Chasse *et al.* for a SI engine in [13]. The robustness of this energy management for a Diesel HEV was studied by Thibault and Leroy in [14]. The method to determine the equivalence factor off-line according is described in [15]. In the next section, we present the influence of the weight between the fuel consumption and the NO_x emissions.

1.4 Results

The simulation results are presented in Figure 3. These simulations are performed from the fuel economy optimization to the NO_x abatement optimization. The weighting parameter k_{fc/NO_x} ranging from 0 (minimum of fuel consumption) to 1 (minimum of NO_x emission). The trends are clear for all the driving cycles considered (FTP, WLTP and ARTEMIS Urban): the weighting

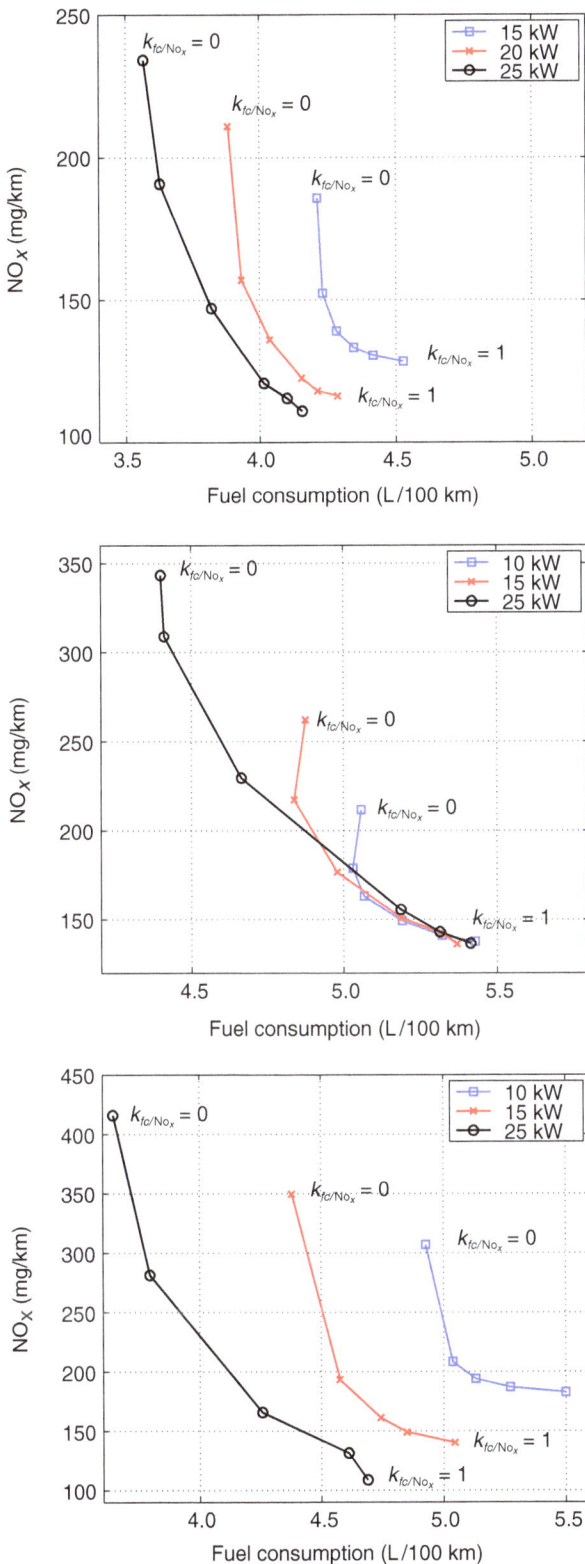

Figure 3

Evolution of the tradeoff betweeen NO_x emissions and fuel consumption for a parametric variation of the electric machine power and the weighting parameter k_{fc/NO_x}. From top to bottom: FTP, WLTP and ARTEMIS Urban.

parameter set the trade-off between FV and NO_x emission. For a pure NO_x optimisation, the Fuel Consumption (FC) penalty can be very high. However a fine tuning of k_{fc/NO_x} provides a great potential for limiting NO_x emission at a price of a very small increase of fuel consumption. The achievable NO_x reduction is close to 40% for FTP. On the other hand, the fuel penalty is small and does not exceed 5%. This behaviour can be observed for several values of the electric machine power.

Figure 4 shows the instantaneous values of some powertrain quantities during the experimental tests. The comparison concerns the HEV with a 20 kW electric motor. Two cases are presented: the fuel-economy optimization and the NO_x reduction optimization. The test results are summarized in Table 2 and in Figure 5. For the optimization of the fuel consumption, the electric motor is mostly used for traction during the urban part of the cycle while the Diesel engine is mostly used at high loads during the extra-urban part to recharge the battery. The ECMS allows to run the ICE mostly for high efficiency setpoints. For the NO_x reduction case, the instantaneous values of torque are limited and the optimal NO_x strategy avoids the high torque values (especially during the extra-urban part of the drive cycle). When the NO_x emissions are not considered ($k_{fc/NO_x} = 0$), the Diesel engine operates mostly at high loads where the engine efficiency is high. The selected operating points are close to the border limit of the LTC region where the amount of EGR is reduced compared to the NO_x optimal region. For the NO_x optimization case ($k_{fc/NO_x} = 0.5$), most of the selected operating conditions are located at middle engine loads. This region of the engine mapping represents a good compromise between NO_x emissions and fuel consumption. The pure engine mode or the recharging mode is limited to the middle torque operating conditions. These results prove that the adapted ECMS is suitable for the Diesel HEV and the optimal behavior is modified compared to the optimization policy applied for gasoline HEV.

2 TRANSIENT TORQUE CONTROL

2.1 Analysis of the Transient NO_x Emisssions

The pollutant emission behaviour of an engine is not purely quasi-static. Thus, the assumption that the emissions can be modeled using static maps can be inadequate. An example of transient NO_x emissions for a portion of a NEDC is displayed in Figure 6. This test was performed on a high dynamic test bench and the NO_x emissions were recorded with a gas analyzer.

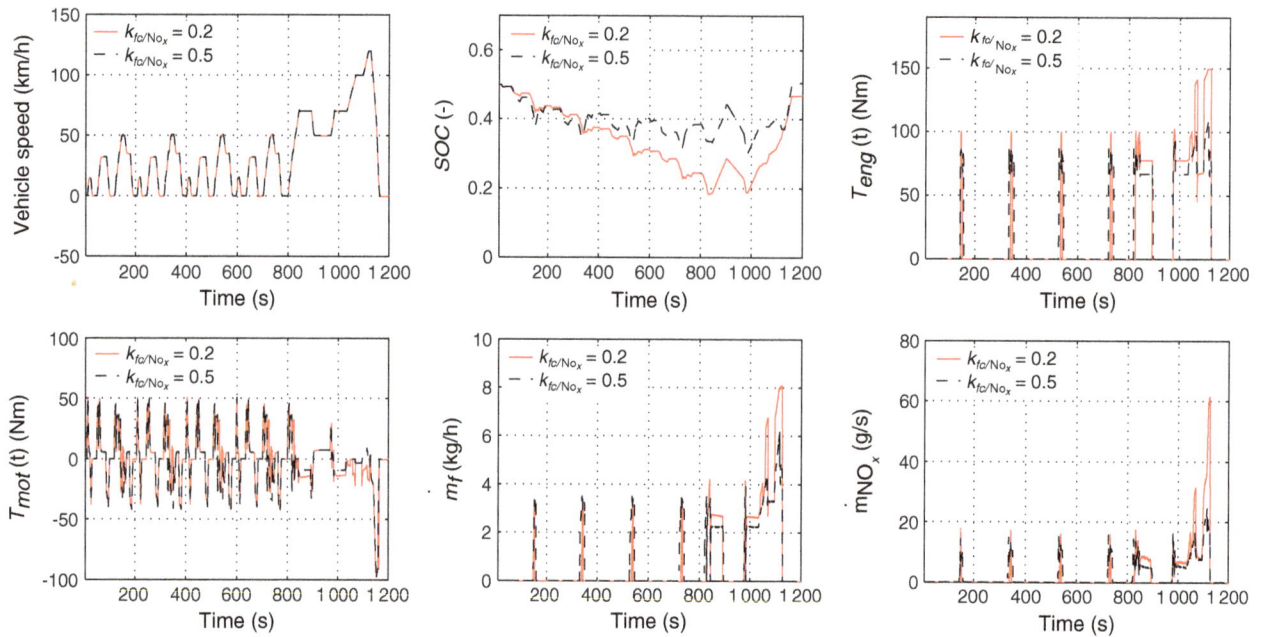

Figure 4

Experimental results for NEDC with the HEV with a 20 kW electric motor. Comparison of two optimisation trade-offs: fuel consumption oriented optimisation with $k_{fc/NO_x} = 0.2$ (red line) and NO_x-oriented optimisation with $k_{fc/NO_x} = 0.5$ (black dotted line).

TABLE 2

Experimental results for a NEDC: comparison of powertrain configuration including FC and NO_x oriented optimization

Vehicle	Baseline	S&S	HEV	HEV
EM power (kW)	-	-	20	20
Mass (kg)	1 470	1 470	1 630	1 630
k_{fc/NO_x}	-	-	0.2	0.5
FC (L/100 km)	4.5	4.16	3.43	3.5
NO_x (mg/km)	110	100	82	67
HC (mg/km)	120	111	20	27

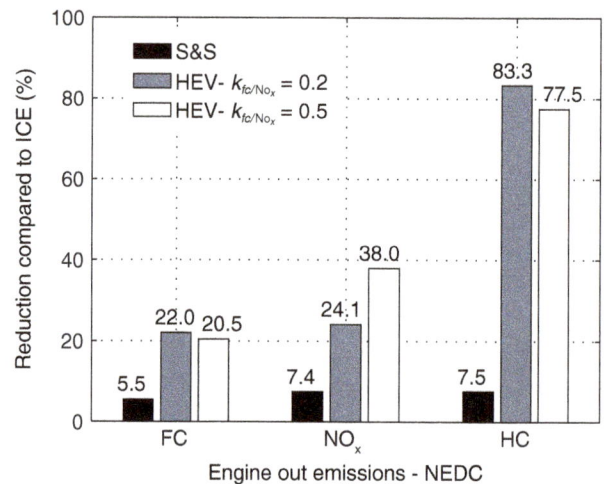

Figure 5

Summary of the experimental results. The gains are relatives to the stand-alone Diesel engine. Three configurations are displayed: Stop and Start, HEV with $k_{fc/NO_x} = 0.2$ and HEV with $k_{fc/NO_x} = 0.5$.

Figure 5 exhibits several peaks in NO_x during the first acceleration of the extra-urban part of the driving cycle. At each gear change, the injection cut-off decreases the exhaust equivalence ratio. This limits the availability of burned gas at the engine intake. During the transient, the intake Burned Gas Ratio (BGR) drops dramatically. When the driver re-accelerates, the intake BGR target cannot be reached instantaneously due to burned gas transport from the exhaust to the intake plenum. During this transient, the intake gas composition and thus the cylinder burned gas ratio are not in steady-state operating conditions leading to spikes in measured NO_x.

The transient part of the pollutant emission may vary according to the engine, its calibration and the driving cycle considered. In the experimental data presented in Figure 5, the instantaneous NO_x levels can be twice the

TABLE 3

Engine performances measured during NEDC and FTP driving cycles. The Transient Part (TP) contribution (in %) of the FC (in L/100 km) and NO_x (in mg/km) emissions are computed from the experimental measures and the engine static maps

	Driving cycle					
	NEDC			FTP		
	Bench	Map	TP	Bench	Map	TP
NO_x	110	100	9%	170	100	42%
FC	4.5	4.3	3%	4.7	4.4	6%

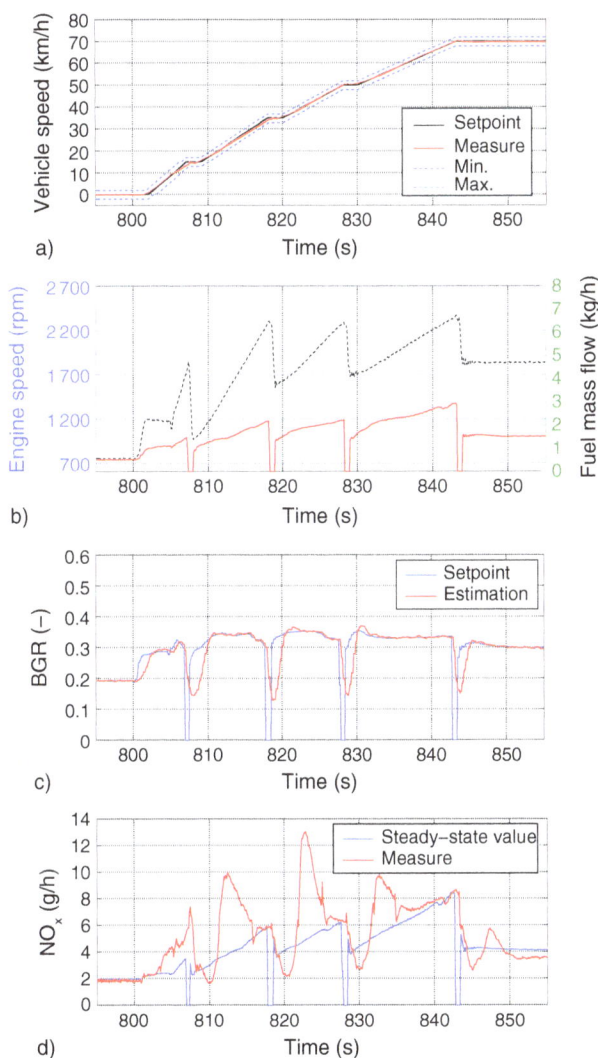

a)

b)

c)

d)

Figure 6

First acceleration of the extra-urban part of the NEDC. a) Vehicle speed, b) engine speed (dotted line) and fuel mass flow, c) BGR and d) NO_x emissions. The steady state NO_x value is calculated from a static map depending on engine speed and torque. The measured NO_x value is delayed and filtered due to the gas analyzer dynamics.

steady state values during transients. Table 3 recaps the transient NO_x contribution on the engine performances during driving cycles. These results correspond to a NECD and a FTP driving cycles performed in conventional engine-driven operations. For the NEDC, the transient NO_x contribution is lower than 9%. However, for the FTP cycle, the transient contribution rises up to 42%.

For the next generation Euro 7 vehicles, a new driving cycle (WLTC) is going to be defined. This latter is more representative of real driving conditions (such as the FTP) and imposes more drastic transient solicitations compared to the actual NEDC. Thus, a higher transient part of the total NO_x emissions is expected. This paper addresses this issue and aims at developing a suitable control strategy for Diesel HEV.

2.2 Transient Torque Control

NO_x peaks occurring during engine transients account for an important part of the total NO_x emissions. The reduction of the transient part of NO_x can be achieved by further improvements of the air system architecture (shorter LP EGR system, combination of HP and LP EGR systems, internal EGR using Variable Valve Actuation (VVA)) or by including a combustion control strategy to adapt injection settings according to air system errors (as proposed by [16]). Here, we consider a Diesel engine with a LP EGR system and without transient combustion control strategy. The goal is to take the advantage of the additional degree of freedom provided by the hybridization only. The idea is to use the electric machine to limit the internal combustion engine dynamics. Some preliminary solutions of transient emission limitation by means of electric boost are proposed in the literature. The limitation of the transient NO_x emissions by an adaptation of the transient torque demand is proposed by [17]. A similar approach, called phlegmatising, was developed by [18]. Recently, Nuësch *et al.* [19] investigate the influence of considering transient emissions in Diesel hybrid electric vehicles. They consider nitrogen oxide and particulate matter using empirical emissions models.

2.2.1 Key Idea and Principle of the Strategy

Figure 7 shows the influence of the torque demand on the amplitude of a NO_x peak obtained with our experimental setup. The NO_x peak amplitude is higher for fast torque transient and these peaks decrease as far as the torque gradient decreases. Also, the NO_x peaks are well correlated with the BGR error ($\varepsilon F_1 = F_1^{sp} - F_1^{est}$). This figure demonstrates that the NO_x emissions are strongly linked with the torque demand and the amplitude of the peak increases with the torque gradient.

These results confirm the conclusion drawn by [17] and [18] who proposed heuristic methods consisting in a limitation of the engine torque setpoint dynamics. In this paper, we apply a similar principle while introducing models of the system to compute the limited engine torque demand. The principle of this strategy is illustrated in Figure 8.

The transient consists in an increasing engine torque step from point A to point B. In this case, the EMS proposes a torque step noted $T_{eng,ss}^{sp}$. This choice is the solution of Equation (15) (respecting (1)) computed from purely static maps without any consideration for the Diesel engine constraints. The principle of the strategy is to keep the same torque setpoint B requested by the EMS but the trajectory followed to reach this value is adapted. The torque setpoint control consists in defining a new torque trajectory $T_{eng,t}^{sp}$ from point A to point B' (in red in *Fig. 8*) such that the transient NO_x peak is avoided or reduced. The torque request at wheel, T_{pwt}^{sp}, depends on the driving conditions (speed and acceleration) and can be evaluated from driver's demand (throttle and brake pedal positions). This demand can be satisfied using the engine, the electric motor or any combination of these two torque sources:

$$T_{pwt}^{sp}(t) = R_1 T_{mot,ss}^{sp}(t) + R_{gb} T_{eng,ss}^{sp}(t) \qquad (18)$$

where R_{gb} is the gear ratio and R_1 is the front axle ratio. The torque setpoint can be positive or negative depending on the vehicle operating conditions (traction or braking). $T_{mot,ss}^{sp}$ is the electric motor torque setpoint and $T_{eng,ss}^{sp}$ is the engine torque setpoint. The steady-state split ratio is chosen by the EMS and the principle of the strategy is to modify its value during the engine transients only. In steady-state, the torque split ratio is maintained. In transient, Equation (18) is not modified but the

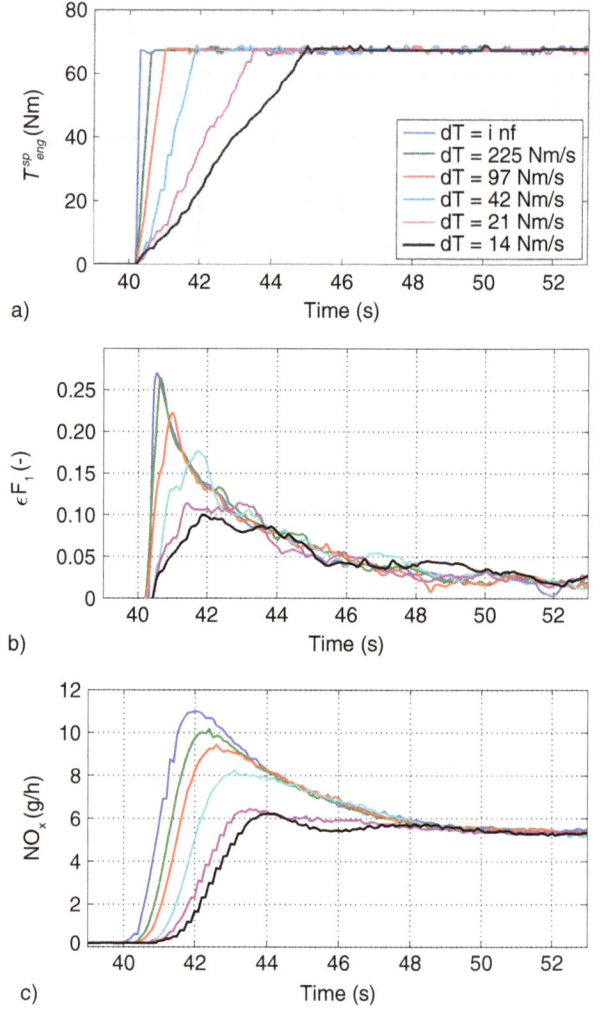

a)

b)

c)

Figure 7

Experimental results of a transient representing an engine cut-off followed by a tip-in with a torque gradient limitation. a) Engine torque, b) BGR error and c) NO_x emissions.

steady-state engine and motor torques become two trajectories:

$$T_{pwt}^{sp}(t) = R_1 T_{mot,t}^{sp}(t) + R_{gb} T_{eng,t}^{sp}(t) \qquad (19)$$

In both cases, the torque request at wheel is not modified. The command $u(t)$ is defined as the motor torque correction $u(t) = \Delta T_{mot}$. Then, the dynamic motor torque request corresponding to the corrected motor torque setpoint is:

$$T_{mot,t}^{sp}(t) = T_{mot,ss}^{sp}(t) + u(t) \qquad (20)$$

The command $u(t)$ is a portion of the motor torque that compensates for the engine torque during transients.

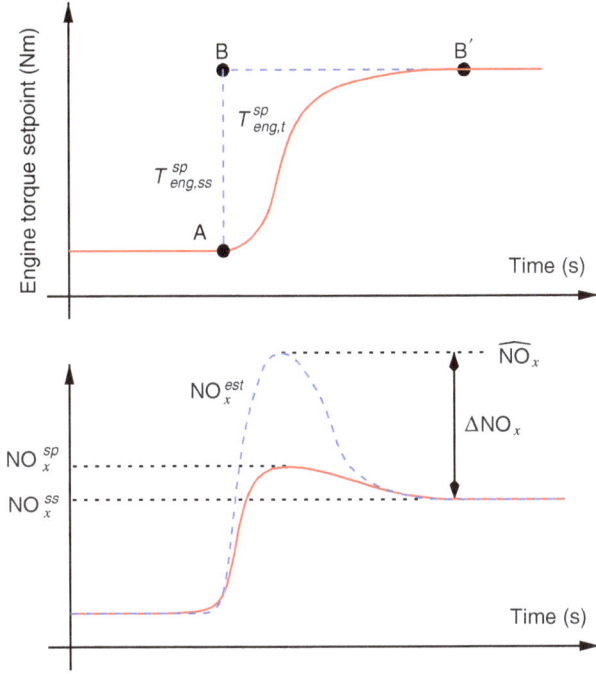

Figure 8

Principle of the control strategy to limit the amplitude of the NO_x peak during a torque step.

The command u is deduced from Equation (19) and (20) and it writes:

$$u(t) = \frac{T_{pwt}^{sp}(t) - R_{gb}T_{eng,t}^{sp}(t)}{R_1} - T_{mot,ss}^{sp}(t) \quad (21)$$

The command $u(t)$ is directly linked to the static motor and powertrain torque setpoints and the dynamic torque setpoint $T_{eng,t}^{sp}$. This latter is a key variable and it is computed such that the NO_x peak generated during a transient is limited. The computation of the corrected engine torque demand $T_{eng,t}^{sp}$ (or trajectory) is explained in the next section. Following this approach, this strategy is applied in cascade with the EMS displayed in Figure 1. The ECMS determines the optimal operating points at some (relatively large) time scale, while a faster controller determines the engine and motor trajectories that minimize the NO_x emissions.

2.2.2 NO_x Modeling

Nitrogen oxides produced by Diesel engines mainly consist of Nitric Oxides (NO) and, to a lesser extent, of nitrogen dioxides (NO_x). In standard Diesel combustion conditions, NO is essentially produced by the extended

Zeldovich mechanism. According to [20], under the equilibrium assumption, the initial NO_x formation rate (in $mol.cm^{-3}.s^{-1}$) may be written:

$$\frac{d[NO]}{d\theta} = \frac{1}{6N_e} \frac{6 \cdot 10^{16}}{\Theta_b^{\frac{1}{2}}} \exp\left(-\frac{69\,090}{\Theta_b}\right) [O_2]_e^{\frac{1}{2}}[N_2]_e \quad (22)$$

where θ represents the crank angle, N_e is the engine speed, Θ_b the temperature in the burned gases and $[X]_e$ refers to the equilibrium concentration of the species X. NO_x formation is thus promoted by high O_2 concentrations and elevated temperatures in the post-combustion gases. Moreover, [21] experimentally shows that the critical time period is when burned gas temperatures are at a maximum. The complexity level of the model employed in the present study to predict NO_x emissions has to be compatible with its integration in the transient torque control strategy, preventing the use of a crank angle resolved thermodynamical model. As the torque trajectory is determined a priori, the use of measured values as input variables of the model is also banned. A semi-physical mean value model proposed by [22], and inspired by NO_x kinetics, has been chosen to estimate NO_x emissions. A detailed analysis of this model is proposed by [23]. NO_x concentration in the output of the cylinders is expressed as a function of the engine speed N_e, the intake manifold burned gas ratio F_1 and the maximum value of in-cylinder temperature $\hat{\Theta}_{cyl}$, according to the following expression:

$$NO_x = \alpha_1\left(\frac{N_e}{\alpha_2}\right)^{\alpha_3}\left(\alpha_4\left(\hat{\Theta}_{cyl} - \alpha_5\right)\right)^{\alpha_6(1-\alpha_7F_1)} \quad (23)$$

The coefficients α_i are calibration parameters learnt on experimental data. The values obtained for the Diesel engine considered in this paper are given in Table 4. In this model, the temperature in the burned gases has been replaced by the maximum value of the mean temperature in the cylinder for sake of simplicity, as the temperature in the burned gases is not measurable for control applications. In spite of this simplification, the model predicts NO_x emissions with a good accuracy over the whole engine operating range as well as for variations of BGR, as shown in [23]. An example of the NO_x model sensitivity according to a BGR variation is displayed in Figure 9.

TABLE 4

NO_x model parameters

α_1	α_2	α_3	α_4	α_5	α_6	α_7
0.37	3 250	-0.56	12.4	0.63	3.06	1.15

MVM model validations in transient are reported in Figures 10 and 11.

During transients, the intake manifold gas compositions are not in steady-state operating conditions. The maximum in-cylinder temperature reaches its steady-state values much faster, and as a result, is considered to be quasi-static. A simplified model is used to represent the dynamics of the intake manifold burned gas ratio (*Eq. 24*). It consists of a delayed first order filter of the BGR static value:

$$\tau(t)\dot{F}_1^{est}(t) + F_1^{est}(t) = F_1^{ss}(t - t_d(t)) \qquad (24)$$

where the time constant τ and the time delay t_d are parametrized as a function of the engine speed:

$$\tau(t) = \frac{\beta_f}{N_e(t)} \qquad (25)$$

and

$$\tau_d(t) = \begin{cases} \frac{\beta_d}{N_e(t)}, & \text{if } T_{eng}^{sp}(t) > 0 \\ 0, & \text{else} \end{cases} \qquad (26)$$

The BGR map F_1 has the general form:

$$F_1 = \vartheta(N_e, T_{eng}) \qquad (27)$$

The burned gas ratio dynamic model is compared with experimental results in Figure 12. The model is in good agreement with the observed test data and can be used as a reference EGR system model into the torque control strategy.

Table 5 compares the estimated cumulated emissions and the NO_x transient contribution on these emissions during driving cycles with the corresponding experimental results given in Table 3. The model correctly predicts both the cumulated NO_x emissions and the transient part. It is employed to validate the NO_x limitation strategy in simulation.

2.2.3 Engine Torque Setpoint Correction

This section explains how to compute the corrected engine torque trajectory. The transient correction of

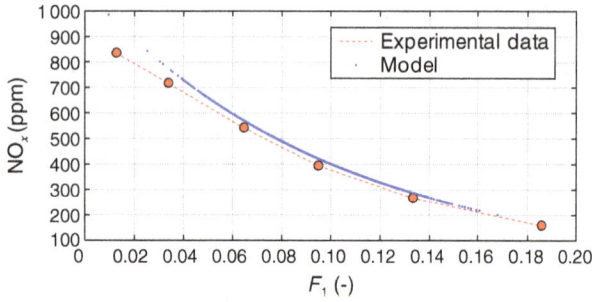

Figure 9

Comparison between the NO_x model and experimental data for a BGR variation.

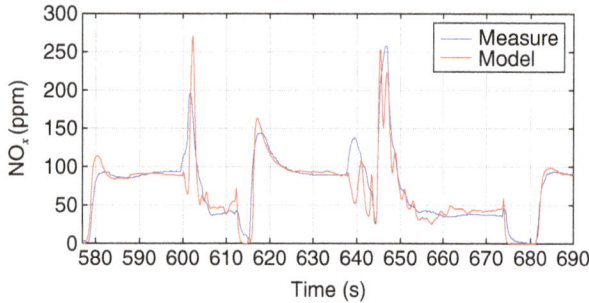

Figure 10

Comparison between NO_x model and experimental data for a portion of NEDC driving cycle.

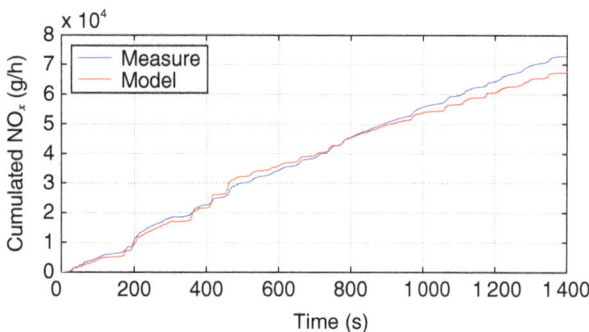

Figure 11

Comparison between the cumulated NO_x model and experimental data for a FTP driving cycle.

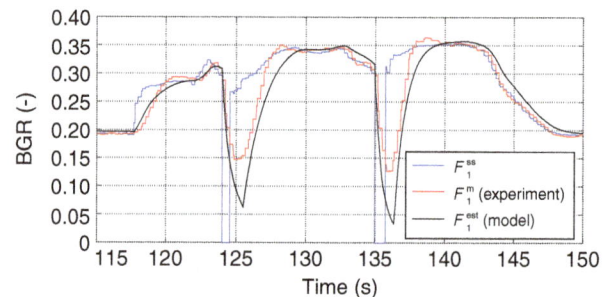

Figure 12

Comparison between measured and estimated intake manifold burned gas fraction (top) and NO_x emission (bottom) during a gear change.

TABLE 5

Comparison between measured and estimated NO_x emissions during NEDC and FTP driving cycle. The experimental Steady-State Part (SSP) of the emissions are computed from the static map. The estimated steady-state emissions are calculated from the model using the intake BGR setpoint F_1^{sp}

| | Driving cycle | | | | | |
| | NEDC | | | FTP | | |
	Total (mg/km)	SSP (mg/km)	TP (%)	Total (mg/km)	SSP (mg/km)	TP (%)
Bench	110	100	9	170	100	42
Model	109	94	14	179	125	30

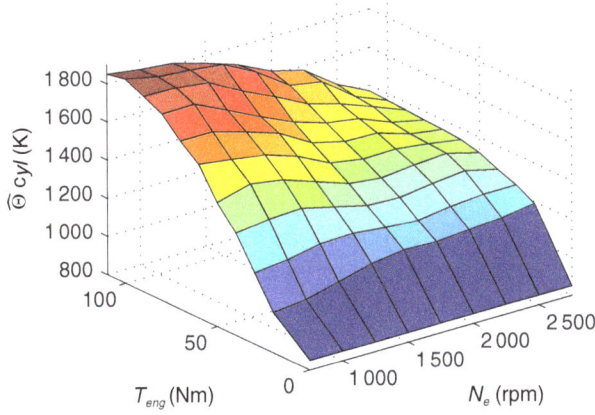

Figure 13

Maximum cylinder temperature map.

the torque setpoint requested by the EMS relies on a Mean Value Model (MVM) of the engine NO_x emissions presented in the previous subsection. This model links the NO_x emissions with the engine speed (N_e), the intake BGR (F_1) and the maximum cylinder temperature ($\hat{\Theta}_{cyl}$). These variables are recognized as first order variables of NO_x formation in compression-ignition engines. The NO_x model, has the general form:

$$NO_x = \phi(N_e, \hat{\Theta}_{cyl}, F_1) \tag{28}$$

The steady-state NO_x level is:

$$NO_x^{ss} = \phi(N_e, \hat{\Theta}_{cyl}^{ss}, F_1^{ss}) \tag{29}$$

The BGR sepoint F_1^{ss} is computed from a static map (27). Temperature setpoint, $\hat{\Theta}_{cyl}^{ss}$ is also mapped according to engine speed and torque (Fig. 13):

$$\begin{cases} F_1^{ss} &= \vartheta(N_e, T_{eng,ss}^{sp}) \\ \hat{\Theta}_{cyl}^{ss} &= \psi_1(N_e, T_{eng,ss}^{sp}) \end{cases} \tag{30}$$

During a transient, the BGR is not equal to its setpoint due to the EGR system lag (as displayed in Fig. 6). For the engine control, an online estimation of the BGR is used [24]. This estimation is valid for the actual BGR value, however, we need to determine the expected BGR value before the transient has occurred. Then, this estimation cannot be used for the high level torque control strategy. Here, the burned gas transport delay from exhaust manifold to intake manifold is modeled using the MVM detailed previously. This model provides the estimated BGR (F_1^{est}) assuming a first order dynamics and a pure delay applied to the BGR map as described in Equation (24). The function ϕ appearing in Equation (28) is invertible and a cylinder temperature setpoint can be computed from this BGR estimation and from the target NO_x^{sp} (Fig. 8):

$$\hat{\Theta}_{cyl}^{sp} = \phi^{-1}(N_e, NO_x^{sp}, F_1^{est}) \tag{31}$$

The NO_x target computation is explained in the next subsection. Knowing $\hat{\Theta}_{cyl}^{sp}$, the cylinder temperature map ψ_1 can be inverted in order to find the corrected value of the engine torque (or torque trajectory):

$$T_{eng}^t = \psi_1^{-1}(N_e, \hat{\Theta}_{cyl}^{sp}) \tag{32}$$

For sake of simplicity, we assume that the static maximum temperature map ψ_1, established for the nominal BGR setpoint, is representative of the actual BGR condition. A dependency of the temperature map according to the actual BGR would provide acceptable results. For that, the cylinder temperature map ψ_1 can be modified as follows:

$$\hat{\Theta}_{cyl} = \psi_2^{-1}(N_e, T_{eng}, F_1) \tag{33}$$

Adding this dependency into a modified function ψ_2 will increase the calibration part of the strategy. We are currently investigating a model-based approach to reduce the experimental tests needed for the calibration of the maximum temperature function ψ_2. This work is not the purpose of this paper and will be reported in a further paper with the justification for the BGR dependency simplification in map ψ_1.

2.2.4 NO$_x$ Target Definition

This section explains how to compute the NO$_x$ trajectory used for torque trajectory computation. The NO$_x$ target definition is purely heuristic and relies on a tunable reduction factor of the maximum NO$_x$ peak amplitude. The achievable NO$_x$ target NO$_x^{sp}$ (*Fig. 8*) is supposed to be included between the actual (or estimated) value and the steady-state value:

$$\mathrm{NO}_x^{est} \geq \mathrm{NO}_x^{sp} \geq \mathrm{NO}_x^{ss} \qquad (34)$$

The target NO$_x$ is tuned empirically with a reduction factor ξ such that:

$$\mathrm{NO}_x^{sp} = \mathrm{NO}_x^{ss} + \Delta\mathrm{NO}_x\left(1 - \frac{\xi}{100}\right) \qquad (35)$$

where $\Delta\mathrm{NO}_x$ is the amplitude of the NO$_x$ peak referred as the steady NO$_x$ value as shown in Figure 8:

$$\Delta\mathrm{NO}_x = \widehat{\mathrm{NO}_x} - \mathrm{NO}_x^{ss} \qquad (36)$$

This method is simple and allows to flexibly tune the level of NO$_x$ reduction. However, the ability for the system to achieve this target is not guaranteed because the system saturation is not considered. Moreover, the reduction factor ξ value is constant for each transient. Including the system saturation to define what can be the reachable NO$_x$ target is a necessary improvement to make the transient torque controller more generic and easy to tune. This will lead to define a limiting factor ξ according to a feasible NO$_x$ target instead of an empirical one. This is the main perspective of this work.

2.2.5 Actuator Limitation and System Saturation

The torque control strategy must account for the actuator limitation that depends on the maximum motor torque T_{mot}^{max} and the static motor torque defined by:

$$T_{mot,ss}^{sp}(t) = \frac{T_{pwt}^{sp}(t) - R_{gb}T_{eng,ss}^{sp}(t)}{R_1} \qquad (37)$$

Then the command $u(t)$ is bounded such that:

$$u(t) \in \left[0(T_{mot}^{max} - T_{mot,ss}^{sp}(t))\right] \qquad (38)$$

From Equations (21, 37) and (38), the minimum and maximum engine torque values write:

$$\begin{cases} T_{eng}^{sp,min} & = \frac{T_{pwt}^{sp} - R_1 T_{mot}^{max}}{R_{gb}} \\ T_{eng}^{sp,max} & = T_{eng,ss}^{sp} \end{cases} \qquad (39)$$

The transient engine torque setpoint is obtained by saturating the engine torque trajectory (32):

$$T_{eng,t}^{sp} = sat\left(\min(T_{eng}^t, T_{eng}^f), T_{eng}^{sp,min}, T_{eng}^{sp,max}\right) \qquad (40)$$

The notation $sat(u, u_m, u_M)$ is used for the function defined by:

$$sat(u, u_m, u_M) = \begin{cases} u_{min} & \text{if} & u(t) \leq u_m \\ u & \text{if} & u_m \leq u(t) \leq u_M \\ u_{max} & \text{if} & u_M \leq u(t) \end{cases} \qquad (41)$$

T_{eng}^f is the feasible torque trajectory that defines the achievable transient NO$_x$ emissions. This latter corresponds to the existing minimum value of the NO$_x$ emissions that can be performed during transient conditions where the cylinder oxygen content (*i.e.* BGR) is not in steady-state condition. The system saturation is not included in this paper and is an ongoing work at IFP Energies nouvelles. Thus, the transient engine torque setpoint writes:

$$T_{eng,t}^{sp} = sat\left(T_{eng}^t, T_{eng}^{sp,min}, T_{eng}^{sp,max}\right) \qquad (42)$$

2.3 Results

This section presents the simulation results of the transient torque control strategy. The objective is to determine whether limiting the Transient Part (TP) of the total NO$_x$ emissions is possible or not. We also would like to characterize the influence of the strategy on the Steady-State Part (SSP) of the total NO$_x$ emissions as well as the Fuel Consumption (FC).

A simulation platform modelling the complete hybrid vehicle has been developed to simulate driving cycles. In this paper, we consider the FTP. The idea is to simulate the system for several values of the NO$_x$ reduction factor ξ ranging from 0% (baseline case) to 100%. This latter case corresponds to a limitation of all the NO$_x$ emissions TP. In order to compare these simulations, the equivalence factor is adapted (dotted line in *Fig. 1*). The value of the equivalence factor was found by dichotomy in order to satisfy the constraint on the final battery state of charge:

$$SOC(t_0) = SOC(t_f) = 50\% \qquad (43)$$

t_0 and t_f are the initial and final times of the considered driving cycle. The simulations are made for three hybridization levels defined by the power of the electric motor:

$$P_{mot} = \{8 \text{ kW}; 14 \text{ kW}; 20 \text{ kW}\} \qquad (44)$$

To be realistic, the battery capacity and vehicle mass are adapted to the electric motor power as described in [15]. Pure electric driving is only enabled in the 20 kW case. The EMS finds the steady-state optimal torque split repartition which minimizes a trade-off between quasi-static NO_x and fuel consumption. The results displayed are obtained for a constant value of the parameter $k_{fc/NO_x} = 0.4$ in Equation (15). The impact of the hybridization level and this setting variable on NO_x emissions and fuel consumption for a Diesel HEV is not the purpose of this paper since it affects only the static balance between NO_x and FC. A sensitivity analysis was done by [25].

The strategy acts on transient phases involving increasing BGR setpoint. Figure 14 illustrates its action on two torque transients belonging to the FTP (*Fig. 15*). $\xi = 0\%$ corresponds to the reference case, without transient strategy; the two others cases (50% and 90%) illustrate the impact of the adjustable transient NO_x reduction parameter. For the second torque transient ($t = 203$ s), the NO_x peak can be completely avoided and the tuning parameter allows a flexible reduction of the peak's amplitude. This reduction is made by limiting the maximal cylinder temperature until enough burned gas is available. In the first torque transient ($t = 199$ s), a saturation on NO_x reduction is observed between the 50% and 90% cases. This saturation can be easily explained looking at the motor torque setpoint which has already reached its maximal value (T_{mot}^{max}) for $\xi = 50\%$. As a consequence, NO_x abatement is not achievable without increasing the electric motor maximum torque (*i.e.* the motor size). We can notice that the strategy does not affect the steady-state engine and motor torque setpoints as claimed in the previous section. They are equal to the one chosen by the EMS once the transient is over ($T_{mot,ss}^{sp} = T_{mot,t}^{sp}$ and $T_{eng,ss}^{sp} = T_{eng,t}^{sp}$).

The impact of the transient strategy on the battery SOC during a complete FTP is illustrated in Figure 16. When the strategy is enabled, the SOC decreases progressively because of the electric energy spent to cut NO_x peaks during each transient phase. The missing energy must be recuperated in order to complete the driving cycle with a final SOC constraint respected. This was done by adjusting the equivalence factor in the energy management strategy which slightly modifies the static operating point choice.

The effect of the strategy on the cumulated NO_x emissions and fuel consumption is illustrated in Figure 17.

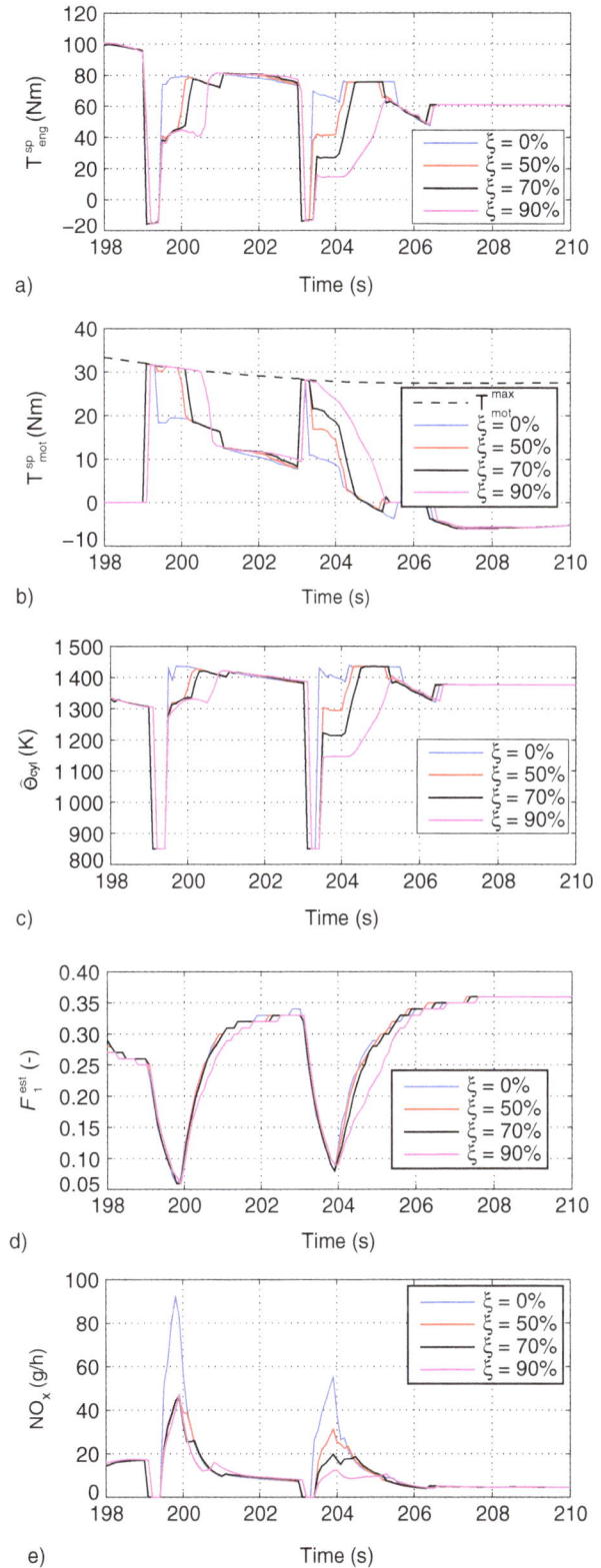

Figure 14

Impact of the transient strategy parameter ξ – FTP Full Hybrid 20 kW. a) Engine torque, b) motor torque, c) maximal cylinder temperature, d) burned gas ratio and e) NO_x emissions.

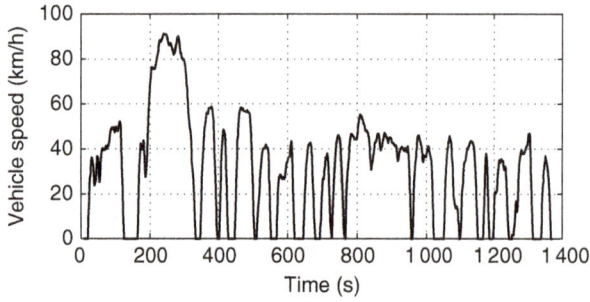

Figure 15

FTP vehicle speed profile.

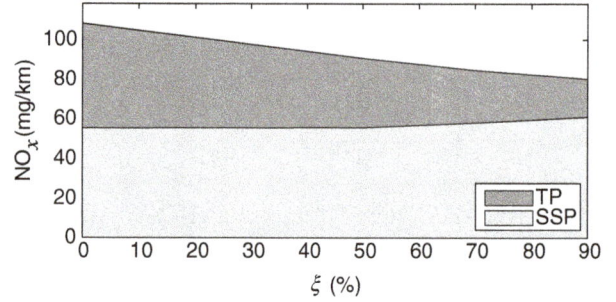

Figure 18

Impact of the transient strategy parameter ξ on the NO_x transient part – FTP Full Hybrid 20 kW.

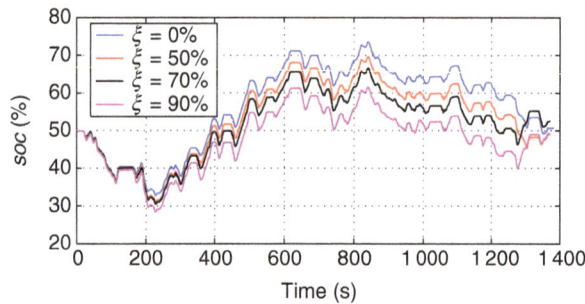

Figure 16

Impact of the transient strategy on the battery SOC – FTP Full Hybrid 20 kW.

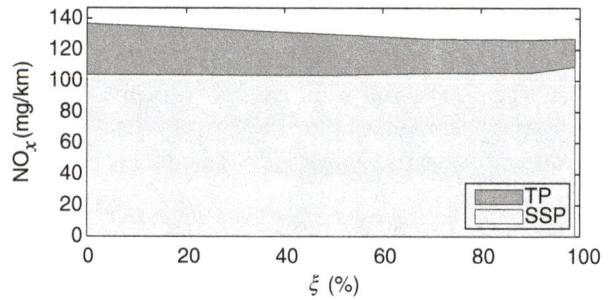

Figure 19

Impact of the transient strategy parameter ξ on the NO_x transient part – FTP Mild Hybrid 14 kW.

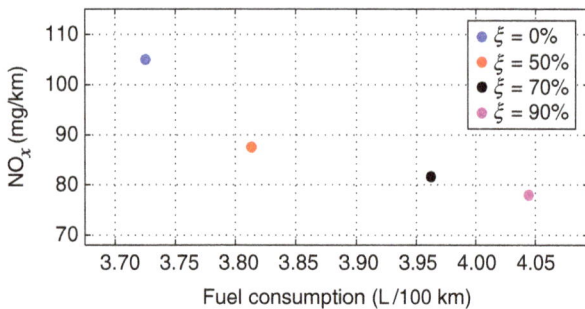

Figure 17

Impact of the transient strategy parameter ξ on the trade-off between NO_x emissions and fuel consumption – FTP Full Hybrid 20 kW.

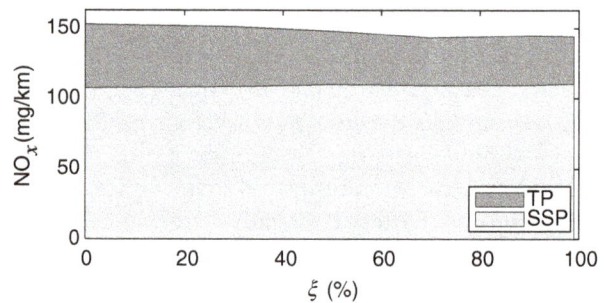

Figure 20

Impact of the transient strategy parameter ξ on the NO_x transient part – FTP Mild Hybrid 8 kW.

Due to transient peak reduction, the total NO_x emissions significantly drop at the price of a slight increase of fuel consumption. This demonstrates the interest of choosing a partial NO_x reduction instead of a total elimination. Indeed in this case, the most interesting case is $\xi = 50\%$ since it allows a global NO_x reduction of 17% with a fuel penalty of only 2%.

The transient part of NO_x emissions can be controlled with the damping factor ξ as illustrated in Figures 18, 19 and 20. The TP of NO_x emissions is defined as the difference between total and steady-state emissions. The total emissions are computed from the model (28):

$$\text{NO}_x^{est} = \phi(N_e, \psi_1(N_e, T_{eng,t}^{sp}), F_1^{est}) \qquad (45)$$

The steady-state NO_x emissions are computed with the same NO_x model. But, instead of using the estimated BGR as input, we use its steady-state value computed from the map (27):

$$\text{NO}_x^{ss} = \phi(N_e, \psi_1(N_e, T_{eng,t}^{sp}), \vartheta(N_e, T_{eng,t}^{sp})) \qquad (46)$$

They represent what the emissions would be if the exhaust gas recirculation loop had an instantaneous settling time. Even if the static part of the cumulated NO_x emissions increases slightly with ξ due to the slight modification of static operating points, the total NO_x emissions are decreasing. The strategy allows to divide the transient part by two (from 49% to 24%) while the cumulated NO_x emissions are reduced by 26%. For $\xi = 90\%$, the remaining transient part corresponds to the saturation of the electric motor which is not able to provide enough torque.

For lower hybridization levels (14 kW and 8 kW), the electric torque available to allow transient NO_x reduction is lower and the emissions can not be reduced as much as in the full hybrid case. In the 8 kW, the electric motor saturation is quickly reached. As a consequence the transient NO_x is limited (*Fig. 20*).

As a conclusion, the transient strategy allows to control the NO_x transient emissions as long as the maximal electric motor torque is not reached. With a 20 kW electric motor, a spectacular NO_x reduction is achievable with a reasonable increase in fuel consumption. The value of the damping factor ξ tunes the trade-off between the NO_x TP and the FC.

CONCLUSION

In Diesel hybrid vehicles, steady-state and transient NO_x emissions must be considered in the supervision strategy. First, we adapted the ECMS-based energy management strategy to account for the steady NO_x. An additional tuning parameter was introduced into the ECMS and allows to tune the trade-off between NO_x abatement and fuel consumption reduction. These results clearly demonstrate that model-based strategy (ECMS) allows an easy integration of the NO_x objective. In addition to the simulation results, the adapted ECMS is validated on the experimental hybrid test bench. The trend observed using the simulator was experimentally reproduced leading to a substantial reduction of the NO_x

emissions. In addition, we introduced a torque control cascaded with the EMS in order to correct the torque split ratio computed by the ECMS. This strategy focuses on transient NO_x emissions and adapts the torque split ratio during transient operations where NO_x peaks are produced. The strategy tends to smooth the engine torque using the motor torque as torque compensator. This controller uses a mean value model for the intake manifold burned gas ratio dynamics and for the NO_x production. Using such models into a vehicle supervision level is the novelty of the approach. The simulation results have shown that the transient part of the NO_x emissions can be reduced. The strategy allows to limit the transient part of the emissions without modifying the steady-state part. Also, the use of the electric motor during transients has a price and the transient NO_x reduction slightly increases fuel consumption. This trade-off can be tuned with the reduction factor. The proposed strategy can be valuable to deal with driving cycles presenting more transient phases. This will be the case for Euro 7 vehicles with the new worldwide harmonized light-duty transient cycle.

REFERENCES

1 Chasse A., Chauvin J., Sciarretta A. (2010) Online optimal control of a parallel hybrid with costate adaptation rule, *Proceedings of the 6th IFAC Symposium Advances in Automotive Control*, Munich, 12-14 July. pp. 318-331.

2 Johnson V.H., Wipke K.B., Rausen D.J. (2009) HEV control strategy for real-time, *SAE Technical Paper* 2009-24-0068.

3 Montazeri-Gh M., Poursamad A., Ghalichi B. (2006) Application of genetic algorithm for optimization of control strategy in parallel hybrid electric vehicles, *Journal of the Franklin Institute* **343**, 4-5, 420-435.

4 Lin C.-C., Peng H., Grizzle J.W., Kang J.-M. (2003) Power management strategy for a parallel hybrid electric truck, *IEEE Transactions on Control Systems Technology* **11**, 6, 839-849.

5 Musardo C., Rizzoni G., Staccia B. (2005) A-ECMS: An adaptive algorithm for hybrid electric vehicle energy management, *Proceedings of the 44th IEEE Conference on Decision and Control*, Seville, 12-15 Dec.

6 Musardo C., Staccia B., Midlam-Mohler S., Guezennec Y., Rizzoni G. (2005) Supervisory control for NO*x* reduction of an HEV with a mixed-mode HCCI/CIDI engine, *Proceedings of the American Control Conference*, Portland, OR, 8-10 June.

7 Dextreit C., Kolmanovsky I.V. (2013) Game theory controller for hybrid electric vehicles, *IEEE Transactions on Control Systems Technology* **PP**, 99, 1-1.

8 Chasse A., Pognant-Gros P., Sciarretta A. (2009) Online implementation of an optimal supervisory control for a parallel hybrid powertrain, *SAE Paper* 2009–01-1868, *SAE Int. J. Engines* **2**, 1, 1630-1638.

9 Del Mastro A., Chasse A., Pognant-Gros P., Corde G., Perez F., Gallo F., Hennequet G. (2009) Advanced hybrid vehicle simulation: from "virtual" to "HyHiL" test bench, *SAE Technical Paper* 2009-24-0068.

10 Ambühl D., Sciarretta A., Onder C., Guzzella L., Sterzing S., Mann K., Kraft D., Küsell M. (2007) A causal operation strategy for hybrid electric vehicles based on optimal control theory, *Proceedings of the 4th Symposium on Hybrid Vehicles and Energy Management*, Stadthalle Branuschweig, Germany, Feb. pp. 318-331.

11 Sciarretta A., Guzzella L. (2007) Control of hybrid electric vehicles - optimal energy management strategies, *IEEE Control Systems Magazine* **27**, 2, 60-70.

12 Sciarretta A., Back M., Guzzella L. (2004) Optimal control of parallel hybrid electric vehicles, *IEEE Transactions on Control Systems Technology* **12**, 3, 352-363.

13 Chasse A., Hafidi G., Pognant-Gros P., Sciarretta A. (2009) Supervisory control of hybrid powertrains: an experimental benchmark of offline optimization and online energy management, *Proceedings of E-COSM'09 - IFAC Workshop on Engine and Powertrain Control, Simulation and Modeling*, Rueil-Malmaison, France, Dec.

14 Thibault L., Leroy T. (2013) Optimal online energy management for Diesel HEV: Robustness to real driving conditions, *SAE Technical Paper* 2013-01-1471.

15 Grondin O., Thibault L., Moulin P., Chasse A. (2011) Energy management strategy for Diesel hybrid electric vehicle, *Proceedings of the 7th IEEE Vehicle Power and Propulsion Conference (VPPC'11)*, Chicago, USA, 6-9 Sept.

16 Hillion M., Chauvin J., Petit N. (2011) Control of highly diluted combustion in Diesel engines, *Control Engineering Practice* **19**, 11, 1274-1286.

17 Predelli O., Bunar F., Manns J., Buchwald R., Sommer A. (2007) Laying out Diesel-engine control systems in passenger-car hybrid drives, *Proceeding of the 4th symposium on Hybrid Vehicle and Energy Management*, Stadthalle Branuschweig, Germany, 14-15 Feb. pp. 131-151.

18 Lindenkamp N., Stöber-Schmidt C.-P., Eilts P. (2009) Strategies for reducing NOx and particulate matter emissions in Diesel hybrid electric vehicles, *SAE Technical Paper* 2009-01-1305.

19 Nüesch T., Wang M., Voser C., Guzzella L (2012) Optimal energy management and sizing for hybrid electric vehicles considering transient emissions, *2012 IFAC Workshop on Engine and Powertrain Control Simulation and Modeling*, Rueil-Malmaison, France, 23-25 Oct.

20 Bowman C.T. (1975) Kinetics of pollutant formation and destruction in combustion, *Progress in Energy and Combustion Science* **1**, 1, 33-45.

21 Heywood J.B. (1988) *Internal Combustion Engine Fundamentals*, McGraw-Hill, New york.

22 Schmitt J.-C., Fremovici M., Grondin O., Le Berr F. (2009) Compression ignition engine model supporting powertrain development, *Proceeding of the E-COSM Symposium*, Rueil-Malmaison, France, 30 Nov. – 2 Dec. pp. 75-82.

23 Quérel C., Grondin O., Letellier C. (2012) State of the art and analysis of control oriented NOx models, *SAE Technical Paper* 2012-01-0723.

24 Grondin O., Moulin P., Chauvin J. (2009) Control of a turbocharged Diesel engine fitted with high pressure and low pressure exhaust gas recirculation system, *Proceedings of the 48th IEEE Conference on Decision and Control & Chinese Control Conference*, Shanghai, China, 16-18 Dec.

25 Thibault L., Grondin O., Quérel C., Corde G. (2012) Energy management strategy and optimal hybridization level for a Diesel HEV, *SAE Technical Paper* 201201-1019.

6

Investigation of Cycle-to-Cycle Variability of NO in Homogeneous Combustion

A. Karvountzis-Kontakiotis and L. Ntziachristos*

Aristotle University of Thessaloniki, Laboratory of Applied Thermodynamics (LAT), GR54125, POB 458, Thessaloniki, Greece
e-mail: akarvout@auth.gr - leon@auth.gr

* Corresponding author

Abstract — Cyclic variability of spark ignition engines is recognized as a scatter in the combustion parameter recordings during actual operation in steady state conditions. Combustion variability may occur due to fluctuations in both early flame kernel development and in turbulent flame propagation with an impact on fuel consumption and emissions. In this study, a detailed chemistry model for the prediction of NO formation in homogeneous engine conditions is presented. The Wiebe parameterization is used for the prediction of heat release; then the calculated thermodynamic data are fed into the chemistry model to predict NO evolution at each degree of crank angle. Experimental data obtained from literature studies were used to validate the mean NO levels calculated. Then the model was applied to predict the impact of cyclic variability on mean NO and the amplitude of its variation. The cyclic variability was simulated by introducing random perturbations, which followed a normal distribution, to the Wiebe function parameters. The results of this approach show that the model proposed better predicts mean NO formation than earlier methods. Also, it shows that to the non linear formation rate of NO with temperature, cycle-to-cycle variation leads to higher mean NO emission levels than what one would predict without taking cyclic variation into account.

Résumé — Enquête de la variabilité cycle-à-cycle du NO dans la combustion homogène — La variabilité cyclique des moteurs à allumage commandé est reconnue comme une dispersion dans les enregistrements des paramètres de combustion lors du fonctionnement réel dans des conditions stables. Des variabilités de combustion peuvent se produire en raison des fluctuations dans le développement précoce du noyau de la flamme et dans la propagation turbulente de la flamme avec un impact sur la consommation de carburant et les émissions. Cette étude présente un modèle chimique détaillé pour prévoir la formation de NO dans des conditions de combustion homogène. Le paramétrage de Wiebe est utilisé pour prévoir le dégagement de chaleur ; les données thermodynamiques calculées sont ensuite intégrées au modèle chimique pour prévoir l'évolution de NO à chaque degré d'angle de rotation du vilebrequin. Les données expérimentales obtenues à partir de l'analyse des publications antérieures ont été utilisées pour valider les niveaux moyens de NO calculés. Le modèle a ensuite été appliqué pour prévoir l'impact de la variabilité cyclique sur le taux moyen de NO formé et l'amplitude de sa variation. La variabilité cyclique a été simulée en introduisant des perturbations aléatoires qui suivent une distribution normale, aux paramètres de la fonction de Wiebe. Les résultats de cette approche montrent que le modèle proposé prédit mieux le taux moyen de formation de NO que les méthodes précédentes. Les résultats montrent également qu'une vitesse de formation non linéaire de NO avec la température et la variation cycle-à-cycle, entraîne une moyenne plus élevée des niveaux d'émission de NO que celle prédite sans prendre en compte la variation cyclique.

NOMENCLATURE

ABDC	After Bottom Dead Centre
ATDC	After Top Dead Centre
BBDC	Before Bottom Dead Centre
BTDC	Before Top Dead Centre
CCV	Cycle to Cycle Variation
CD	Combustion Duration
CFD	Computational Fluid Dynamics
COV	Coefficient Of Variation
EVC	Exhaust Valve Close
EVO	Exhaust Valve Open
imep	indicated mean effective pressure
IVC	Intake Valve Close
IVO	Intake Valve Open
m	Wiebe shape coefficient
MC	Mean Cycle
PL	Partial Load
SD	Standard Deviation
SI	Spark Ignition
SOI	Start Of Ignition
WOT	Wide Open Throttle

INTRODUCTION

Combustion in engines evolves differently in each operation cycle even at steady state operating conditions. Experimentally, Cycle-to-Cycle Variability (CCV) is best observed by the scatter of the measured cylinder pressure around the mean pressure curve. Such fluctuations of the cylinder pressure have an impact on engine performance [1], fuel consumption [2] and pollutant emissions [3, 4], while in some extreme cases such as highly diluted lean mixtures could result in misfiring or knocking [2]. The Coefficient Of Variation of the indicated mean effective pressure (COV_{imep}) is used for the classification of CCV [5]. In general, COV_{imep} should be limited to up to about 10% in order to avoid vehicle drivability problems [5, 6].

There are several reasons that may cause CCV. These may include variations in the early flame kernel development due to corresponding spark variance in each cycle or the turbulence conditions in the spark neighbourhood. The kernel development affects flame propagation, which by turn results to different macroscopic combustion parameters. The spark discharge

characteristics [2], the local equivalence ratio of the mixture and its inhomogeneity close to the spark plug [2, 7, 8], turbulence in the vicinity of spark plug at the ignition time [8], and mixture temperature and pressure at the time of ignition [8] are all related with the variations of the early flame kernel development. On the other hand, the overall equivalence ratio [9], the extent of mixture homogeneity [10, 11], the percentage of the residual gas fraction of the mixture [10] and the averaged turbulence intensity [12-16] are factors that affect the main flame propagation.

Combustion CCV also leads to variability in the combustion products. NO_x formation in particular shows a strong dependence on combustion duration. NO_x emissions decrease as combustion time decreases and this dependence becomes stronger as air-fuel ratio becomes leaner [17]. In other studies, it was found that the variance of NO_x is higher compared with the variance of imep and the maximum combustion pressure [3, 18].

There have been several model approaches aiming at simulating combustion development and pollutants formation in SI engines. The Wiebe function [19] has been applied in most studies for the approximation of heat release due to fuel consumption. However this empirical function does not have a physical meaning and its predictability is not always satisfactory. Zero-dimensional phenomenological models may better approach the actual physics, taking into account different temperature zones and compositions. However the turbulence conditions in the combustion chamber cannot be modeled with this kind of models [20], hence they cannot be used to simulate CCV. As a result, CFD models (1D/3D) are mainly used for the simulation of CCV, because they are able to precisely simulate both the rate of the early flame development and the flame propagation [12, 13, 21, 22]. Their disadvantage is their high computational cost and the difficulty in setting up a satisfactory combustion CFD model [20].

In SI modeling, NO emissions are usually simulated by applying the extended Zeldovich mechanism, also known as the thermal mechanism [23, 24]. However, in stoichiometric and slightly rich mixtures, the prompt (also known as Fenimore) mechanism could be responsible for up to fifteen percent of the total nitric oxide emissions [25].

The objective of this study is the investigation of the combustion CCV in nitric oxide emissions, using a detailed chemical mechanism. The simple two-zone

Wiebe model is used for the description of the mixture temperature and pressure during combustion. The thermodynamic parameters for each cycle are used as input in the detailed chemical mechanism for the prediction of NO formation as a function of degree of crank angle. The model is then used to predict the impact of CCV on NO emission levels.

1 MODEL APPROACH

The model presented in this paper consists of a detailed chemical mechanism, coupled to a two zone Wiebe model [19, 23]. For the aim of this study, the three parameters of the Wiebe function were individually perturbated around central values to simulate CCV, thus having an impact on the burning rate and the NO formation.

1.1 Thermodynamic Model

The commercial engine simulation package AVL BOOST was used for the simulation of the heat release rate and the in-cylinder thermodynamic properties. The combustion submodel used for the prediction of heat release was a two-zone Wiebe model. The Wiebe function describes the burned gas mass fraction at a given crank angle:

$$\frac{Q_f(\phi)}{Q_{f,total}} = 1 - e^{-\alpha\left(\frac{\phi - \phi_{SOI}}{\Delta\phi_{CD}}\right)^{m+1}} \qquad (1)$$

In Equation (1), ϕ_{SOI} is the degree of crank angle where ignition starts, $\Delta\phi_{CD}$ is the duration of combustion in crank angle degrees, m is a shape parameter for the Wiebe function, and α is a combustion efficiency parameter.

The two-zone approach consists of a burned zone with a temperature for the combustion products and an unburned zone with a different temperature for the unburned mixture and any residuals from the previous combustion cycle. A uniform pressure for both zones is assumed. Although the Wiebe model is an empirical model and it is not recommended for the investigation of the CCV origins, variation of its parameters provides a good approximation in simulating combustion exothermy variability.

1.2 Emission Model

The chemistry model used to predict NO formation was based on SENKIN, a FORTRAN based code developed in *Sandia Laboratories* [26] that has been later evolved

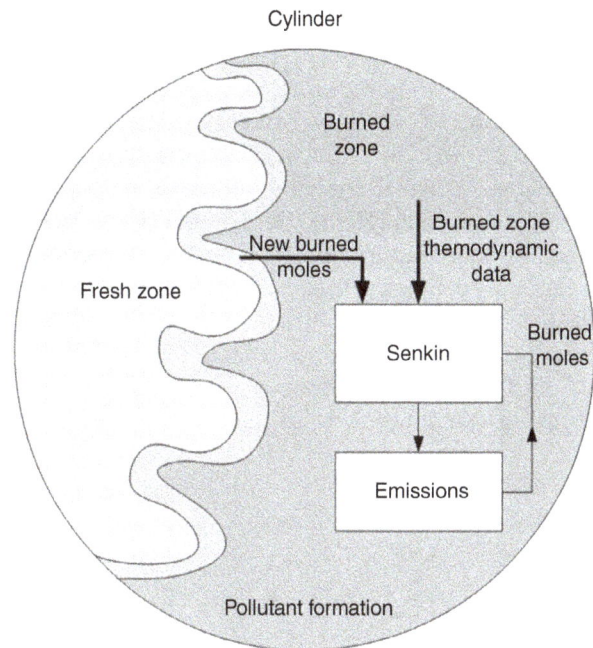

Figure 1

Schematic of the emission modeling approach.

into the CHEMKIN software package. SENKIN calculates combustion evolution in homogeneous gas phase mixtures. The code solves the chemical kinetics differential equations and predicts the formation rate of products. This solution can refer either to constant pressure, constant volume, or constant temperature conditions. The default reaction scheme of SENKIN v1.8 that was used in this study, consisted of 53 species and 325 chemical reactions [26, 27]. The reaction scheme involved of a number of carbon-nitrogen species and radicals which are relevant in the NO formation chemistry, including HCN, H_2CN, CN, HCNO and HOCN.

Figure 1 illustrates the coupling between the thermodynamic and the chemical kinetic modeling developed in the current study. The thermodynamic data for the hot zone are imported in the converter at each crank angle. The burned mass fraction from the Wiebe function defines the newly burned moles which enter from the flame front to the burned zone. The newly burned moles are calculated from the oxidation rate of the fuel, according to the stoichiometry of the combustion shown in Table 1.

The newly burned moles and the composition from the previous step are imported as the initial input composition of the burned zone in SENKIN. SENKIN calculates as an output the new composition of the burned zone which will be again imported in the next crank angle.

When the thermodynamic model calculates the end of combustion, no new moles are assumed in the SENKIN input scenario. The loop therefore ends and kinetics are thereafter considered frozen. In earlier typical two zone models [18, 22, 23, 28] only the thermal mechanism was considered, while the other necessary species for the thermal mechanism (H_2, H, O_2, O, OH, H_2O) were calculated assuming equilibrium. The proposed emission model uses a detailed chemical mechanism which includes the thermal and the prompt mechanism, while the other necessary species are calculated from detailed kinetics. This improves the precision in NO_x prediction, with a cost in computational time. A reduced detailed chemical mechanism but with explicit kinetics for intermediate species could serve as a compromise between accuracy and computational time.

1.3 Modeling of NO CCV

The modeling of NO CCV was performed by introducing perturbations into the Wiebe function parameters, regarding the ignition timing (SOI), the Combustion Duration (CD) and the parameter m. Each of these three parameters was described by a normal distribution, characterized by a mean value and a standard deviation. The mean value of each distribution was the Wiebe value of the mean-cycle model, while the range of perturbations was taken from experimental data, as it will be later discussed. Finally the CCV thermodynamic data were imported in the detailed chemical mechanism. This procedure was modeled in MATLAB.

1.4 Modeling Assumptions

The following assumptions were considered for the simplification of the emission modeling and the CCV analysis:
- uniform pressure in the cylinder (burned and unburned zone at the same pressure);

TABLE 1

The new species imported in the emission model at each iteration, assuming C_aH_b as the chemical formula of the hydrocarbon fuel

Species	$\lambda \geq 1$	$0.9 < \lambda < 1$
CO_2	a	$a \cdot (2\lambda - 1) + b/2 \cdot (\lambda - 1)$
H_2O	$b/2$	$b/2$
CO	0	$(2a + b/2) \cdot (1 - \lambda)$
O_2	$(2a + b/2) \cdot (\lambda - 1)$	0
N_2	$3.76 \cdot \lambda \cdot (a + b/4)$	$3.76 \cdot \lambda \cdot \alpha s$

- a complete combustion of hydrocarbon fuel with air;
- uniform composition in the burned zone;
- NO_x emissions solely consisting of NO.

The validity and impact of these assumptions in the final results is investigated in the results section.

2 EXPERIMENTAL DATA

Experimental data are necessary for the validation of the model developed in this study. In most CCV analysis, only the thermodynamic data are measured, without considering the emission data. Ball *et al.* (1998) [4] used experimental data from a *Rover* K4 optical engine to investigate cycle-to-cycle variation in combustion and NO emissions. The fuel used in those experiments was methane. That engine from the Ball *et al.* (1998) [4] work was simulated in the present study, as many engine specifications necessary for the modeling are contained in that publication and are summarized in Table 2. The model was applied to this engine and the results of the

TABLE 2

The *Rover* K4 optical engine characteristics (Ball *et al.*, 1998) [4]

Main specification		
Bore	80	(mm)
Stroke	89	(mm)
ConRod length	160	(mm)
Compression ratio	10	(-)
Cam timing		
IVO	12	BTDC
IVC	52	ABDC
Peak lift inlet	8.8	(mm) at 70 BBDC
EVO	52	BBDC
EVC	12	ATDC
Peak lift exhaust	8.8	(mm) at 70 ABDC
Cylinder head		
Type		Rover K16 1.4 MPI
Pent angle	45	
Inlet valve seat diameter	24	(mm)
Inlet valve seat diameter	19.6	(mm)
Number of valves	4	(-)

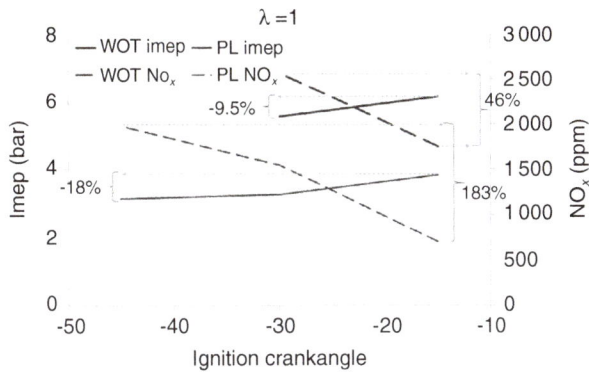

Figure 2

Effect of the ignition timing on imep and NO$_x$ for stoichiometric conditions.

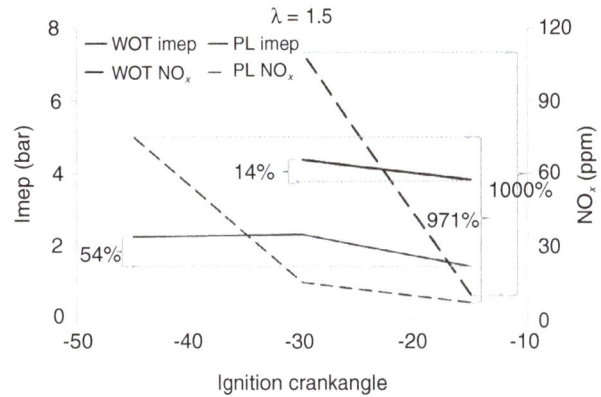

Figure 3

Effect of ignition timing on imep and NO$_x$ for lean conditions.

simulations were compared with the experimental data for validation.

This optical engine was measured under partial load and Wide Open Throttle conditions (WOT), for different crank angle ignition durations and lambda values. Information about the engine performance and the engine emissions (NO$_x$, HC) was also available for each measured engine point.

3 EFFECT OF ENGINE OPERATION PARAMETERS ON EMISSIONS

Based on the experimental data presented in the previous section, Figure 2 presents a graph of imep and NO$_x$ concentration for stoichiometric combustion, during Partial Load (PL) and WOT operation. It is observed that while the Start Of Ignition (SOI) changed from 15° BTDC to 45° BTDC partial load imep only differed by 18%, the WOT imep differed by 9.5%, while NO$_x$ concentrations changed by 183% and 46%, respectively. This shows how much more sensitive NO formation is than the thermodynamic properties of the engine when combustion parameters change.

The corresponding graph for lean operation ($\lambda = 1.5$) is shown in Figure 3. The impact of the variation in combustion parameters on NO$_x$ formation is even more magnified in this case compared to the stoichiometric combustion.

By comparing the two cases, it is observed that in lean operating conditions, the impact of the ignition timing on the indicated mean effective pressure is higher compared to the stoichiometric mixture, an observation which is in agreement with other studies [2-9]. From these data it seems that cycle-to-cycle combustion

variability is more pronounced in lean and highly diluted mixtures, even with slight modification of the combustion parameters. In addition such conditions lead to high NO$_x$ formation, hence the CCV effect is magnified in this case as well.

This non-linearity of NO$_x$ formation is not easy to simulate in detail with a simplified mechanism. Hence, a detailed and more precise chemical mechanism is applied in this study, in order to simulate this non-linearity and high sensitivity in NO$_x$ formation. The model presented in this study can be used to predict the amplitude of variation of NO$_x$ emissions due to CCV and, in this way, to more accurately predict the compliance of an engine with a given emission limit target.

4 RESULTS

For validation, the proposed emission model is first used to predict the *Rover* K4 NO$_x$ measured emissions at both stoichiometric and lean conditions. First, the measured data of *Rover* K4 are used regarding NO$_x$ emissions and engine performance characteristics to relate the tendency between performance and emissions. Second, the comparison between simulated and experimental cycle-averaged NO data is presented for the validation of the simulation. Measurement and simulation are discussed and the importance of the prompt NO formation mechanism is justified. Last but not least, the NO CCV is investigated.

4.1 Mean Cycle NO Modeling

The *Rover* K4 was simulated with the AVL BOOST model for mean cycle Wiebe parameters and the results

TABLE 3
Validation of the thermodynamic model

Case	λ	θign (BTDC)	Imep (bar)	Pmax (bar)	CaPmax (ATDC)	10% MFB (ATDC)
Part load						
P1015 exp	1	15	3.84	18.05	20	5
P1015 sim	1	15	3.84	16.45	20.11	4.94
P1030 exp	1	30	3.27	25.92	7	−10
P1030 sim	1	30	3.27	20.47	10.92	−11.42
P1045 exp	1	45	3.15	29.34	2	−18
P1045 sim	1	45	3.14	32.37	3.07	−18.32
P1515 exp	1.5	15	1.46	10.23	0	19
P1515 sim	1.5	15	1.44	12.2	1.34	19.37
P1530 exp	1.5	30	2.33	15.13	12	−2
P1530 sim	1.5	30	2.33	14.96	11.84	−2.24
P1545 exp	1.5	45	2.25	21.44	6	−16
P1545 sim	1.5	45	2.23	25.48	6.26	−15.91
WOT						
W1015 exp	1	15	6.14	33.02	18	3
W1015 sim	1	15	6.15	35.76	19.99	2.27
W1030 exp	1	30	5.55	45.68	5	−11
W1030 sim	1	30	5.55	55.34	7.45	−11.18
W1515 exp	1.5	15	3.83	17.33	10	11
W1515 sim	1.5	15	3.82	21.07	3.51	10
W1530 exp	1.5	30	4.38	29.2	12	−3
W1530 sim	1.5	30	4.38	29.67	13.83	−3.23

were compared with the experimental cycle-averaged engine data found in Ball *et al.* (1998) [4]. The comparison between experimental and simulated data refers to the imep, the maximum pressure during the combustion phase, the crank-angle degree where maximum pressure occurs, and the crank angle degree where 10% of the fuel mass is burned (MFR). All these data are presented in Table 3. The designation of each point in Table 3 is done with the P and W initials corresponding to partial load or wide open throttle operation, respectively, followed by two digits corresponding to the lambda value (10 corresponding to $\lambda = 1$ and 15 corresponding to $\lambda = 1.5$), followed by two digits of crank angle degree where ignition starts before top dead centre.

The predicted thermodynamic data of the ten simulated operating points were used as an input for the NO prediction. The simulated NO emissions are compared with the experimental NO emissions in Figure 4 for stoichiometric combustion and in Figure 5 for the lean combustion. In the stoichiometric combustion, NO emissions are presented with and without the effect of the prompt mechanism. The prompt mechanism has been switched off by zeroing the HCN radicals in the chemical mechanism.

The model appears to have a rather good accuracy over a wide NO_x range, that is from NO_x concentrations of less than 10 ppm (P1515) to more than 2 000 ppm (W1030). For these cases where large differences can be seen (*e.g.* W1015), one should also observe related

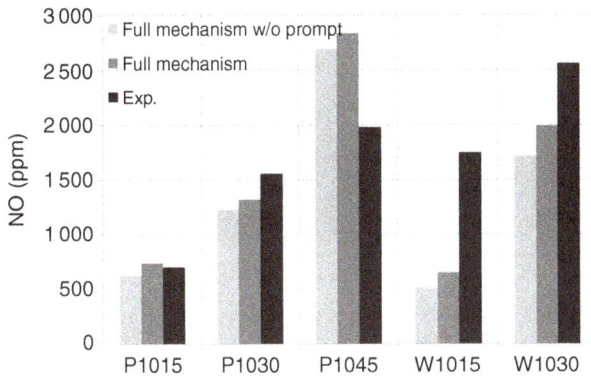

Figure 4

Comparison of measured and simulated NO molar fractions for stoichiometric combustion. Results without prompt mechanism are also included.

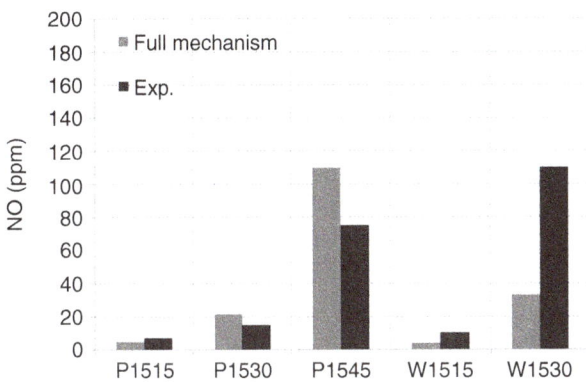

Figure 5

Comparison of measured and simulated NO molar fractions for lean ($\lambda = 1.5$) operating engine conditions.

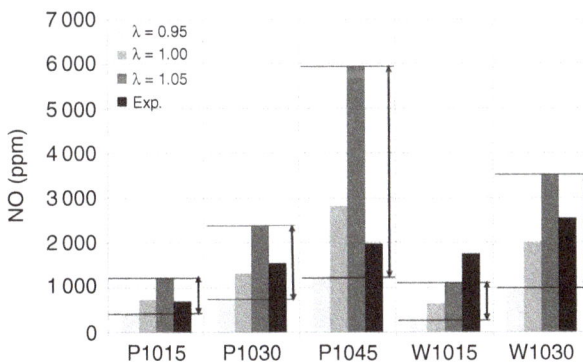

Figure 6

Impact of slight stoichiometry variation on NO_x formation.

differences in the thermodynamic data and not only in the reaction modeling. Cases with lower thermodynamic error show better prediction in NO_x results (example P1015). By using a more sophisticated combustion model [21, 22], the burning rate prediction could be improved with significant improvement in NO_x prediction as well.

As one might expect, the availability of oxygen is the key variable affecting NO_x prediction. This may be an additional reason of difference between measured and experimental data. Within a typical stoichiometric window of [$0.95 < \lambda < 1.05$] that appears in actual engines during stoichiometric operation, slight differences in lambda could affect the total amount of NO_x formed during combustion. The stoichiometric cases of the experimental data were also simulated with a slightly rich ($\lambda = 0.95$) and slightly lean ($\lambda = 1.05$) mixture. The results are presented in Figure 6. It is observed that the measured NO_x concentration is almost always between these slightly lean and rich simulated values. Hence, slight departures from the set lambda in the experimental data may be a significant reason for the difference between experiment and simulation.

Figure 4 also shows that the "prompt" mechanism increases the total amount of NO_x concentrations by 10%-15% in case of stoichiometric combustion. Including the prompt formation one can increase the accuracy of the chemical mechanism. Bachmaier et al. (1973) [25] used an experimental configuration to define the equivalence ratio in which the prompt mechanism becomes significant in terms of total NO_x formation for various hydrocarbon mixtures. They found that the prompt NO formation starts to become significant as the mixture moves towards stoichiometry from $\lambda = 1.33$, in the case of methane. The prompt mechanism was negligible for leaner ($\lambda \geq 1.5$) conditions. Our results confirm the significance of the prompt mechanism in addition to the thermal one, even for stoichiometric combustion.

The thermodynamic input scenario is also important in lean conditions; however the oxygen availability does not affect the final results as much as in the stoichiometric case. In lean combustion, it seems that non-homogeneities in the burned zone can become important for accurately predicting final NO emissions. Multi-zoning is mostly used in 0D-engine models to take into account mixture stratification. In multi-zone modeling, different lambda and temperatures are assumed in each zone. Including multi-zones is a development we are currently working on in our model.

Another reason for differences between the simulated and experimental results could be the uncertainty in the high concentration of hydrocarbons (HC) that this engine emits (up to 9 000 ppm). By assuming the measured concentration of HC in the model, the prompt mechanism appears very significant, even in the lean case. As this engine is an optical and not a production one, these HC were assumed to be generated from crevices in the piston/cylinder interface and oil oxidation, rather than from fuel combustion itself. Although these HC do not participate in combustion, they could have an effect in a cold outer zone of a multi-zone model.

4.2 Cycle-to-Cycle NO Variability

The detailed chemical mechanism was then used for the investigation of NO CCV. From the various engine points in Figure 4, four engine points were chosen for the CCV analysis; two in partial load (P1015, P1030) and two at wide open throttle operation (W1015, W1030). All engine points were selected in stoichiometric conditions, to also include the effect of the prompt mechanism in NO formation.

NO variability was investigated using a statistical analysis. Wiebe combustion parameters such as the ignition timing (SOI), the CD and the Wiebe shape coefficient (m) were randomly varied within limits, assuming that these parameters follow a normal distribution. The mean values for these distributions were equal to the values used in the case of mean cycle modeling. The range of the variation considered was taken from a relevant analysis in the framework of the FP6 LESS-CCV research project [29] and differed for partial load and WOT operation. Full load points correspond to higher CCV than low load engine points [2]. One hundred engine cycles were simulated in each engine point and the results of imep and NO_x concentrations are presented in distributions. Differences between mean cycle indices and CCV values are discussed.

4.2.1 Results of Cycle-to-Cycle Variation

Cycle-to-cycle variation of pressure and temperature are illustrated in Figures 7 and 8, respectively, for the engine point of partial load and ignition timing of 15° BTDC (P1015). The mean value of maximum pressure is 16.6 bar and the standard deviation is 0.98 bar, while the peak temperature has a mean value of 2 172 K and a standard deviation of 20 K.

Figures 9 and 10 illustrate the distributions of imep and NO concentration, respectively, for the same engine point (P1015), due to the variation of the combustion parameters. Both figures include the statistical characteristics of the distributions such as the mean value and the Standard Deviation (SD). Mean Cycle imep (MC imep)

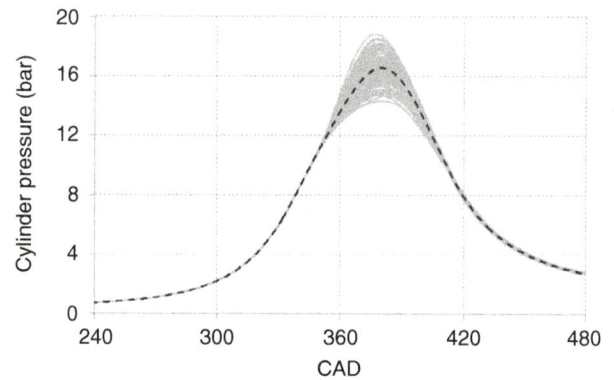

Figure 7

CCV of in-cylinder pressure (P1015 point).

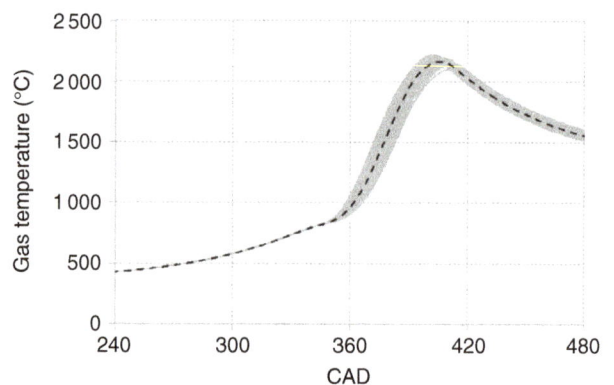

Figure 8

CCV of temperature evolution (P1015 point).

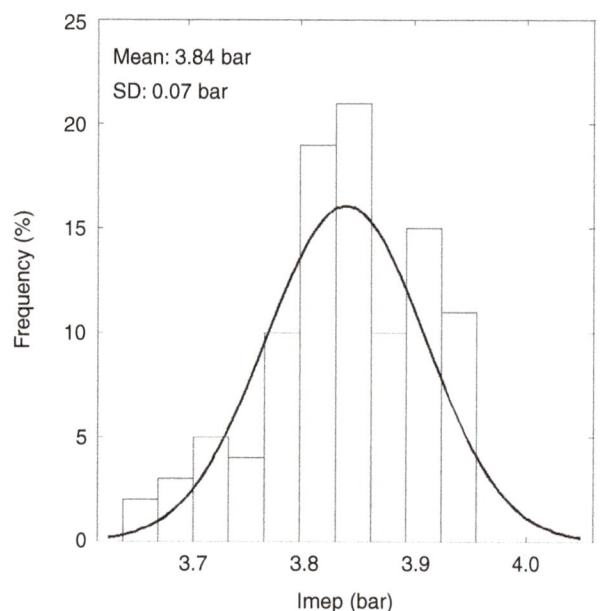

Figure 9

CCV of imep (P1015 point).

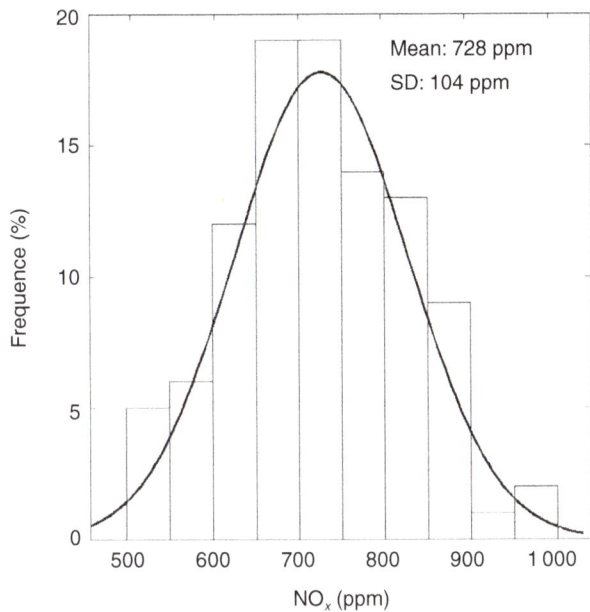

Figure 10

CCV NO_x (P1015 point).

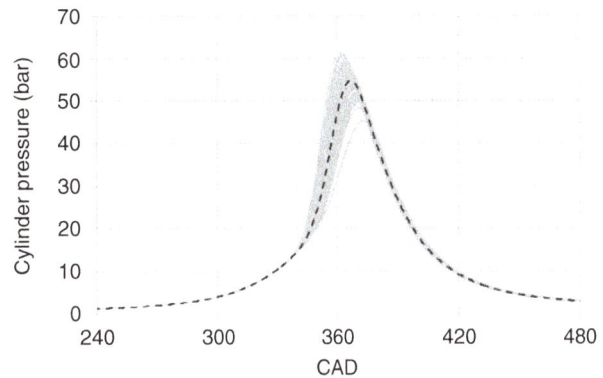

Figure 11

CCV of in-cylinder pressure (W1030 point).

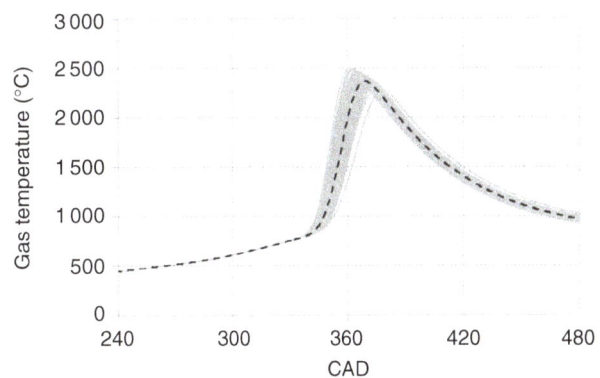

Figure 12

CCV of temperature evolution (W1030 point).

and the mean imep value of the CCV analysis coincide perfectly, while MC NO_x value and CCV mean NO_x value seems to have a slight deviation.

The same approach was also followed for the operation point at partial load (P1030) with ignition timing 30° BTDC. The peak pressure distribution has a mean value of 20.6 bar and the standard deviation is 1.05 bar. The peak temperature has a mean of 2 173 K and 30 K respectively. Distributions of this engine point for imep present no difference for the MC imep and the CCV imep value. In the case of NO, a small difference between MC NO_x and CCV NO_x values is again shown.

In the case of WOT, the same approach with a higher range of Wiebe parameters was used for the CCV analysis. Pressure and temperature plots of the engine point of 30° BTDC (W1030) are presented in Figures 11 and 12, respectively. Pressure and temperature peak values have a higher range as a result of higher range in the combustion parameters. In the case of W1015, the mean value of the maximum pressure is 35.7 bar and SD is equal with 3.1 bar, while the peak temperature varies from 2 121 K to 2 310 K with a mean value and standard deviation equal with 2 200 K and 39 K, respectively. Same order of magnitude differences are noticed for the case of W1030, where peak pressure varies from 45.5 bar to 61.7 bar (mean 55.5 bar, standard deviation

2.9 bar) and peak temperature varies from 2 281 K to 2 503 K (mean 2 410 K, standard deviation 42 K).

The distributions of imep and NO are illustrated in Figures 13 and 14. Due to higher CCV, MC imep and CCV imep are slightly different in both cases. Thus, MC NO_x value and CCV NO_x values present a higher deviation compared with the partial load. This also indicates that deviation between MC and CCV NO values is affected by the range of change of the combustion parameters.

4.2.2 Contribution of Prompt Mechanism on the Cycle-to-Cycle NO Variation

The impact of the prompt mechanism on NO_x CCV has been also investigated. In the case of mean cycle modeling, it was observed that the prompt mechanism accounts for an additional 10% to 15% in the final NO_x concentration. Therefore, it is expected that the

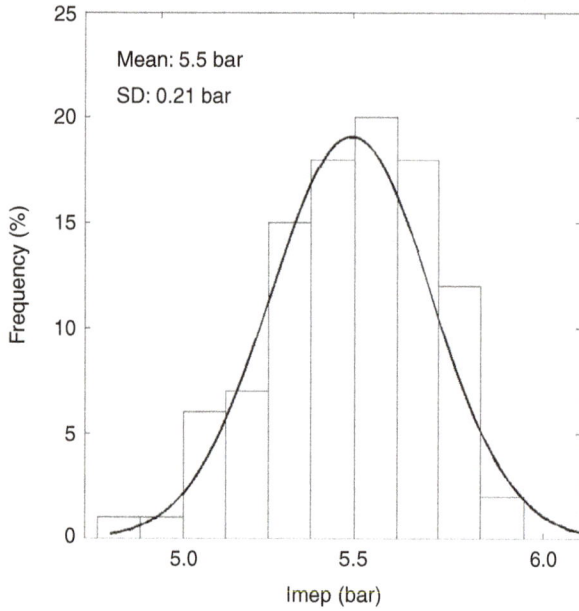

Figure 13
CCV of imep (W1030 point).

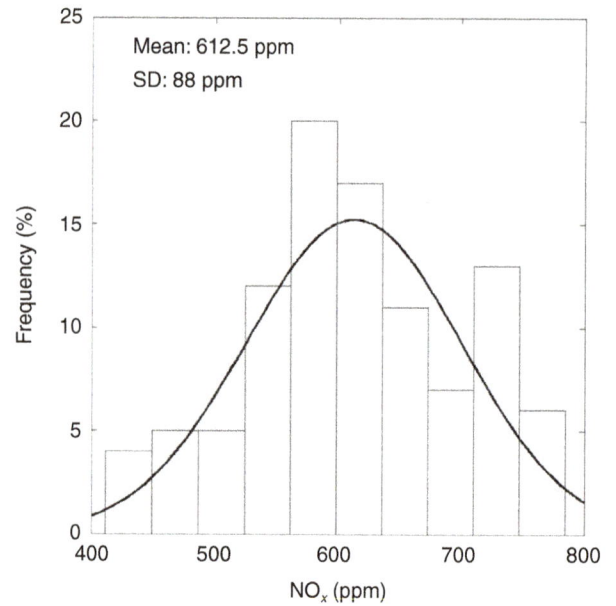

Figure 15
CCV NO_x without the prompt mechanism (P1015 point).

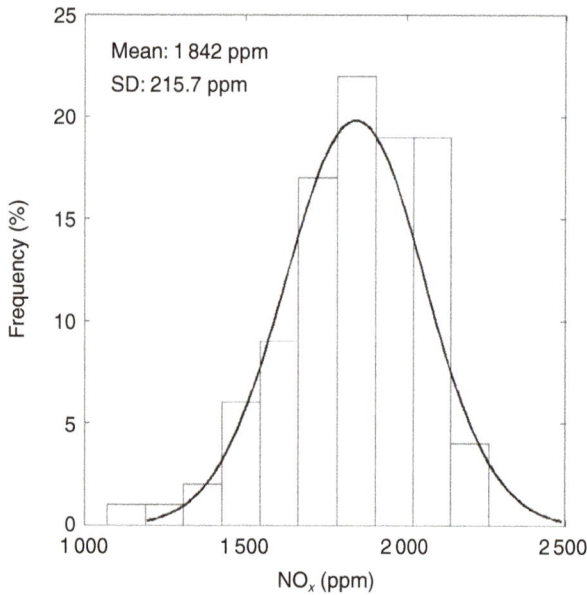

Figure 14
CCV NO_x (W1030 point).

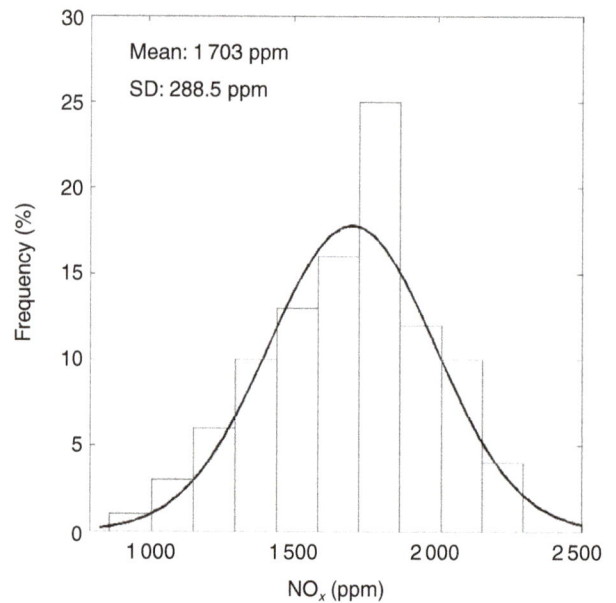

Figure 16
CCV NO_x without the prompt mechanism (W1030 point).

prompt mechanism should have a corresponding effect on NO CCV.

Figures 15 and 16 demonstrate the distributions of NO for a partial load and a full load engine point of CCV analysis, considering only the thermal mechanism.

Mean values of NO distribution show a decrease compared to the mean CCV NO values using the full mechanism. In addition, for the case of using the detailed chemical mechanism without the prompt one, a slight difference between MC NO values and CCV NO values

TABLE 4

Comparison of mean cycle values (MC) and CCV values for imep and NO

Engine point	Case	Imep (bar)	Diff (%)	NO full model (ppm)	Diff (%)	NO w/o prompt (ppm)	Diff (%)
P1015	MC	3.84	0.00	735	0.99	623	1.71
	CCV	3.84		728		613	
P1030	MC	3.27	0.00	1 320	0.65	1 230	4.14
	CCV	3.27		1 311		1 181	
W1015	MC	6.15	0.49	646	2.02	516	1.78
	CCV	6.12		633		507	
W1030	MC	5.55	0.90	1 990	8.05	1 720	0.97
	CCV	5.5		1 842		1 703	

is also observed. However, the prompt mechanism has an additional impact in the statistic characteristics of the NO distribution, which is described in the next section.

5 DEVIATION BETWEEN MEAN CYCLE VALUES AND MEAN CCV VALUES

CCV does not only result in a range of values for NO_x emissions but, due to the non-linearity of NO_x formation with combustion parameters and primarily with temperature, it may also have an impact on the average NO_x emitted. Hence, comparison between cycle-averaged values and the mean CCV value is important.

Partial load and full load are two cases that exhibit different variability for the combustion parameters. In partial load, the mean value of imep CCV distribution is almost the same to the mean cycle imep value. On the other hand, the CCV imep values are always lower than the mean cycle imep values in full load operation (Tab. 4). This means that the impact of CCV on average is a degradation of the engine performance.

NO formation is also affected by the variability in the combustion parameters. Both in full and partial load CCV NO_x values are always less than MC NO_x values, which reflects the CCV impact in NO formation (Tab. 4). In WOT operation, this impact is higher than in partial load. This result is related with the non-linearity of NO formation and for this reason it can not quantitatively correlated with imep variation. As shown in Table 4, the higher the difference between CCV imep and MC imep is, the higher is this difference between CCV NO_x and MC NO_x, too.

TABLE 5

Comparison of COV values for imep and NO with and without the prompt mechanism

Case	Imep COV (%)	NO full model COV (%)	NO w/o prompt COV (%)
P1015	1.81	14.30	14.35
P1030	0.84	14.50	13.77
W1015	0.81	26.03	26.89
W1030	3.81	11.71	16.94

The coefficient of variation is used as a metric of the intention of the NO CCV in Table 5. The impact of the prompt mechanism is also separated in this table. NO in general presents higher variability due to CCV than imep does. Also, the results show that it is not possible to establish a direct link between imep CCV and NO CCV. The latter is dependant on both the operation point and the CCV of imep. Finally, the impact of the prompt mechanism on CCV is also specific to the engine point considered. In one of the WOT conditions examined, the prompt mechanism led to a significant increase in NO CCV, that is not obvious in the other cases. This means that the combination of heat release rate with reaction kinetics is unique for each engine point that results to a behaviour which cannot be generalized at this stage. Simulations with other engines and further refinements in the model may lead to a more consistent behaviour of CCV NO with CCV in other combustion parameters.

CONCLUSIONS

In this study, a detailed chemical mechanism was used for the prediction of homogeneous engine-out NO emissions. Literature experimental data were used for the validation of the simulated values. The model satisfactorily predicts NO emissions, ranging from a few ppm to a couple of thousand of ppm of NO molar fraction, in both stoichiometric and lean conditions. Then, the model was used for the simulation of NO variation due to combustion CCV. It was found that CCV NO distributions exhibit a higher COV compared to the imep distributions. In addition, mean CCV NO values are always lower than the average cycle NO values. The impact of prompt mechanism in NO result was also investigated. In the case of average cycle emissions, it was found that the prompt mechanism increases the accuracy of the prediction, especially in stoichiometric conditions by up to 15%. In CCV, the prompt mechanism has an impact in the COV and mean value of NO distributions, although the impact was dependant on the engine operation point considered.

ACKNOWLEDGMENTS

This study was performed within the framework of the FP7 LESSCCV research project (Grant agreement 233615).

REFERENCES

1 Brehob D.D., Newman C.E. (1992) Monte Carlo Simulation of Cycle by Cycle Variability, *SAE Paper* 922165.

2 Ozdor N., Dulger M., Sher E. (1994) Cyclic Variability in Spark Ignition Engines. A Literature Survey, *SAE Paper* 940987.

3 Karvountzis-Kontakiotis A., Ntziachristos L. (2012) A detailed chemical mechanism to predict NO cycle-to-cycle variation in homogeneous engine combustion, *IFAC Workshop on Engine and Powertrain Control, Simulation and Modeling*, IFP Energies nouvelles, France, 23-25 Oct.

4 Ball J.K., Raine R.R., Stone C.R. (1998) Combustion analysis and cycle-by-cycle variations in spark ignition engine combustion Part 2: A new parameter for completeness of combustion and its use in modelling cycle-by-cycle variations in combustion, *Proceeding of the Institution of Mechanical Engineers, Part D: Journal of Automobile Engineering June 1* **212**, 6, 507-523.

5 Heywood J.B. (1988) *Internal Combustion Engine Fundamentals*, McGraw-Hill, Singapore.

6 Young M.B. (1981) Cyclic Dispersion in the Homogeneous-Charge Spark-Ignition - A Literature Survey, *SAE Paper* 810020.

7 Stone C.R., Brown A.G., Beckwith P. (1996) Cycle-by-Cycle Variations in Spark Ignition Engine Combustion – Part II: Modelling of Flame Kernel Displacements as a Cause of Cycle-by-Cycle Variations, *SAE Paper* 960613.

8 Johansson B. (1996) Cycle-to-Cycle Variations in S.I. Engines – The Effects of Fluid Flow and Gas Composition in the Vicinity of the Spark Plug on Early Combustion, *SAE Paper* 962084.

9 Whitelaw J.H., Xu H.M. (1995) Cyclic Variations in a Lean-Burn Spark Ignition Engine Without and With Swirl, *SAE Paper* 950683.

10 Fox W.J., Cheng K.W., Heywood B.J. (1993) A Model for Predicting Residual Gas Fraction in Spark-Ignition Engines, *SAE Paper* 931025.

11 Hamai K., Kawajiri H., Ishizuka T., Nakai M. (1988) Combustion Fluctuation Mechanism Involving Cycle-to-Cycle Spark Ignition Variation Due to Gas Flow Motion in S.I. Engines, *21st Int. Symposium on Combustion* **21**, 505-512.

12 Vermorel O., Richard S., Colin O., Angelberger C., Benkenida A., Veynante D. (2009) Towards the understanding of cyclic variability in a spark ignited engine using multi-cycle LES, *Combustion and Flame* **156**, 1525-1541.

13 Lacour C., Pera C. (2011) An Experimental Database Dedicated to the Study and Modelling of Cyclic Variability in Spark-Ignition Engines with LES, *SAE Paper* 2011-01-1282.

14 Martin J., Witze P., Borgnakke C. (1985) Combustion Effects on the Preflame Flow Field in a Research Engine, *SAE Paper* 850122.

15 Matekunas F. (1983) Modes and Measures of Cyclic Combustion Variability, *SAE Paper* 830337.

16 Shen H., Hinze P., Heywood J. (1996) A Study of Cycle-to-Cycle Variations in SI Engines Using a Modified Quasi-Dimensional Model, *SAE Paper* 961187.

17 Watson C.H., Goldsworthly C.L., Milkins E.E. (1976) Cycle-by-Cycle Variations of HC, CO, and NOx, *SAE Paper* 760753.

18 Ball K.J., Bowe J.M., Stone C.R., Collings N. (2001) Validation of a Cyclic NO Formation Model with Fast NO Measurements, *SAE Paper* 2001-01-1010.

19 Jante A. (1960) Das Wiebe-Brenngesetz; ein Fortschritt in der Thermodynamik der Kreisprozesse von Verbrennungsmotoren, *Kraftfahrzeugtechnik* **9**, 340-346.

20 Stiesch G. (2003) *Modeling Engine Spray and Combustion Processes*, Springer, Berlin.

21 Duclos J.-M., Zolver M., Baritaud T. (1999) 3D Modeling of Combustion for DI-SI Engines, *Oil & Gas Science and Technology* **54**, 2, 259-264.

22 Richard S., Bougrine S., Font G., Lafossas F.-A., Le Berr F. (2009) On the Reduction of a 3D CFD Combustion Model to Build a Physical 0D Model for Simulating Heat Release, Knock and Pollutants in SI Engines, *Oil & Gas Science and Technology* **64**, 3, 223-242.

23 Heywood J.B., Higgins J.M., Watts P.A., Tabaczynski R.J. (1979) Development and Use of a Cycle Simulation to Predict SI Engine Efficiency and NOx Emissions, *SAE Paper* 790291.

24 Pattas K., Häfner G. (1973) Stichoxidbildung bei der ottomotorichen Verbrennung, *MTZ*, Nr. 12.

25 Bachmaier F., Eberius K.H., Just T.H. (1973) The formation of Nitric Oxide and the Detection of HCN in Premixed Hydrocarbon – Air Flames at 1 Atmosphere, *Combustion Science and Technology* **7**, 77-84.

26 Lutz A.E., Kee R.J., Miller J.A. (1988) Senkin: A FORTRAN program for predicting homogeneous gas phase chemical kinetics with sensitivity analysis, SAND87-8248.

27 Glassman I., Richard A.Y. (2008) *Combustion*, Academic Press, California, USA.

28 Kergin U. (2002) Study on the prediction of the effects of design and operating parameters on NOx emissions from a leanburn natural gas engine, *Energy Conversion and Management* **44**, 907-921.

29 Shuemie A., Fairbrother R., Pötsch Ch, Tatschl R. (2011) *LESSCCV Project Meeting*, Milano, WP4.

30 Merker G.P., Schwarz C., Stiesch G., Otto F. (2004) *Simulation of combustion and pollutant formation for engine-development*, Springer.

7

Combustion Noise and Pollutants Prediction for Injection Pattern and Exhaust Gas Recirculation Tuning in an Automotive Common-Rail Diesel Engine

Ivan Arsie[1]*, Rocco Di Leo[1], Cesare Pianese[1] and Matteo De Cesare[2]

[1] Dept. of Industrial Engineering, University of Salerno, Fisciano (SA) 84084 - Italy
[2] Magneti Marelli Powertrain, Bologna 40134 - Italy
e-mail: iarsie@unisa.it - rodileo@unisa.it - pianese@unisa.it - matteo.decesare@magnetimarelli.com

* Corresponding author

Abstract — *In the last years, emissions standards for internal combustion engines are becoming more and more restrictive, particularly for NO_x and soot emissions from Diesel engines. In order to comply with these requirements, Original Equipment Manufacturers (OEM) have to face with innovative combustion concepts and/or sophisticate after-treatment devices. In both cases, the role of the Engine Management System (EMS) is increasingly essential, following the large number of actuators and sensors introduced and the need to meet customer expectations on performance and comfort. On the other hand, the large number of control variables to be tuned imposes a massive recourse to the experimental testing which is poorly sustainable in terms of time and money. In order to reduce the experimental effort and the time to market, the application of simulation models for EMS calibration has become fundamental. Predictive models, validated against a limited amount of experimental data, allow performing detailed analysis on the influence of engine control variables on pollutants, comfort and performance.*

In this paper, a simulation analysis on the impact of injection pattern and Exhaust Gas Recirculation (EGR) rate on fuel consumption, combustion noise, NO and soot emissions is presented for an automotive Common-Rail Diesel engine. Simulations are accomplished by means of a quasi-dimensional multi-zone model of in-cylinder processes. Furthermore a methodology for in-cylinder pressure processing is presented to estimate combustion noise contribution to radiated noise.

Model validation is carried out by comparing simulated in-cylinder pressure traces and exhaust emissions with experimental data measured at the test bench in steady-state conditions. Effects of control variables on engine performance, noise and pollutants are analyzed by imposing significant deviation of EGR rate and injection pattern (i.e. rail pressure, start-of-injection, number of injections). The results evidence that quasi-dimensional in-cylinder models can be effective in supporting the engine control design toward the optimal tuning of EMS with significant saving of time and money.

Résumé — **Prédiction du bruit de combustion et des polluants pour le réglage des paramètres d'injection et de l'EGR (*Exhaust Gas Recirculation*) dans un moteur Diesel *Common-Rail* pour l'automobile** — Ces dernières années, les normes d'émissions pour les moteurs à combustion interne sont de plus en plus restrictives, en particulier pour les émissions des NO_x et la production de suie par les moteurs Diesel. Afin de se conformer à ces exigences, les équipementiers (OEM, *Original Equipment Manufacturer*) doivent faire face à des concepts innovants de combustion et/ou à des dispositifs de post-traitement sophistiqués. Dans les deux cas, le rôle du système de gestion du moteur (EMS, *Engine Management System*) est de plus en plus essentiel, considérant le grand nombre de capteurs et d'actionneurs introduits et l'exigence pour répondre aux attentes de performance et de confort des clients. Par ailleurs, le grand nombre de variables de contrôle à optimiser, impose un recours massif à l'essai expérimental qui est peu rentable pour le rapport temps/argent. Afin de réduire l'effort expérimental et le délai de mise sur le marché, l'application des modèles de simulation pour le calibrage de l'EMS est fondamentale. Les modèles prédictifs, validés par rapport à un nombre limité de données expérimentales, permettent de réaliser des analyses détaillées sur l'influence des variables de contrôle du moteur sur les polluants, le confort et les performances. Cet article présente une analyse de simulation sur l'impact des paramètres d'injection et du taux de recirculation des gaz d'échappement (EGR, *Exhaust Gas Recirculation*) sur la consommation de carburant, le bruit de combustion, les émissions de NO et la production de suie par le moteur automobile Diesel *Common-Rail*. Les simulations sont effectuées par un modèle multi-zone, quasi-dimensionnel des phénomènes internes au cylindre. Cet article présente également une méthode pour le traitement de la pression interne au cylindre afin d'évaluer le bruit de combustion par rapport au bruit émis. La validation du modèle est effectuée en comparant la pression interne au cylindre et les émissions d'échappement simulées aux données expérimentales traitées au banc d'essai en régime stationnaire. Les effets des variables de contrôle sur les performances du moteur, le bruit et les polluants sont analysés en imposant la variation du taux de l'EGR et les paramètres d'injection (par exemple la pression du rail, le début d'injection et le nombre d'injections). Les résultats démontrent que les modèles quasi-dimensionnels des phénomènes internes au cylindre peuvent être efficaces à la conception du contrôle du moteur pour optimiser le réglage de l'EMS avec un gain de temps et d'argent.

NOMENCLATURE

b_i	Injection calibration parameter (-)
BMEP	Break Mean Effective Pressure (bar)
C_1	Calibration factor for spray impingement (-)
C_2	Parameter of the turbulence model (-)
C_3	Empirical parameter for ignition delay (-)
C_4	Proportional factor of eddy turnover (-)
E	Activation energy (J/mol)
EID	Effective Injection Duration (ms)
ET	Energizing Time (ms)
E_{ith_zone}	Internal energy in the i-th zone (J)
h_{ith_zone}	Specific enthalpy in the i-th-zone (J/kg)
IMEP	Indicated Mean Effective Pressure (bar)
ISD	Injection Start Delay (ms)
L	Current number of radial parcels (-)
L_I	Integral length scale (m)
L_h	Maximum number of radial parcels (-)
m_{ae}	Mass of entrained air (kg)
m_b	Mass of burned fuel (kg)
m_e	Mass of evaporated fuel (kg)
$m_{f,inj}$	Mass of injected fuel (kg)
m_{ith_zone}	Fuel mass in the i-th zone (kg)

m_v	Mass of fuel vapor (kg)
N_{rad}	Maximum number of radial parcels (-)
n_{fv}	Molar fraction of fuel vapor (-)
n_{O_2}	Molar fraction of oxygen (-)
PCCI	Premixed Combustion Compression Ignition
p_{rail}	Common rail pressure (bar)
p	In-cylinder pressure (bar)
p_{eff}	Sound pressure (Pa)
p_0	Pressure hearing threshold (Pa)
Q_{ith_zone}	Heat transfer rate to the i-th zone (J)
R_0	Universal gas constant (J/(mol.K))
S	Penetration of the generic spray core parcel (m)
S_L	Radial penetration of the L-th parcel of spray (m)
SOI	Start Of Injection (°ATDC)
SOC	Start Of Combustion (°ATDC)
T	In-cylinder temperature (K)
T_b	Temperature of burned gas (K)
T_{eff}	Time period for sound pressure evaluation (s)
u'	Turbulence intensity (m/s)
U_f	Initial spray velocity (m/s)
U_{mp}	Mean piston velocity (m/s)
V_a	Air-zone volume (m^3)

V_{cyl}	Instantaneous cylinder volume (m^3)
V_{finj}	Volume of injected fuel (m^3)
V_{ith_zone}	Instantaneous volume of the i-th zone (m^3)
W_{ith_zone}	Work transfer rate out of the i-th zone (J)

Greek Symbols

γ	Weight factor (-)
ρ	Air zone density (kg/m^3)
τ_b	Characteristic combustion time (s)
$\tau_{b,lam}$	Characteristic time of laminar combustion (s)
$\tau_{b,turb}$	Characteristic time of turbulent combustion (s)
τ_{id}	Ignition delay (s)

INTRODUCTION

The interest in Diesel engines for automotive application has dramatically grown in the last decade, due to the benefits gained with the introduction of common-rail system and electronic control. A strong increase in fuel economy and a remarkable reduction of emissions and combustion noise have been achieved, thanks to both optimized fuelling strategy and advanced fuel injection technology. Namely, the improvement of injector time response, injection pressure and nozzle characteristics have made feasible the operation of multiple injections and have enhanced the fuel atomization. The actuation of early pilot and pre injections enhances the occurrence of a smoother combustion process with benefits on noise. Improved fuel atomization enhances the air entrainment making the combustion cleaner and more efficient, thus reducing both particulate emissions and fuel consumption but with a negative impact on NO$_x$ emissions (Tennison and Reitz, 2001). On the other hand, the recourse to Exhaust Gas Recirculation (EGR) lowers in-cylinder peak temperature and NO$_x$ emissions but with a negative impact on particulate emissions.

In the last years, many efforts are addressed towards new combustion concepts, in order to face with the soot/NO$_x$ trade-off and the increasingly restrictive emission standards. Earlier injections and large EGR rate promote premixed combustion and lead to lower peak temperature, with benefits on both particulate and NO$_x$ emissions. The drawback is the increase of combustion noise, due to the large delay of premixed combustion up to the Top Dead Center (TDC) that results in a dramatic and sharp increase of in-cylinder pressure (Torregrosa et al., 2011). In this context, it is clear that a suitable design of engine control strategies is fundamental in order to overcome with the simultaneous and opposite impact of combustion law on NO$_x$/soot

emissions and combustion noise. Nevertheless the large number of control variables (i.e. injection pattern, EGR, Variable Geometry Turbine (VGT) rack position) makes the experimental testing extremely expensive in terms of time and money. Massive use of advanced mathematical models to simulate engine and system components (mechanical and electronic devices) is therefore recommended to speed up the design and optimization of engine control strategies.

The complexity of Diesel engine combustion, which is governed by the turbulent fuel-air-mixing, causes an unresolved trade-off between computational time and accuracy. Single zone models based on empirical heat release laws, largely used to simulate SI engine performance and emissions, are inadequate to simulate the heterogeneous character of Diesel combustion (Barba et al., 2000). This problem is particularly felt for emissions prediction; in that case a huge effort has to be spent for parameters identification to reach a satisfactory accuracy. Therefore, in order to achieve suitable precision, most of the studies in the field of Diesel engine modeling have focused on the basic phenomena involved into fuel injection/evaporation, air entrainment, combustion and emission formation, with particular emphasis on particulate matter (mainly soot). On the other hand, many advanced models are available in the literature, based on the complete 3D description of turbulent, multi-phase flow field inside the cylinder (Gang and Tao, 2010; López et al., 2013; Javadi Rad et al., 2010). Despite their accuracy, these models present a large computational demand and are indeed oriented to engine design (combustion chamber shaping, fuel jet/air interaction, swirl) rather than to control design application.

Phenomenological two-zone or multi-zone combustion models have been proposed in literature to meet the requirements for engine control design. Such models are accurate enough to predict fuel evaporation, air entrainment, fuel-air distribution and thermal stratification with a reasonable computational demand (Kouremenos et al., 1997; Arsie et al., 2006). Particularly, the identification analysis of the main model parameters enhances the development of predictive tools for efficient and accurate simulation of the effects of control injection variables on combustion process and exhaust emissions formation (Rakopoulos et al., 2003; Arsie et al., 2007, 2012).

In the present paper, a multi-zone model is applied to simulate engine combustion and predict noise and pollutant emissions depending on injection pattern and EGR rate. The novelties with respect to previous works presented by the authors (Arsie et al., 2006, 2007, 2012) are the development of improved models for fuel injection and ignition delay and a methodology to estimate the combustion noise.

1 MULTI-ZONE MODEL

Simulation of in-cylinder pressure is accomplished by a thermodynamic model, which is based on the energy conservation for an open system and on the volume conservation of the total combustion chamber (Assanis and Heywood, 1986; Hiroyasu and Kadota, 1983; Bi et al., 1999; Arsie et al., 2006):

$$\dot{E}_i = \dot{Q}_i - \dot{W}_i + \sum_{j,i \neq j} \dot{m}_{i,j} \times h_{i,j} \qquad (1)$$

$$V_{cyl} = V_a + \sum_i V_i \qquad (2)$$

The combustion chamber is divided into several zones, with homogeneous pressure and different temperature and chemical composition. In each zone, the gas is assumed ideal and the thermodynamic properties are function of temperature, pressure and composition (Ferguson, 1986). During the compression stroke, only one homogeneous zone containing air and residual gas (air zone, a) is considered as shown in Figure 1.

When the injection takes place, the fuel jet forms a number of sprays, depending on the number of injection nozzle holes. Each spray is divided into several parcels along both axial and radial direction. For each parcel, a burned zone composed by combustion products and an unburned zone composed by fuel, entrained air and residual gas, are considered. This process is repeated for each injection, neglecting interactions among the sprays and energy or mass transfer among the parcels (Arsie et al., 2006). The model simulates temperature and chemical composition in each parcel thus enhancing prediction of NO and soot engine emissions.

1.1 Fuel Injection

Fuel injection strongly affects the heat release rate and its modeling is a critical issue to deal with. This is also due to the lack of experimental data collected at the flow test bench on the injection rate shape which inhibits the development of data-driven models. Nowadays several multi-dimensional commercial codes are available to model mechanical, hydraulic and electromagnetic phenomena, thus taking into account the inertia and the dynamics of every component inside the injector. Nevertheless these approaches involve a huge computational effort, not suitable for the current model application.

In order to overcome this issue, in the model presented herein the injection rate shape is simulated by an empirical formulation derived from a set of experimental data measured at the flow test bench. Figure 2 shows the

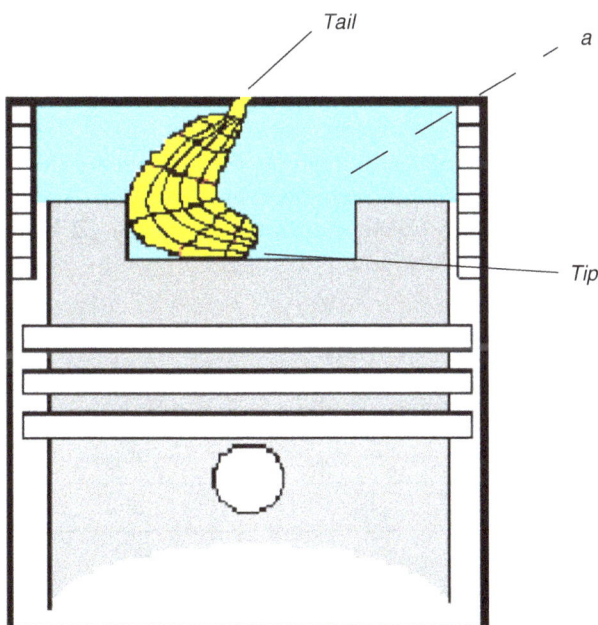

Figure 1

Scheme of in-cylinder stratification with air zone (a) and spray discretization in axial and radial direction.

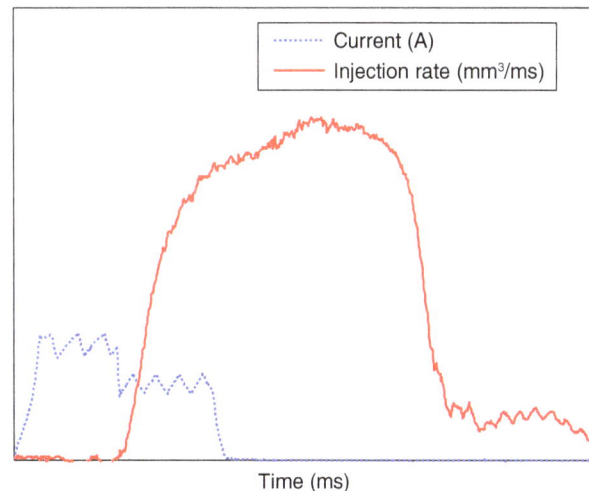

Figure 2

Experimental injection rate shape. P_{rail} = 1 600 bar, ET = 730 µs. The scales are omitted for confidential issues.

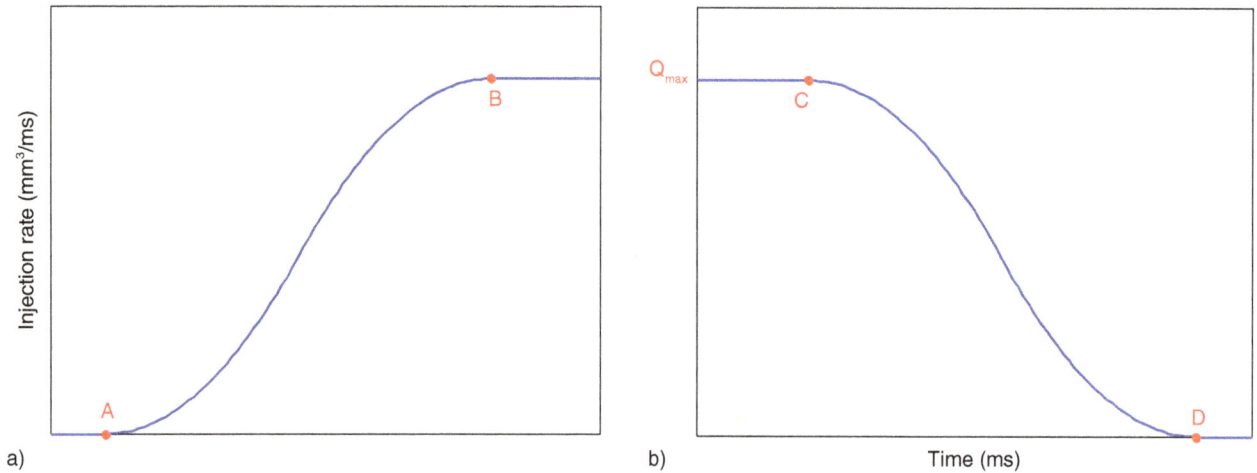

Figure 3

a) *S*- and b) *Z*- functions defined to describe the injection rate trajectory.

injection rate shape experimentally detected for a Common-Rail injector in case of rail pressure (P_{rail}) and Energizing Time (*ET*) set at 1 600 bar and 730 µs, respectively. Figure 2 evidences that the experimental injection rate trajectory does not correspond to a regular geometrical shape; it usually shows fluctuations around the maximum flow rate, due to the wave effects inside the injector pipes. However, at least for the main injections, such fluctuations can be neglected without significant lack of accuracy. The mentioned maximum flow rate, which depends on common rail and combustion chamber pressure and on the characteristics of the injector, is calculated from the static flow rate provided by the manufacturer. Nevertheless, it is worth noting that in case of short injection timing, the static flow rate might not be reached and the maximum flow rate has to be evaluated differently, as it will be described later.

Following the experimental trajectory, the injection rate shape was modeled by coupling two symmetric curves, namely a *S*-function and a *Z*-function which account for needle lift and closing, respectively, as shown in Figure 3.

The *S*-function can be defined by means of three parameters corresponding to:
– the Injection Start Delay (ISD) (point A in *Fig. 3*);
– the time at which the maximum flow rate is reached (point B);
– the maximum flow rate reached (Q_{max}).

Being symmetric, the *Z*- function was derived consequently. It is worth noting that the ISD is defined as the time it takes from the start of injector Energizing Time (*ET*) and

the effective fuel flowing through the nozzle (Coppo and Dongiovanni, 2007). Moreover, the time range defined by the points A and D in Figure 3 corresponds to the Effective Injection Duration (EID). The three parameters ISD, EID and Q_{max} are expected to be dependent on needle inertia, rail pressure and back-pressure into the combustion chamber (*i. e.* in-cylinder pressure).

The parameters identification was accomplished making use of a set of experimental injector rate trajectories, ranging the *ET* and P_{rail} as reported in Table 1. Particularly, from the analysis of the experimental data, the ISD resulted to be almost constant and it was set to 0.335 ms for all the operating conditions. On the other hand, the EID was identified for each operating condition and finally expressed as function of rail pressure and *ET* by the following relationship:

TABLE 1

Values of rail pressure (P_{rail}) and Energizing Time (*ET*) experimentally investigated for the injection rate identification

Test case	P_{rail} (bar)	*ET* (µs)
1	740	285
2	740	525
3	800	345
4	800	650
5	1 600	315
6	1 600	750

$$EID = b_1 + b_2 \times P_{rail} + b_3 \times ET \qquad (3)$$

with the parameters b_1, b_2 and b_3 equal to -0.574, -1.60×10^{-5} and 2.60×10^{-3}, respectively, for the current injection system.

Finally, the maximum flow rate Q_{max} was identified by recursive processing of S- and Z- functions, in order to achieve the target mass of injected fuel.

The comparison between measured and estimated injection rate trajectories for the six test cases considered is shown in Figure 4, which exhibits a good accuracy of the developed injector rate model.

1.2 Fuel Spray and Evaporation

The injected fuel moves as a liquid column, until the break-up time elapses. Then it is assumed that the fuel atomizes to fine droplets which move into the combustion chamber decreasing their velocity while entraining the surrounding air (Hiroyasu and Masataka, 1990; Jung and Assanis, 2006). The break-up time is calculated using the correlations proposed by Hiroyasu and Kadota (1983). The spatial development of the spray is simulated using the Naber correlation (Naber and Siebers, 1996). This quasi-dimensional approach allows estimating the spray penetration along the central axial direction. The radial discretization is defined generalizing the correlation proposed by Hiroyasu and Kadota (1983), as follows:

$$S_L = S \cdot \exp\left[-8.557 \cdot 10^{-3} \cdot \frac{(L_h - 1)^2}{(N_{rad} - 1)^2} \cdot (L - 1)^2\right] \qquad (4)$$

where S is the penetration of the generic spray core parcel, S_L is the penetration of the L-th parcel of spray in radial direction, $L_h = 10$ is the maximum number of radial parcels considered by Hiroyasu and Kadota (1983) and Hiroyasu and Masataka (1990), N_{rad} is the current number of radial parcels.

The air entrainment model is derived from the momentum conservation law:

$$\dot{m}_{ae} = -C_1 \frac{m_{f,inj} \times U_f}{\left(\frac{dS}{dt}\right)^2} \times \frac{d^2S}{dt^2} \qquad (5)$$

where the parameter C_1 accounts for the influence of air swirl and the effects of the spray impingement on piston bowl and/or cylinder wall. For the current study, the parameter C_1 was identified by fitting measured and simulated in-cylinder pressure for five engine operating conditions at constant engine speed and increasing load.

After the break-up time, due to the high surrounding gas temperature, the droplets evaporate and mix with the entrained air. For a complete description of the fuel evaporation model the reader is addressed to a previous paper (Arsie et al., 2006). Its approach relies on the equations of the mass diffusion and heat transfer for a spherical droplet with initial diameter equal to the Sauter Mean Diameter (SMD) (Hiroyasu and Kadota, 1983; Kuo, 1986; Jung and Assanis, 2006). Moreover the model assumes the heat transfer to the cylinder wall as sum of radiative and convective heat transfer, following the Woschni formulation (Ramos, 1989). The total heat transfer is shared among the zones according with their mass and temperature.

1.3 Turbulence Model

The turbulence model is based on the k-ε approach. The values of the turbulent kinetic energy (k) and its dissipation rate (ε) have been assumed homogeneous in the combustion chamber and they have been computed by the two following equations (Ramos, 1989):

$$\frac{dk}{dt} = \frac{2}{3}\frac{k}{\rho}\frac{d\rho}{dt} - \varepsilon \qquad (6)$$

$$\frac{d\varepsilon}{dt} = \frac{4}{3}\frac{\varepsilon}{\rho}\frac{d\rho}{dt} - \frac{2\varepsilon^2}{k} \qquad (7)$$

These equations do not consider the combustion influence on the turbulence.

The initial condition of k at Intake Valve Closing (IVC) is estimated considering its definition for isotropic homogeneous turbulence and assuming that the initial value of the turbulence intensity (u') depends on the mean piston velocity:

$$k = \frac{3}{2}(u')^2 \qquad (8)$$

$$k(IVC) = C_2 \frac{3}{2}\left(U_{mp}\right)^2 \qquad (9)$$

where $C_2 = 0.10$.

The initial value of ε is estimated assuming the equilibrium between production and dissipation of turbulence kinetic energy (Ramos, 1989):

$$\varepsilon(IVC) = \frac{[k(IVC)]^{3/2}}{L_I(IVC)} \qquad (10)$$

where L_I is the integral length scale, whose value at IVC was set to 10 mm, corresponding to the maximum intake

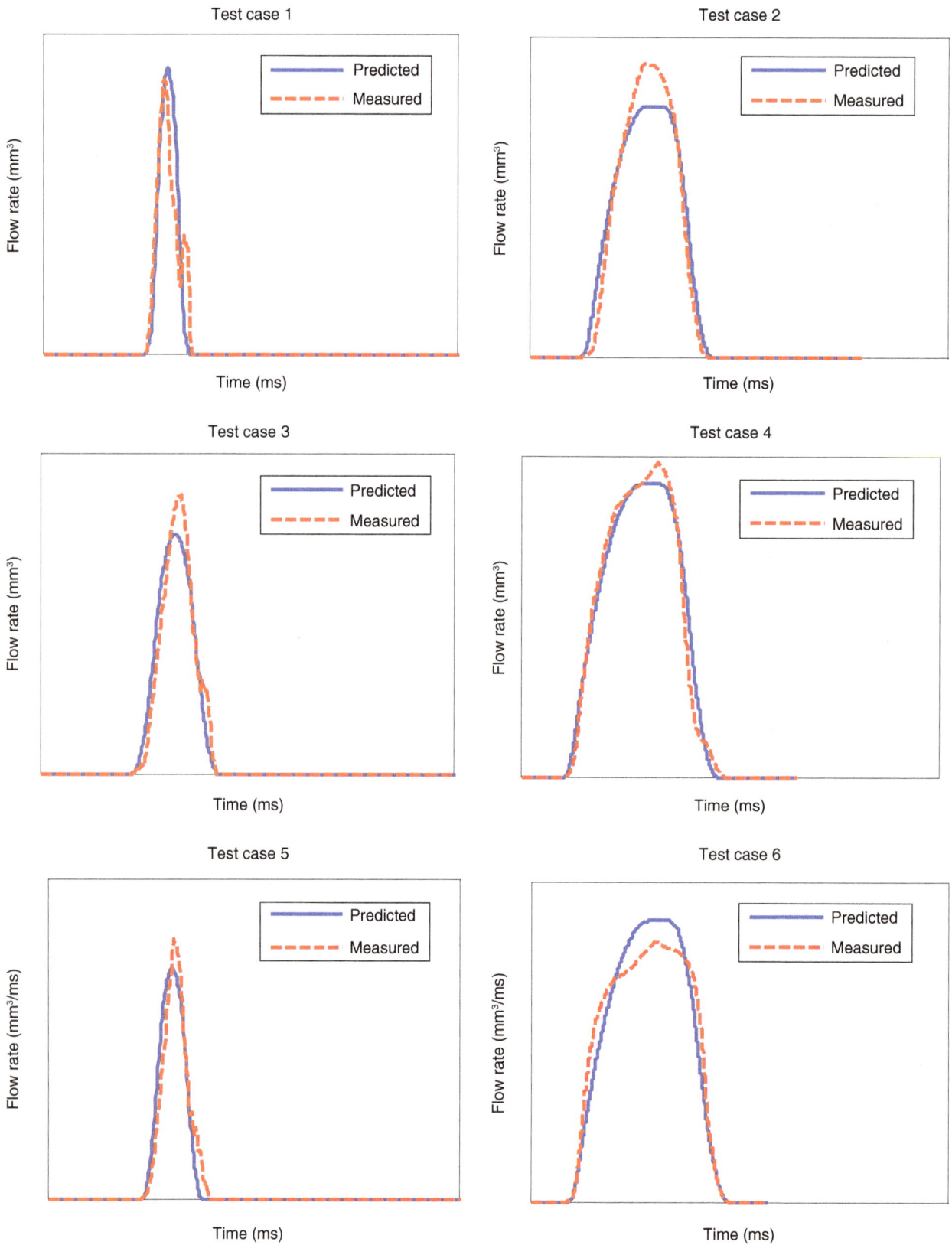

Figure 4

Comparison between measured and predicted injection rate trajectories.

valve lift. At IVC, Equations (9) and (10) are used to calculate u' and L_I to fix the initial condition of Equations (6) and (7).

1.4 Ignition Delay

The ignition delay is due to the combustion kinetics which depends on the cylinder pressure and temperature at the injection timing through an Arrhenius correlation (Hiroyasu and Kadota 1983):

$$\tau_{id} = C_3 p^{-1.02} \phi^{-0.1} \exp\left(\frac{1600}{T}\right) \qquad (11)$$

where p and T are the in-cylinder pressure and temperature, respectively, ϕ is the equivalent ratio of the mixture. C_3 is an empirical parameter and was set to 3.45 according with literature data (Heywood, 1988).

1.5 Combustion

The combustion model is based on the laminar and turbulent characteristic-time approach (Kong et al., 1995; Tanner and Reitz, 1999; Ayoub and Reitz, 1995). The fuel combustion in the burning region is described by the following equation (Ramos, 1989):

$$\frac{dm_b}{dt} = \frac{m_e - m_b}{\tau_b} \qquad (12)$$

where the characteristic time τ_b is the same for each chemical reactant. In order to account for the effects of turbulence on the chemical reactions, the characteristic time is calculated as the weighted sum of the laminar timescale ($\tau_{b,lam}$) and the turbulent timescale ($\tau_{b,turb}$):

$$\tau_b = \tau_{b,lam} + \gamma \tau_{b,turb} \qquad (13)$$

with the weight γ given as:

$$\gamma = \frac{1 - e^{-x}}{0.632} \qquad (14)$$

where x is the burned fuel fraction defined as:

$$x = \frac{m_b}{m_v} \qquad (15)$$

Zeroing the concentration of fuel at equilibrium, the laminar time scale is computed as:

$$\tau_{b,lam} = \left(7.68 \times 10^8 [n_{fv}]^{-0.75} \\ \times [n_{O_2}]^{1.5} \exp\left(-\frac{E}{R_0 T_b}\right)\right)^{-1} \qquad (16)$$

where $E = 77.3 \times 10^3$ J/mol and $R_0 = 8.3144$ J/(mol.K).

Finally the turbulent combustion time is function of the eddy turnover:

$$\tau_{b,turb} = C_4 \frac{k}{\varepsilon} \qquad (17)$$

the proportional factor C_4 was set to 0.142 according with literature data (Kong et al., 1995; Tanner and Reitz, 1999; Ayoub and Reitz, 1995).

1.6 Nitrogen Oxide Emissions

NO_x emissions from Diesel engines are mainly due to the thermal NO formation for dilute (lean mixture and EGR) operation (Heywood, 1988). The thermal NO formation process is modeled making use of the well-known extended Zeldovich mechanism applied to the mixing zone, which considers three reactions with seven species as main responsible for NO production (Heywood, 1988; Ramos, 1989; Arsie et al., 2006, 2007).

More detailed models have been proposed, as the super extended Zeldovich mechanism by Miller et al. (1998), which accounts for 13 species and up to 67 reactions and can led to a significant improvement of model accuracy. On the other hand, this approach could thwart the benefits of phenomenological models because of its higher computational complexity.

According with the well known assumptions on steady state nitrogen formation and equilibrium concentration for the reactants (Heywood, 1988), the Zeldovich mechanism holds the following rate of variation for the NO concentration:

$$\frac{1}{V_b}\frac{dn_{NO}}{dt} = \frac{2R_1\left\{1 - \left(\frac{[NO]}{[NO]_{eq}}\right)^2\right\}}{1 + \left(\frac{[NO]}{[NO]_{eq}}\right)\frac{R_1}{R_2 + R_3}} \qquad (18)$$

where n_{NO} is the number of NO moles in the burned gas volume V_b, while R_1, R_2 and R_3 are computed as follows:

$$R_1 = k_1^+ [O]_e [N_2]_e \quad k_1^+ = 7.6 \times 10^{13} \exp\left(\frac{-38\,000}{T}\right)$$

$$R_2 = k_2^- [NO]_e [O]_e \quad k_2^- = 1.5 \times 10^9 \exp\left(\frac{-19\,500}{T}\right)$$

$$R_3 = k_3^- [NO]_e [H]_e \quad k_3^- = 2 \times 10^{14} \exp\left(\frac{-23\,650}{T}\right)$$

The temperature T is in Kelvin, the concentrations are in mol/cm^3 and the subscript e denotes chemical equilibrium.

The indicated reaction rate constants k_1, k_2 and k_3 are the most frequently used in the literature (Heywood, 1988; Ramos, 1989) and they could present some uncertainty depending on actual temperature and pressure. Several studies have been proposed in order to identify the optimal parameters at different engine operation. Among the others, Miller et al. (1998) proposed a correction factor for the constant k_1 as function of the instantaneous in-cylinder pressure; at high engine load and pressure, the reaction rate is reduced up to 80% of the original value, with a significant reduction of the NO prediction. The authors themselves have proposed an identification method based on a decomposition approach for estimating the optimal parameters as function of the engine operating conditions, with a significant improvement of model accuracy on a wide set of reference data (Arsie et al., 1998).

1.7 Soot Emissions

The mechanism of particulate formation is one of the most critical tasks in Diesel engine modeling. The basic phenomena that characterize the formation, the growth and the oxidation of the soot particles are not completely understood yet. The attempts performed for estimating soot emissions have led to the development of a wide variety of models ranging from phenomenological to empirical (black-box).

The most widely adopted modeling approach is the one originally proposed by Hiroyasu, which describes the soot formation and oxidation processes as kinetically controlled by two Arrhenius equations (Hiroyasu and Kadota, 1983). Thus the net soot mass rate is given by the difference between the mass formation rate and the mass oxidation rate (Patterson et al., 1994):

$$\frac{dm_s}{dt} = \frac{dm_{sf}}{dt} - \frac{dm_{so}}{dt} \qquad (19)$$

The mass formation rate m_{sf} and the mass oxidation rate m_{so} are estimated as:

$$\frac{dm_{sf}}{dt} = A_f m_{fv} P^{0.5} \exp\left(-E_f/RT\right) \qquad (20)$$

$$\frac{dm_{so}}{dt} = A_o m_s Y_{O_2} P^{1.8} \exp\left(-E_o/RT\right) \qquad (21)$$

where m_{fv} and m_s are the mass of fuel vapour and the net mass of soot, respectively, P is the in-cylinder pressure, Y_{O_2} is the oxygen molar fraction, T is the temperature. The pre-exponential coefficients A_f and A_o are model parameters to be identified in order to fit the experimental measurements; for the current analysis the identification was performed with respect to one operating point, corresponding to engine operation at medium load with EGR.

The activation energies E_f and E_o are assumed equal to 12 500 cal/mol and 14 000 cal/mol, as suggested by Hiroyasu and Kadota (1983).

The model given by Equations (19, 20) and (21) has been widely implemented in the framework of multi-zone combustion models (Patterson et al., 1994); the soot and oxidation kinetic equations are solved independently for each zone, which is characterized by uniform pressure, temperature and chemical composition. The total soot emissions are then estimated considering the contributions of all the zones. A different approach was proposed by Bayer and Foster (2003) who developed a detailed spray model and solved the soot formation and oxidation equations (20) and (21) for the whole region bounded by the fuel diffusion flame. This assumption is based on the hypothesis that the soot formation is mainly due to the fuel pyrolysis in the rich core, which is characterized by uniform temperature and composition.

2 MODEL VALIDATION

The present section is devoted to analyze model accuracy by comparing the simulations against a set of experimental data measured at the test bench. The reference engine is a light-duty, 4 cylinders, EURO 5 Diesel engine, equipped with common-rail injection system, high pressure EGR and Variable Geometry Turbine (VGT), whose main characteristics are described in Table 2.

Model accuracy was evaluated via comparison between predicted and measured in-cylinder pressure, NO and soot emissions at 34 different engine operating conditions, with engine speed ranging from 1 000 to 4 500 rpm, BMEP ranging from min to max, EGR rate ranging from 0 to 35%. Furthermore operations with single, double or multiple fuel injections were investigated.

Figure 5-9 show the comparison between predicted and measured in-cylinder pressure traces for five engine

TABLE 2

Engine technical data

Cylinders	4 in line
Displaced volume	1 248 cc
Valves per cylinder	4
Max. power	70 kW @ 4 000 rpm
Max. torque	210 Nm @ 1 750 rpm
Fuel injection system	Common rail solenoid injectors
Compression ratio	16.8:1

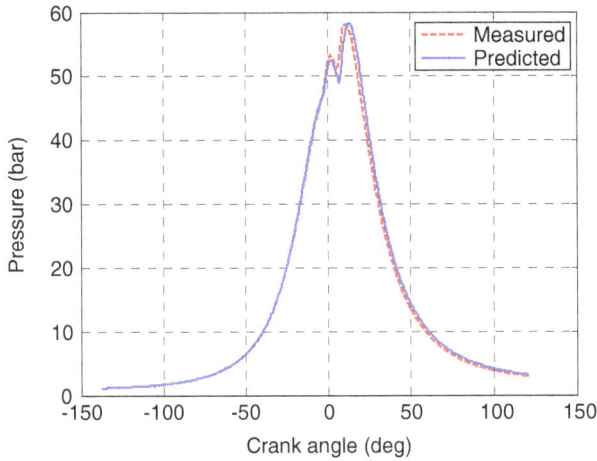

Figure 5

Comparison between measured and predicted in-cylinder pressure. Test case 1.

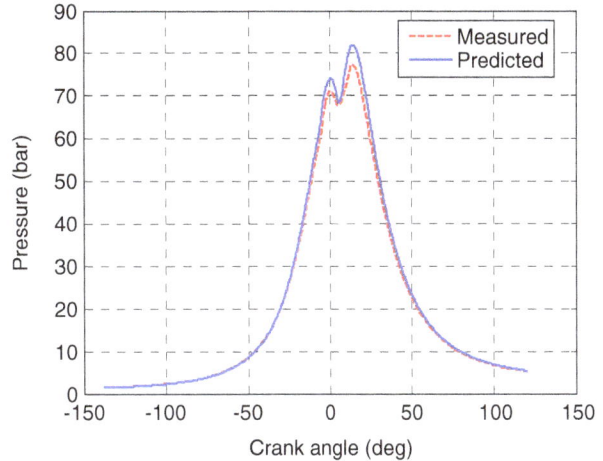

Figure 7

Comparison between measured and predicted in-cylinder pressure. Test case 3.

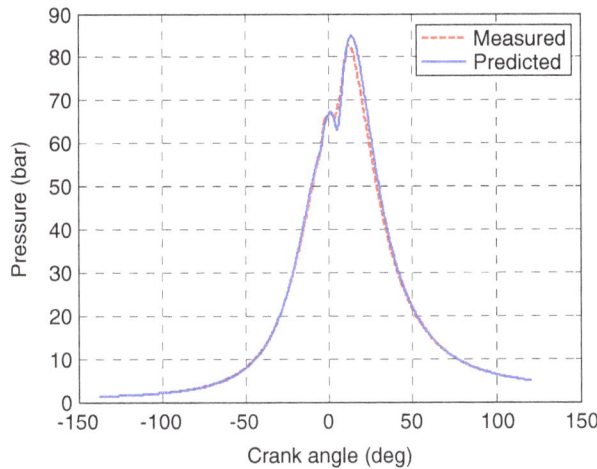

Figure 6

Comparison between measured and predicted in-cylinder pressure. Test case 2.

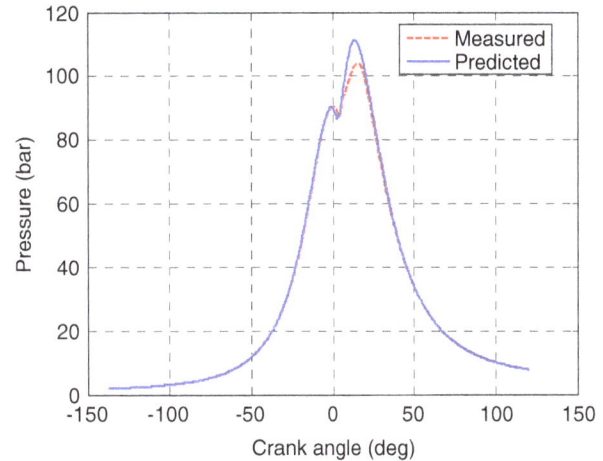

Figure 8

Comparison between measured and predicted in-cylinder pressure. Test case 4.

operating conditions, with different engine speed, load, fuel injections patterns and EGR rate, as reported in Table 3. In all cases, the model exhibits a good accuracy in predicting the engine cycle, even in the most critical conditions in case of high EGR rate (*i.e. Fig. 5, 9*). The model accuracy on the whole data set (34 cases) is shown in Figure 10 where the comparison between measured and predicted gross IMEP is shown. The figure evidences a good agreement with a correlation index R^2 equal to 0.995.

Figures 11-14 show model accuracy in estimating soot and NO emissions, respectively, by a comparison of predicted and measured data. The results refer to ten operating conditions at 2 000 rpm and 2 500 rpm, with increasing BMEP and rail pressure and different EGR rates.

Figures 11 and 12 apparently show poor validation results for the soot model with a quite large error. Nevertheless the model catches the main trends *versus* engine operating conditions, with the initial rise due to load increase and the final reduction due to the strong EGR reduction. An opposite trend is observed as BMEP increases from 4 to 8 bar at 2 000 rpm and EGR is reduced from 30% to 20%. This different

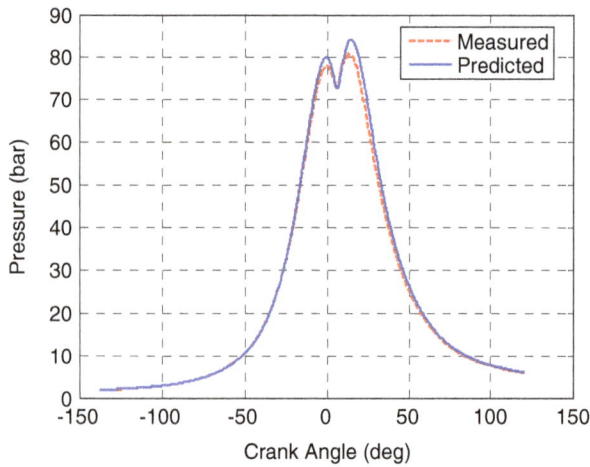

Figure 9

Comparison between measured and predicted in-cylinder pressure. Test case 5.

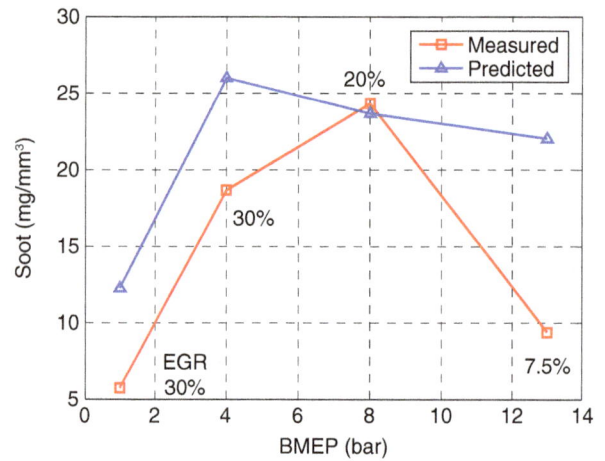

Figure 11

Comparison between measured and predicted engine soot emissions, *versus BMEP* at engine speed = 2 000 rpm. The numerical values in the plot indicate the EGR rate for each operating condition.

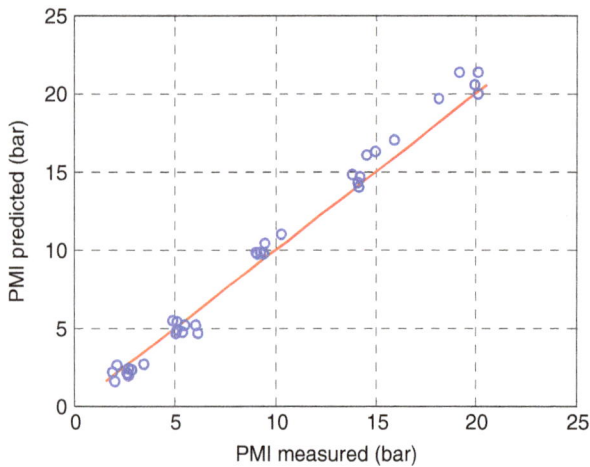

Figure 10

Comparison between measured and predicted IMEP for the whole set of experimental data. $R^2 = 0.99554$.

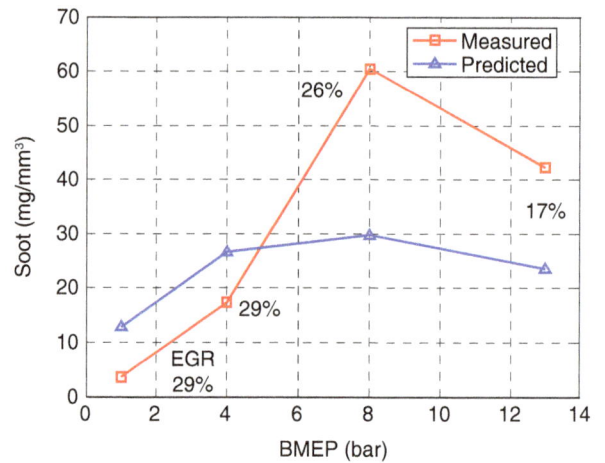

Figure 12

Comparison between measured and predicted engine soot emissions, *versus BMEP* at engine speed = 2 500 rpm. The numerical values in the plot indicate the EGR rate for each operating condition.

TABLE 3

Test cases considered for model validation

Test case	Speed (rpm)	BMEP (bar)	EGR (%)	P_{rail} (bar)	SOI (°ATDC) pil/pre/main
1	1 500	4	32	450	−24/−12/−2.5
2	1 500	8	16	615	−20/−9/−2
3	2 000	8	20	700	−30/−16/−3
4	2 500	13	17	1 000	−29/−7
5	3 000	8	25	910	−23/−6

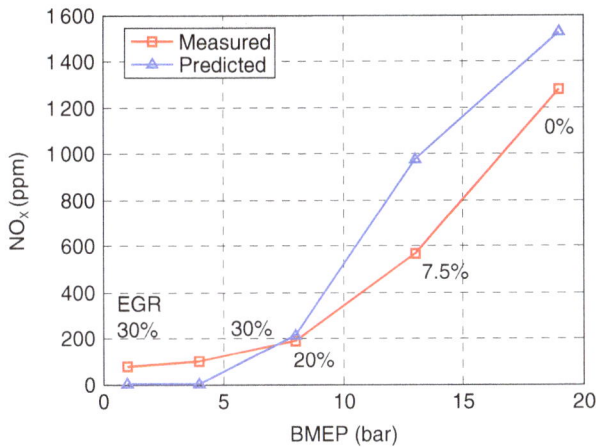

Figure 13

Comparison between measured and predicted engine NO emissions *versus* *BMEP* at engine speed = 2 000 rpm. The numerical values in the plot indicate the EGR rate for each operating condition.

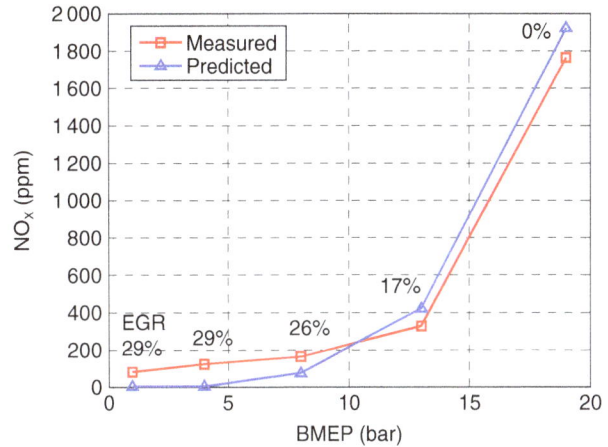

Figure 14

Comparison between measured and predicted engine NO emissions *versus* *BMEP* at engine speed = 2 500 rpm. The numerical values in the plot indicate the EGR rate for each operating condition.

behaviour may be due the superposition of the following effects:
– overestimation of the increased soot oxidation due to greater in-cylinder temperature;
– underestimation of increased soot formation due to greater mass of fuel. These effects also explain the underestimation detected at 2 500 rpm and BMEP equal to 8 and 13 bar.

It is worth noting that soot measurement is very often affected by large uncertainty due to the poor reliability of the instruments used, which are frequently based on empiric laws. Recently, more sophisticated and reliable instruments are coming up but they were not available for the current analysis. This is one of the motivations why physical models, even more complex that this (Patterson *et al.*, 1994; Hountalas, 2008), rarely exhibit a mean relative error below 50% in the whole engine operating domain. Few simulation results showing higher accuracy focus the analysis on three/four engine operating conditions very close each other (Jung and Assanis, 2006).

Finally, regardless to the entity of the validation error, model worthiness can be assessed by simulating soot emissions with perturbation of injection pattern and EGR and verifying whether the results are in accordance with the trends expected from experimental investigation. This parametric analysis is presented in Section 4.

Figures 13 and 14 evidence the good model results in predicting NO emissions with respect to measurements.

The figures show the expected increasing trend of NO with the load, due to the higher in-cylinder temperature following increased injected fuel mass and reduced EGR rate. Poor accuracy is reached at low load, because the Zeldovich mechanism only accounts for thermal NO formation thus lacking accuracy when low in cylinder temperature is reached. Nevertheless it is worth noting that the proposed model is intended to support the EMS tuning in compliance with NO_x/soot regulations. Therefore model accuracy and sensitivity is requested particularly in the most critical operating conditions corresponding to medium-high load, rather than at low load. In such conditions, the model exhibits a mean validation error below 23%, which is comparable to the accuracy achieved by physical models, even more complex that this, presented in the literature (Jung and Assanis, 2006; Patterson *et al.*, 1994; Hountalas, 2008; Miller *et al.*, 1998).

3 COMBUSTION NOISE

Noise is a critical issue for automotive engines and its main source is the in-cylinder pressure gradient generated during combustion. The in-cylinder pressure acts as exciting force on the engine block, causing its vibration and finally resulting in radiated noise (Payri *et al.*, 2005). The combustion noise generated by the sharp increase of in-cylinder pressure is strongly affected by the heat release rate (*i.e.* fuel burning rate) which in turn depends on injection pattern and mixture composition (*i.e.* air, fuel and inert gases). The presented methodology is aimed at predicting the impact of these control variables on combustion noise.

Mechanical noise, generated by the mechanical forces related to moving components (*i.e.* camshafts, connecting rods, pistons, etc.), also concurs to block vibration and noise radiation. Nevertheless it is not affected by engine control and its analysis was neglected, being beyond the scope of the present work.

The proposed approach is based on the estimation of the Sound Pressure Level (SPL), defined as:

$$SPL = 20 \times \log_{10}\left(\frac{p_{eff}}{p_0}\right) \quad (22)$$

The reference value p_0 corresponds to the hearing threshold at a frequency of 1 kHz and is set to 2×10^{-5} Pa. The sound pressure p_{eff} represents the root mean square of the time domain pressure signal and is given by:

$$p_{eff} = \sqrt{\frac{1}{T_{eff}}\int_0^T p(t)^2 dt} \quad (23)$$

Equation (22) is supposed to be applied for pure tones. In case of complex signals, as it is the case for the in-cylinder pressure, decomposition in elementary harmonics has to be accomplished by means of FFT analysis.

The methodology was applied to the simulated in-cylinder pressure traces corresponding to the test cases 1, 3, and 4 previously considered for model validation (*Tab. 3*). This allows analyzing engine working conditions with different speed, load and EGR rate. SPL estimation was performed considering the in-cylinder pressure contribution of all cylinders, as it is shown in Figure 15, to better describe the excitation of the engine structure. The resulting SPL frequency spectra are shown in Figure 16. As expected, the test cases with higher speed and load show a greater SPL in almost the whole frequency band. The maximum values are located at the lowest frequencies corresponding to the combustion period and are dependent on the magnitude of the in-cylinder pressure peak. On the other hand, at medium frequencies (*i.e.* 1 KHz), the SPL of the three test cases almost superimpose due to the impact of the higher pressure gradient following the enhanced premixed combustion for earlier injections and greater EGR rate.

In order to indicate the overall noise generated by the in-cylinder pressure signal, a synthetic index is introduced by the following equation, corresponding to the law of level summation (Möser, 2009):

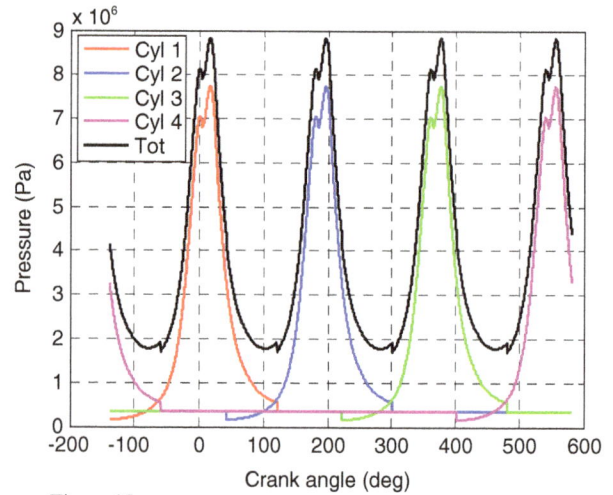

Figure 15

Superposition of the in-cylinder pressure in the four cylinders for the test case 1. The abscissa window corresponds to one engine cycle.

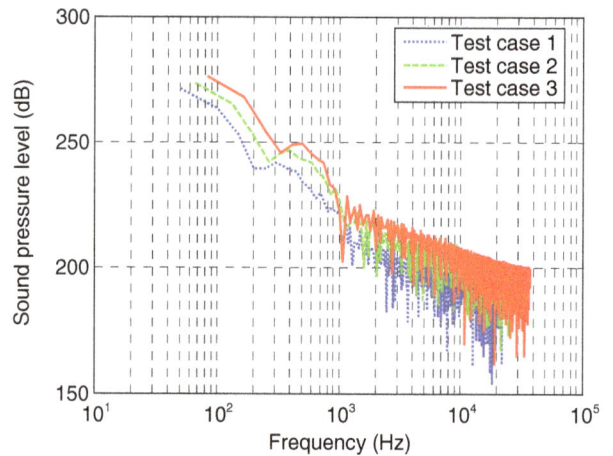

Figure 16

Frequency spectra of SPL for the three considered working conditions.

$$SPL_{tot} = 10 \times \log_{10}\sum_{i=1}^{N} 10^{\frac{SPL_i}{10}} \quad (24)$$

This approach allows estimating the total (or global) sound pressure level in case of more noise sources, as it is the case of the complex in-cylinder pressure signal that exhibits different harmonic components.

Figure 17 shows the estimation of the total SPL for the three working conditions considered. As expected from the SPL spectra analysis shown in Figure 16, the total SPL increases with load as it is mostly influenced

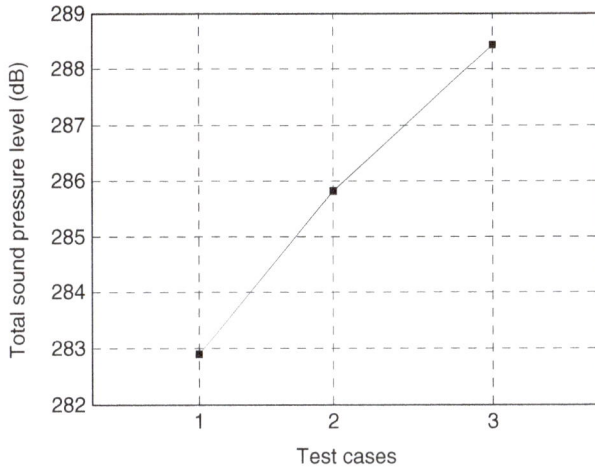

Figure 17

Total SPL for the three test cases considered for model validation.

TABLE 4

Set-points of the combustion control variables investigated to analyze the impact on performance and emissions

P_{rail} (bar)	EGR (%)	SOI (°ATDC)
700	20	Pilot from −30 to −60 by steps of −10
	30	Pre from −16 to −46 by steps of −10
	40	Main from −3 to −33 by steps of −10
1 000	30	Pilot from −30 to −60 by steps of −10
		Pre from −16 to −46 by steps of −10
		Main from −3 to −33 by steps of −10

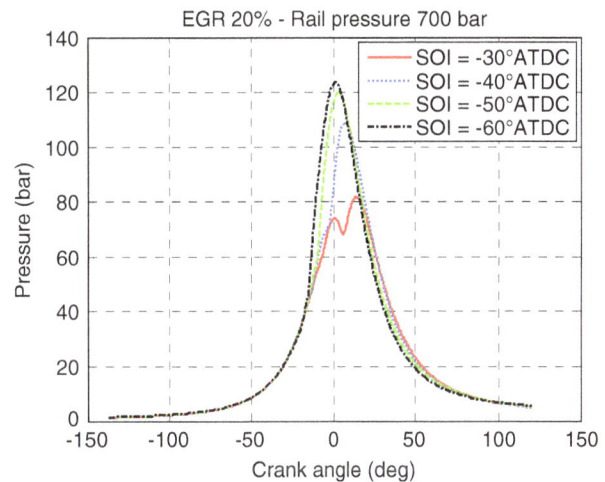

Figure 18

Simulated in-cylinder pressure at different pilot SOI and fixed EGR and rail pressure.

by the greatest values at low frequencies which are related to in-cylinder peak pressure magnitude.

4 RESULTS AND DISCUSSION

The present section analyzes the impact of combustion control variables, namely fuel injection pattern and EGR rate, on heat release rate, in-cylinder pressure and, consequently, noise and pollutants emissions of NO and soot. The analysis is based on the multi-zone model simulations coupled with the methodology for combustion noise prediction. Simulations were carried out at fixed engine speed (*i.e.* 2 000 rpm) and overall amount of injected fuel (*i.e.* 20 mg/cycle), imposing variation of EGR rate, rail pressure and Start of Injection (SOI), as reported in Table 4. A multiple injection strategy with pilot, pre and main injections was applied in all cases.

4.1 Start of Injection

The impact of SOI was investigated by imposing a variation from the baseline values, set to −30/−16/−3°ATDC (for pilot, pre and main injections, respectively), towards BDC up to −60° ATDC for the pilot injection. Fuel delivered for each injection and dwell times were kept constant, consequently as pilot SOI was advanced, pre and main injection were shifted accordingly.

Figures 18 and 19 show the superposition of pressure cycles and heat release rate profiles simulated at fixed

EGR and rail pressure and variable SOI. According to Figure 18, as the SOI is advanced the in-cylinder pressure exhibits a significant increase. This behavior is explained by the heat release rate profiles shown in Figure 19. As the SOI is advanced, the ignition delay is increased, due to the lower in-cylinder temperature (*Eq. 11*), particularly for the pilot and pre injections. The figure evidences that when SOI advance is greater than 40° the heat release of pilot, pre and main injection take place simultaneously, reducing the benefits of multiple injection. As a consequence of the increased ignition delay, the in-cylinder pressure exhibits a greater pressure rise due to the enhanced air-fuel mixing and the larger fraction of fuel burning in premixed mode.

It is worth noting that further advancing SOI towards BDC would amplify these phenomena, promoting a

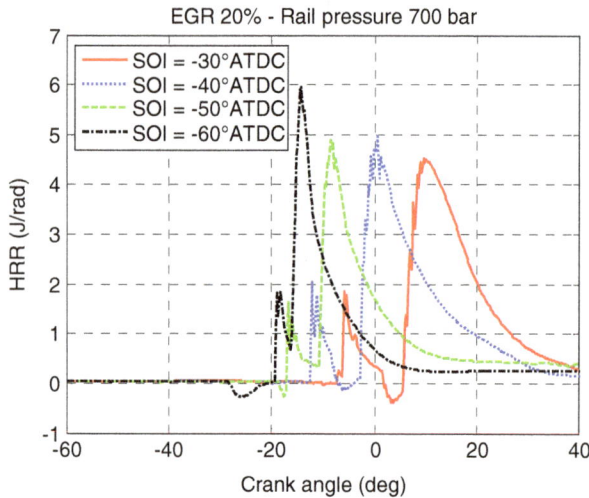

Figure 19

Simulated heat release rate at different pilot SOI and fixed EGR and rail pressure.

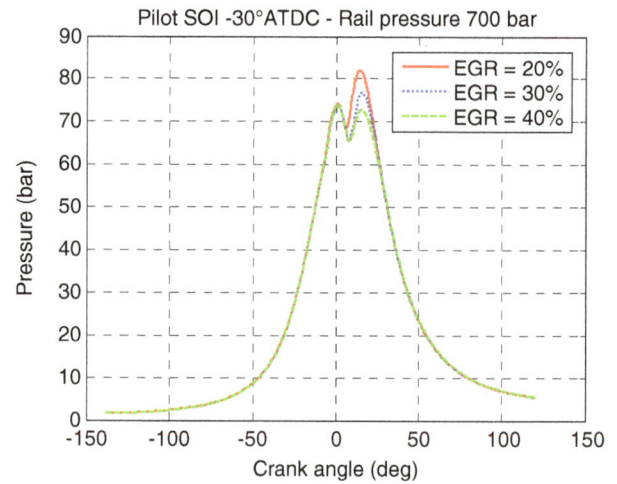

Figure 20

Simulated in-cylinder pressure at different EGR rates and fixed SOI and rail pressure.

complete premixed combustion (*i.e.* Premixed Combustion Compression Ignition – PCCI) in place of the conventional one. Nevertheless, though innovative combustion concepts, such as PCCI, have experimentally proved to be promising in reducing both NO and soot emissions, they were not investigated in the current analyses. The motivation is that advancing injection may result in combustion deterioration and fuel impingement on cylinder or piston walls and none of these effects is actually taken into account by the in-cylinder model.

4.2 Exhaust Gas Recirculation

The impact of inert gases and oxygen concentration in the intake charge was analyzed by considering three EGR rates, corresponding to 20% (*i.e.* baseline setting), 30% and 40%. Figure 20 exhibits that as EGR is increased the in-cylinder pressure presents a lower peak and smoother rise. According to Figure 21 this is due to the less abrupt combustion, due to the lower temperature and oxygen content in the mixing zone (*Eq. 16*).

4.3 Rail Pressure

Two values of injection pressure were considered for the present analysis, corresponding to 700 bar (*i.e.* baseline setting) and 1 000 bar. The increase of injection pressure results in better fuel atomization and improved air-fuel mixing due to the greater flux momentum. The resulting in-cylinder pressure exhibits a greater rise following the enhanced air-fuel mixing. This is evidenced by

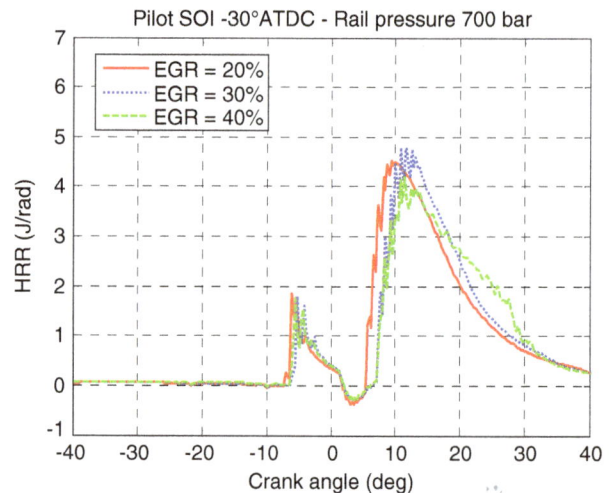

Figure 21

Simulated heat release rate at different EGR rates and fixed SOI and rail pressure.

Figures 22 and 23 that show in-cylinder pressure and heat release rate for the two considered values of injection pressure with fixed SOI and EGR.

4.4 Engine Performance and Emissions

The impact of combustion control variables on engine performance and emissions is shown in the following

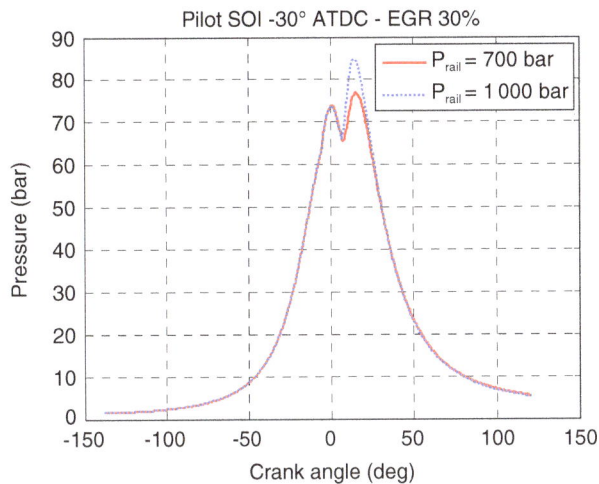

Figure 22

Simulated in-cylinder pressure at different rail pressure and fixed SOI and EGR rate.

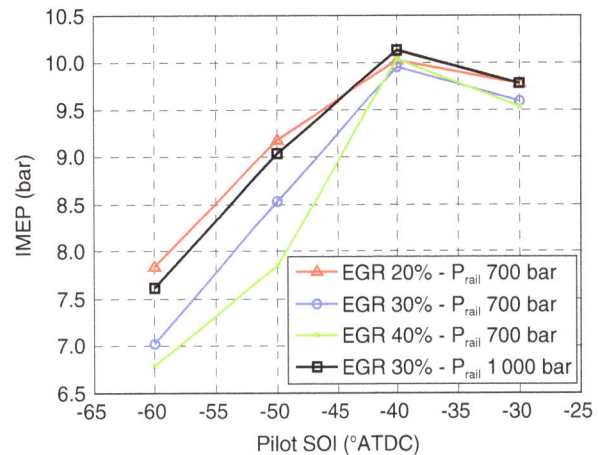

Figure 24

Simulation results: effect of SOI, EGR and P_{rail} on IMEP.

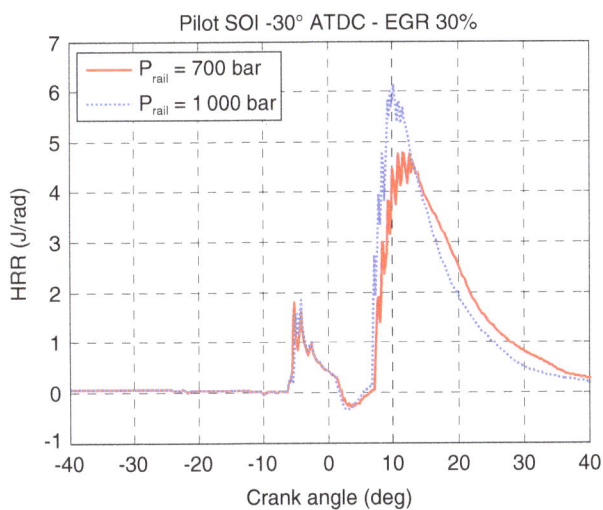

Figure 23

Simulated heat release rate at different rail pressure and fixed SOI and EGR rate.

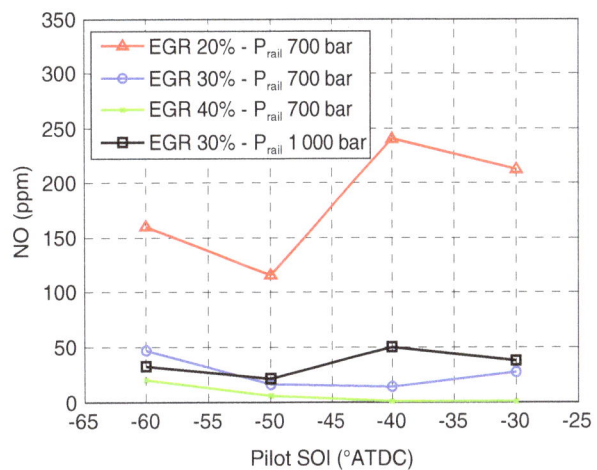

Figure 25

Simulation results: effect of SOI, EGR and P_{rail} on NO emissions.

Figures 24-27 that illustrate the prediction of IMEP, NO, soot and combustion noise in the operated conditions investigated (Tab. 4).

Figure 24 shows that as SOI is advanced the IMEP initially increases due to in-cylinder pressure rise, until pilot SOI reaches approx. −40°. As SOI is further advanced towards TDC, IMEP decreases due to the higher compression work. The opposite impact of EGR and rail pressure on IMEP reflects the behavior of pressure cycle and heat release rate, previously commented in Figures 20-23. It is worth noting that following the assumption of constant mass of injected fuel per cycle, an increase of IMEP corresponds to higher combustion efficiency, with lower specific fuel consumption and CO_2 emissions.

Figure 25 shows that as SOI is advanced, NO emissions initially increase for the higher in-cylinder temperature following the sharp heat release rate (Fig. 19). Further SOI advance results in a reduction of NO due to more uniform air-fuel mixing and reduced local

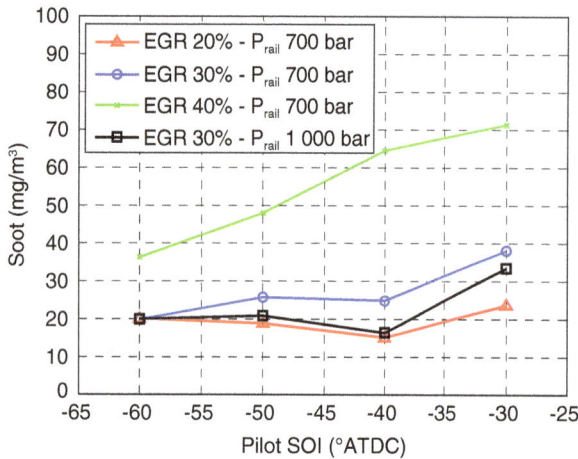

Figure 26

Simulation results: effect of SOI, EGR and P_{rail} on soot emissions.

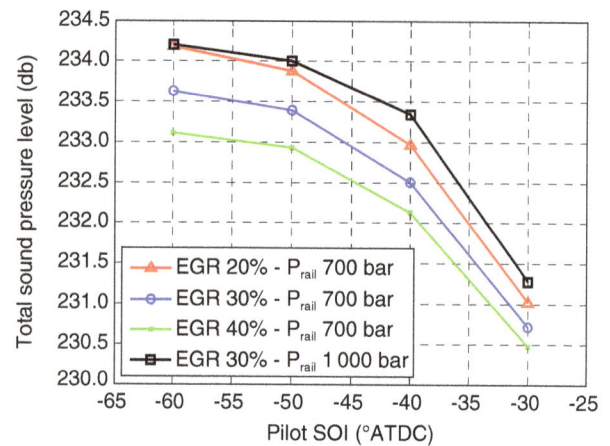

Figure 27

Simulation results: effect of SOI, EGR and P_{rail} on combustion noise.

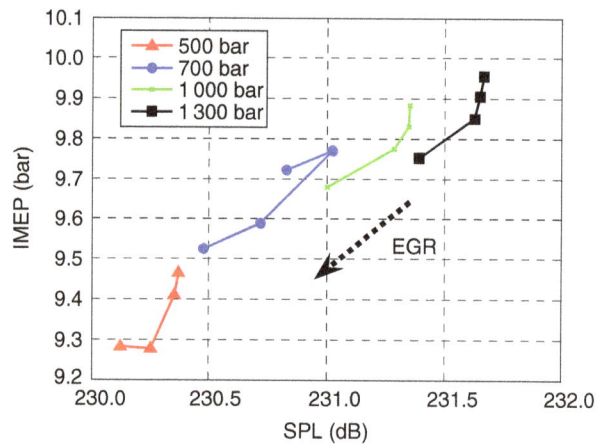

Figure 28

Simulation results: trade-off between acoustic emissions (SPL) and Indicated Mean Effective Pressure (IMEP).

temperature. Both these effects are enhanced by the rail pressure, due to the mentioned influence on fuel atomization and air-fuel mixing. Concerning the EGR rate, Figure 25 exhibits the expected strong impact on NO reduction, due to the significant temperature decrease.

The prediction of soot emissions confirms the expected trade-off with NO emissions. Figure 26 shows that as SOI is advanced soot emissions decrease due to the enhanced air-fuel mixing caused by the longer ignition delay. This phenomenon is even amplified by the higher rail pressure that promotes fuel atomization and air entrainment. On the other hand, high EGR rate results in an increase of soot due to the lower temperature that inhibits soot oxidation.

The impact on combustion noise is shown in Figure 27 and reflects the heat release rate profiles (*Fig. 19, 21, 23*). Advanced SOI and high injection pressure promote premixed combustion, resulting in a sharp heat release rate and a greater sound pressure level. This effect is mitigated by the low oxygen concentration in case of high EGR that makes combustion rate smoother.

In order to highlight the opposite effects of combustion control variables on engine performances and emissions, Figures 28 and 29 show the simulated trade-off of IMEP *versus* SPL and NO *versus* soot emissions, respectively. The simulations were performed imposing constant engine speed (*i.e.* 2 000 rpm), mass of injected fuel (*i.e.* 20 mg/cycle) and SOI ($-30°/-16°/-3°$ATDC) while ranging EGR rate and rail pressure as reported in Table 5.

Particularly, the figures evidence that increasing the rail pressures results in higher IMEP (*i.e.* lower specific fuel consumption) and lower soot emissions with a slight impact on NO emissions. Nevertheless a strong increase of combustion noise is observed. On the other hand, increasing the EGR rate results in a strong reduction of both NO and noise and an increase of specific fuel consumption (*i.e.* reduction of IMEP) and soot.

The presented results evidence that the quasi-dimensional multi-zone modeling approach applied for in-cylinder simulation allows predicting the expected trends of pressure cycle and heat release rate *versus*

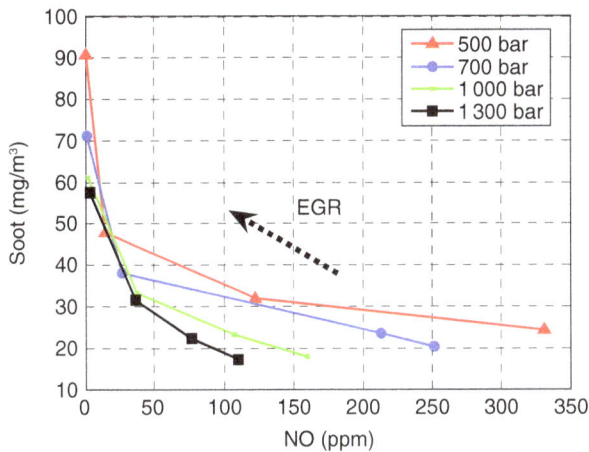

Figure 29

Simulation results: trade-off between soot and NO emissions.

TABLE 5

Operating conditions investigated for the trade-off analysis

P_{rail} (bar)	500 / 700 /1 000 / 1 300
EGR (%)	10 / 20 / 30 / 40

injection pattern and EGR rate. Consequently the effects on engine performance, noise and pollutants are in accordance with those expected from the experimental analyses in the literature (Tennison and Reitz, 2001; Torregrosa *et al.*, 2011). Particularly, the simulation results confirm the complex interaction and the opposite effects of injection timing, injection pressure and EGR on fuel burning rate and pollutants formation and evidence the valuable contribution of simulation models for EMS tuning.

CONCLUSIONS

The impact of injection pattern and EGR on noise and pollutants emissions in a light-duty Diesel engine was investigated *via* simulation analysis. A Multi-zone quasi dimensional model was applied for combustion simulation and NO/soot estimation and successfully validated against experimental data. Furthermore a methodology based on in-cylinder pressure processing was applied for combustion noise prediction.

The simulated effects of injection timing, rail pressure and EGR on engine performance, noise and pollutants were in accordance with those expected from the experimental analyses proposed in the literature. Particularly, the simulation results confirmed the complex interaction

and the opposite effects on fuel burning rate and pollutants formation and evidenced the valuable contribution of the proposed modeling approach for EMS tuning.

Future work will be aimed at improving the fuel jet model to account for combustion deterioration and fuel wall impingement which usually take place in case of advanced injection timing characteristic of premixed combustion (*i.e.* PCCI).

ACKNOWLEDGMENTS

The present research has been funded by Magneti Marelli S.p.A. Powertrain and University of Salerno.

REFERENCES

Arsie I., Pianese C., Rizzo G. (1998) Models for the Prediction of Performance and Emissions in a Spark Ignition Engine - A Sequentially Structured Approach, *SAE Technical Paper* 980779.

Arsie I., Di Genova F., Mogavero A., Pianese C., Rizzo G., Caraceni A., Cioffi P., Flauti G. (2006) Multi-Zone Predictive Modeling of Common-Rail Multi-Injection Diesel Engines, *SAE Technical Paper* 2006-01-1384.

Arsie I., Pianese C., Sorrentino M. (2007) Effects of Control Parameters on Performance and Emissions of HSDI Diesel Engines: Investigation *via* Two Zone Modelling, *Oil & Gas Science and Technology* **62**, 4, 457-469.

Arsie I., Di Leo R., Pianese C., De Cesare M. (2012) Combustion Noise and Pollutants Prediction for Injection Pattern and EGR Tuning in an Automotive Common-Rail Diesel Engine, *IFAC Workshop on Engine and Powertrain Control, Simulation and Modeling (E-COSM'12)*, Rueil-Malmaison (France), 23-25 Oct.

Assanis D.N., Heywood J.B. (1986) Development and Use of a Computer Simulation of the Turbocompounded Diesel System for Engine Performance and Component Heat Transfer Studies, *SAE Technical Paper* 860329.

Ayoub N.S., Reitz R. (1995) Multidimensional Computation of Multicomponent Spray Vaporization and Combustion, *SAE Technical Paper* 950285.

Barba C., Burkhardt C., Boulouchos K., Bargende M. (2000) A Phenomenological Combustion Model for Heat Release Rate Prediction in High-Speed DI Diesel Engines With Common Rail Injection, *SAE Technical Paper* 2000-01-2933.

Bayer J., Foster D.E. (2003) Zero-Dimensional Soot Modeling, *SAE Technical Paper* 2003-01-1070.

Bi X., Yang M., Han S., Ma Z. (1999) A Multi-Zone Model for Diesel Spray Combustion, *SAE Technical Paper* 1999-01-0916.

Coppo M., Dongiovanni C. (2007) Experimental Validation of a Common-Rail Injector Model in the Whole Operation Field, *ASME J. of Engineering for Gas Turbines and Power* **129**, 596-608.

Ferguson C.R. (1986) Internal Combustion Engine, *Applied Thermosciences*, John Wiley.

Gang L., Tao B. (2010) Multiple-Cylinder Diesel Engine Combustion CFD Simulation with Detailed Chemistry Based IPV-Library Approach, *SAE Technical Paper* 2010-01-1495.

Heywood J.B. (1988) *Internal Combustion Engine Fundamentals*, MC Graw Hill.

Hiroyasu H., Masataka A. (1990) Structures of Spray in Diesel Engines, *SAE Technical Paper* 900475.

Hiroyasu H., Kadota T. (1983) Development and Use of a Spray Combustion Modeling to Predict Diesel Engine Efficiency and Pollutant Emission, *Bulletin of the ASME* **26**, 214.

Hountalas D.T., Mavrapoulos G.C., Binder K.B. (2008) Effect of Exhaust Gas Recirculation (EGR) temperature for various EGR rates on heavy duty DI Diesel engine performance and emissions, *Energy* **33**, 272-283.

Javadi Rad G., Gorjiinst M., Keshavarz M., Safari H., Jazayeri A.A. (2010) An Investigation on Injection Characteristics of Direct-Injected Heavy Duty Diesel Engine by Means of Multi-Zone Spray Modeling, *Oil & Gas Science and Technology* **65**, 893-901.

Jung D., Assanis D.N. (2006) Quasidimensional Modeling of Direct Injection Diesel Engine Nitric Oxide, Soot, and Unburned Hydrocarbon Emissions, *ASME J. of Engineering for Gas Turbines and Power* **128**, 388-396.

Kong S.C., Han Z., Reitz R.D. (1995) The Development and Application of a Diesel Ignition and Combustion Model for Multidimensional Engine Simulation, *SAE Technical Paper* 950278.

Kouremenos D.A., Rakopoulos C.D., Hountalas D.T. (1997) Multi-Zone Combustion Modeling for the Prediction of Pollutant Emissions and Performance of DI Diesel Engine, *SAE Technical Paper* 970635.

Kuo K.K. (1986) *Principles of Combustion*, John Wiley.

López J.J., Novella R., García A., Winklinger J.F. (2013) Investigation of the ignition and combustion processes of a dual-fuel spray under Diesel-like conditions using Computational Fluid Dynamics (CFD) modelling, *Mathematical and Computer Modeling* **57**, 1897-1906.

Miller R., Davis G., Lavoie G., Newman C., Gardner T. (1998) A Super-Extended Zel'dovich Mechanism for NO_x Modeling and Engine Calibration, *SAE Technical Paper* 980781.

Möser M. (2009) *Engineering acoustics – An introduction to Noise Control*, Springer.

Naber J.D., Siebers D.L. (1996) Effects of gas Density and Vaporization on Penetration and Dispersion of Diesel Spray, *SAE Technical Paper* 960034.

Patterson M.A., Kong S.C., Hampson G.J., Reitz R.D. (1994) Modeling the Effects of Fuel Injection Characteristics on Diesel Engine Soot and NO_x Emissions, *SAE Technical Paper* 940523.

Payri F., Broatch A., Tormos B., Marant V. (2005) New methodology for in-cylinder pressure analysis in direct injection Diesel engines – Application to combustion noise, *Meas. Sci. Technol.* **16**, 540-547.

Rakopoulos C.D., Rakopoulos D.C., Kyritsis D.C. (2003) Development and Validation of a Comprehensive two-zone Model for Combustion and Emissions Formation in a DI Diesel Engine, *International Journal of Energy Research* **27**, 1221-1249.

Ramos J.I. (1989) *Internal Combustion Engine Modeling*, Hemisphere Publishing Corporation.

Tanner F.X., Reitz R.D. (1999) Scaling Aspects of the Characteristic Time Combustion Model in the Simulation of Diesel Engines, *SAE Technical Paper* 1999-01-1175.

Tennison P.J., Reitz R.D. (2001) An Experimental Investigation of the Effects of Common-Rail Injection System Parameters on Emissions and Performance in a High-Speed Direct-Injection Diesel Engine, *ASME J. of Engineering for Gas Turbines and Power* **123**, 167-178.

Torregrosa A.J., Broatch A., Novella R., Mónico L.F. (2011) Suitability analysis of advanced Diesel combustion concepts for emissions and noise control, *Energy* **36**, 825-838.

8

Development of Look-Ahead Controller Concepts for a Wheel Loader Application

Tomas Nilsson[1]*, Anders Fröberg[2] and Jan Åslund[1]

[1] Department of Electrical Engineering, Linköping University - Sweden
[2] Volvo Construction Equipment, Eskilstuna - Sweden
e-mail: tnilsson@isy.liu.se - anders.froberg@volvo.com - jaasl@isy.liu.se

* Corresponding author

Abstract — *This paper presents two conceptual methods, based on dynamic programming, for one-step look-ahead control of a Continuously Variable Transmission (CVT) in a wheel loader. The first method developed, designated Stochastic Dynamic Programming (SDP), uses a statistical load prediction and stochastic dynamic programming for minimizing fuel use. The second method developed, designated Free-Time Dynamic Programming (FTDP), has vehicle speed as a state and introduces a fixed 0.1 s delay in the bucket controls in a combined minimization of fuel and time. The methods are evaluated using a set of 34 measured loading cycles, used in a 'leave one out' manner.*

The evaluation shows that the SDP method requires about 1/10th of the computational effort of FTDP and has a more transparent impact of differences in the cycle prediction. The FTDP method, on the other hand, shows a 10% lower fuel consumption, which is close to the actual optimum, at the same cycle times, and is able to complete a much larger part of the evaluation cycles.

Résumé — Développement de concepts d'une commande prédictive, destinée à une application pour chargeur sur pneus — Ce document présente deux méthodes de conception, basées sur la programmation dynamique, pour la commande à un pas de prédiction d'une transmission continûment variable (*Continuously Variable Transmission*, CVT) d'un chargeur sur pneus. La première méthode développée, appelée programmation dynamique stochastique (*Stochastic Dynamic Programming*, SDP) utilise une prédiction statistique de la charge et la programmation dynamique stochastique pour minimiser l'utilisation de carburant. La seconde méthode développée, appelée programmation dynamique à temps libre (*Free-Time Dynamic Programming*, FTDP), établit la vitesse du véhicule en tant qu'état et introduit un retard de 0,1 s dans les commandes du godet pour minimiser à la fois l'utilisation de carburant et le temps nécessaire à l'opération.

Les méthodes sont évaluées en s'appuyant sur 34 cycles de chargement mesurés, utilisés selon la méthode de validation croisée « *leave-one-out* ».

L'évaluation montre que la méthode SDP requiert environ 1 dixième de l'effort de calcul de la méthode FTDP, et qu'elle a un impact plus transparent sur les écarts dans la prédiction du cycle. D'un autre côté, avec la méthode FTDP on obtient une réduction de 10 % de la consommation de carburant, ce qui est proche de l'optimum réel, pour les mêmes durées de cycle, et elle permet de réaliser une plus grande partie des cycles d'évaluation.

INTRODUCTION

Background

Wheel loader operation is often highly transient and contains episodes of low speed and high tractive effort, while the engine has to deliver power to both the transmission and the working hydraulics. The most common general transmission layout of heavy wheel loaders is presented in Figure 1. The engine is connected to the hydraulics through a variable displacement pump and to the drive shaft through a hydrodynamic torque converter and an automatic gearbox.

In this setup, the torque converter is a crucial component, since it provides some disconnection between the engine and vehicle speeds. This disconnection makes the system mechanically robust but the solution is also prone to high losses. High thrust is achieved by high torque converter slip, which produces losses. High hydraulic flow requires high engine speed, which also produces transmission torque which, if increased speed is not desired, is balanced by the brakes, causing losses in both the torque converter and the brakes. This lack of efficiency is the reason for a desire to find other transmission concepts for wheel loaders.

On the Choice of a Hydraulic Multi-Mode CVT

Any alternative transmission has to enable increased efficiency in the typical operation conditions mentioned. The low speeds at which the machine often operates makes it impractical to use a stepped gearbox without a torque converter. One alternative is to consider infinitely variable transmissions, such as the Diesel-electric used in [1] or the hydrostatic used in [2]. The drawback with this type of transmission is that the repeated power conversions reduce the peak efficiency. This is addressed by power-split constructions such as those described by [3, 4], in which some part of the power is mechanically transmitted. Multi-mode Continuously Variable Transmissions (CVT) are constructed so that several power-split layouts can be performed with the same device, thus enabling high efficiency at widely spaced gear ratios. In this paper, just as in [5], the transmission is based on a hydrostatic CVT since this solution has a favorable cost and torque rating.

CVT Control in a Wheel Loader

The introduction of a CVT increases both the possibility of fuel saving and the risk of poor operability. The performance depends to a high degree on the implemented controller. Some work has been done on CVT control in wheel loader applications [6, 7]. The focus is often on actuator control though, and there is a lack of work on higher level control, including the choice of the engine operating point. This choice is highly complicated by the operation often being extremely transient.

The most common operating pattern for wheel loaders is the short loading cycle. In this cycle, the loader approaches a pile and fills the bucket, everses, approaches a load receiver and empties the bucket, reverses and starts over. The operation is described in detail in [8, 9]. This easily described and highly repetitive operation may form the basis of a rough prediction of the future load. Because of the extremely transient operation, the benefits of utilizing the prediction in the controller can be expected to become high. Look-ahead control for on-road vehicles has been implemented [10-12]. In the wheel loader application, the potential benefit has been explored [13], but so far there has been no implemented look-ahead controller for wheel loaders. The main difficulties, as compared with on-road applications, are the increase in system complexity and the uncertainties in the future load prediction. This paper introduces and evaluates two different conceptual look-ahead controller implementations for this system, both of which are based on dynamic programming.

Problem Formulation

The goal of this paper is to develop and test, through simulations, conceptual dynamic programming-based look-ahead controllers for use in a multi-mode CVT wheel loader. The controllers should be focused on the short loading cycle, and may therefore use future load predictions derived from data collected during measurements in a number of loading cycles. The aim should be to minimize, or at least to reduce, the fuel consumption without having a negative impact on drivability or performance of the machine.

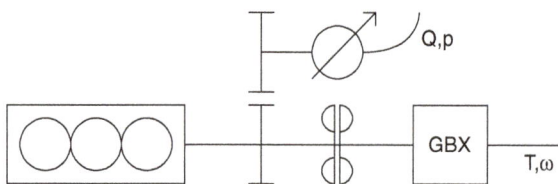

Figure 1

Reference vehicle drivetrain setup.

1 MODELS

1.1 Machine Operation

One of the most common operating patterns for wheel loaders is the Short Loading Cycle (SLC), as described in [8, 9]. This cycle is also the basis for the prediction used in this work.

In the SLC definition used here, and referring to the position designations in Figure 2, the cycle starts at position (2) and consists of four separate phases. In the first phase the machine drives forward to position (1), and during the final part of this phase the bucket is filled. The filling of the bucket often requires high tractive force combined with tilting and some lifting of the bucket. The second phase is reversing back to position (2) and the third is forward driving to the load receiver at position (3). During these two phases, the bucket is raised, and at the end of the third phase it is emptied. The fourth and final phase is reversing back to position (2) while lowering the bucket. In a typical cycle the total duration is around 30 s and the distances between the driving direction changes are around 10 m.

In this paper, a measurement sequence which includes 34 short loading cycles is used. The measurement setup is presented in Figure 3. The basic load components, related to the load components used in the system description in Figure 1, are vehicle speed v_w, tractive force F_w, hydraulic pressure p_H and hydraulic flow Q_H. The main difference from the description in Figure 1 is

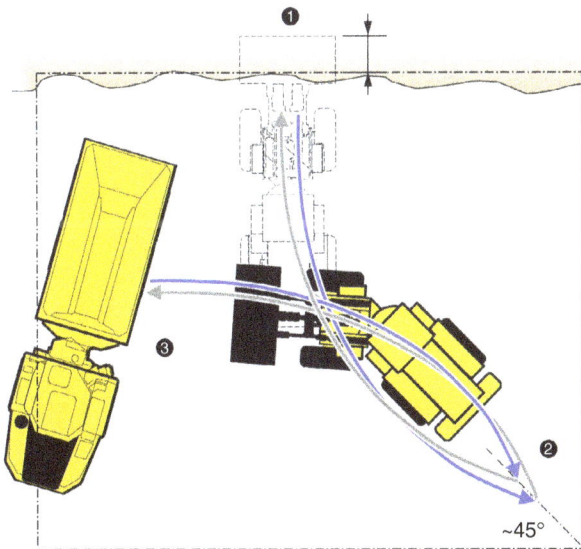

Figure 3

A view of the measurement setup indicating the signals available. Solid lines are mechanical connections and dashed lines are hydraulic connections. The system setup corresponds to that presented in Figure 1.

that F_w does not include inertia forces. These load components are derived as follows. The hydraulic pressure p_H is assumed to be equal to the measured hydraulic pump pressure p_{Ls}. The hydraulic flow Q_H is calculated from the volumes in the lift and tilt cylinders, which are calculated from the lift and tilt angles θ_1 and θ_2. Lowering the bucket generally does not require pressurized hydraulic fluid, and this is therefore not supplied through the pump. The vehicle speed v_w is derived from the torque converter output speed ω_{ct} and the selected gear r_c, which include the selected driving direction. The tractive force F_w during the bucket filling is calculated from the torque converter output torque T_{ct} and the selected gear r_c. The torque converter output torque is calculated from the torque converter input and output speeds, ω_{cp} and ω_{ct}, according to:

$$v_c = \frac{\omega_{ct}}{\omega_{cp}} \tag{1a}$$

$$T_{cp} = M_P(v_c)\left(\frac{\omega_{cp}}{\omega_{cp,ref}}\right)^2 \tag{1b}$$

$$T_{ct} = \mu(v_c)T_{cp} \tag{1c}$$

in which M_P and μ are scalable maps that have been measured at the reference speed $\omega_{cp,ref}$. The tractive force when not filling the bucket is modeled as a constant rolling resistance according to:

$$F_w = \text{sign}(v_w)mgc_r \tag{2}$$

These basic load components are used in constructing the load cases $w(t)$ or $w(s)$, according to the requirements of each dynamic programming implementation.

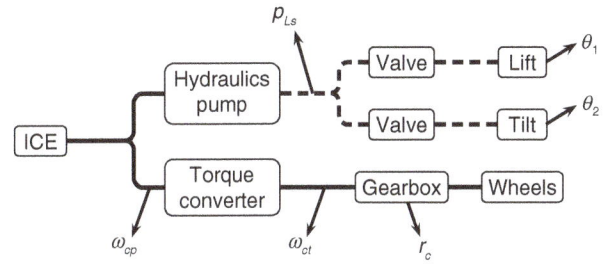

Figure 2

A view of a short loading cycle [8].

One of the measured SLC, as described by the four presented load components, is displayed in Figure 4.

Due to adjustments made in the following load case creations, it is of interest to view the times and distances in the measured cycles. These are displayed in Figure 5. The average unadjusted cycle time is 26.5 s and the average unadjusted distance driven is 35 m.

1.2 Vehicle Model and System Layout

The vehicle is modeled as a mass m, for which the speed dynamics depend on the propulsive torque T_W, the brake torque T_b and the tractive force F_w. The factor r includes the final gear ratio and the wheel radius:

$$\frac{dv_w}{dt} \cdot m = r^{-1}T_W - r^{-1}T_b - F_w \qquad (3)$$

The layout of the system is presented in Figure 6. The main components, which are described in the following sections, are the engine, the multi-mode CVT transmission and the variable displacement hydraulics pump.

1.3 Engine Model

The engine is modeled as an inertia I_e which is affected by the engine torque T_e, the transmission torque T_T and the hydraulic pump torque T_H:

$$\frac{d\omega_e}{dt} \cdot I_e = T_e - T_T - T_H \qquad (4)$$

The relation between fuel use and engine torque is described by a quadratic Willan's efficiency model, as presented in [14], expanded with a torque loss due to lack of intake manifold pressure:

$$T_e = e(\omega_e, m_f) \cdot \frac{q_{lhv}n_{cyl}}{2\pi n_r} \cdot m_f - T_L(\omega_e) - T_{pt} \qquad (5)$$

in which m_f is fuel mass per injection, ω_e is engine speed, e and T_L are efficiency functions, q_{lhv}, n_{cyl} and n_r are constants, and T_{pt} is torque loss due to lack of air intake pressure $p_{off} = p_t - p_{set}(\omega_e, m_f)$. Here, p_t is the actual pressure and p_{set} is a static setpoint map. The turbocharger speed dynamics is assumed to be a first-order system. The dynamics model is expressed in the corresponding intake air pressure:

$$\frac{dp_t}{dt} \cdot \tau(\omega_e) = -p_{off}(\omega_e, m_f) \qquad (6)$$

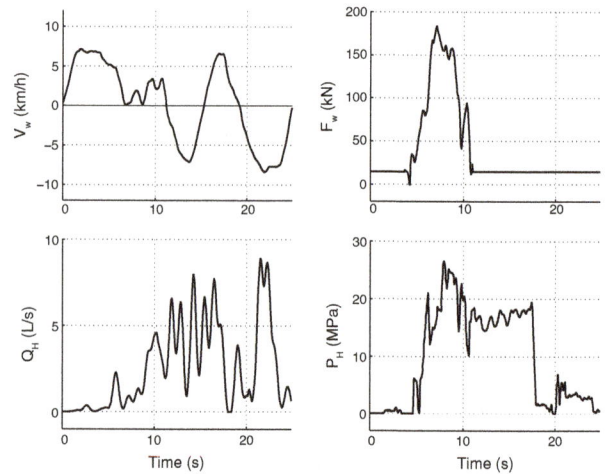

Figure 4

An example of a short loading cycle expressed in the four basic load components vehicle speed v_w, tractive force F_w, hydraulic flow Q_H and hydraulic pressure p_H.

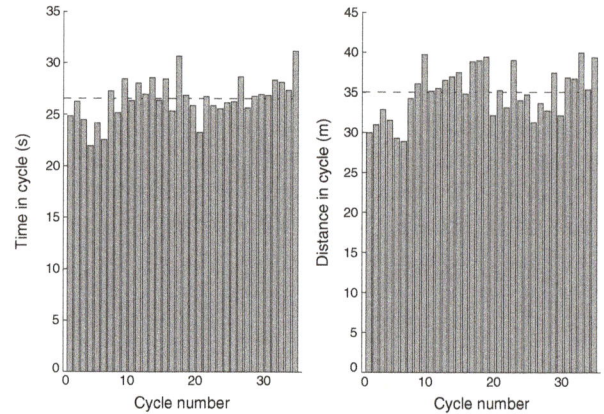

Figure 5

Overview of cycle durations expressed as time in cycle and as distance in cycle. The mean values are indicated by the dashed lines.

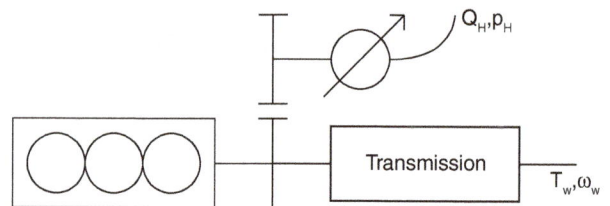

Figure 6

Overview of the layout of the system studied. The transmission is presented in detail in Figure 8.

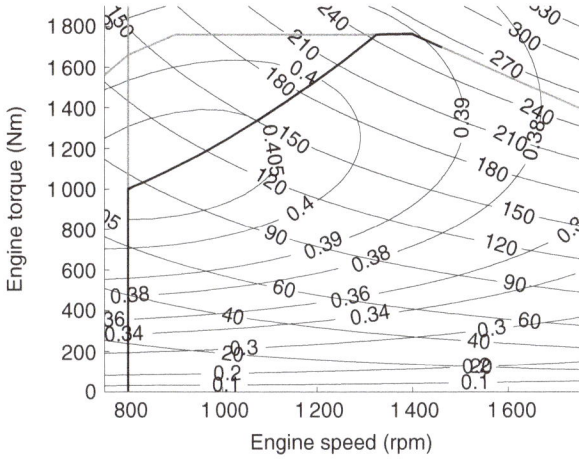

Figure 7

Engine map with static optimal operation line (black), speed and torque limits (thick gray), efficiency curves and output power lines with *kW* markings (thin gray).

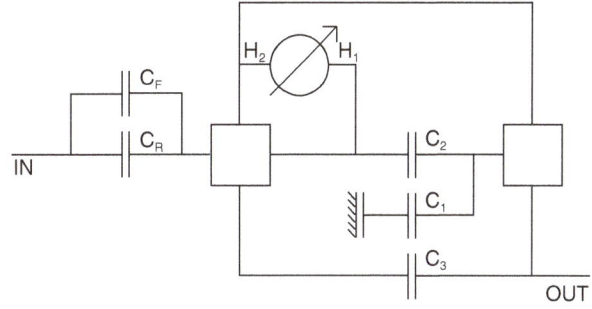

Figure 8

Layout of the multi-mode CVT. The box to the left is a Ravigneaux planetary gearset and the box to the right is a regular planetary gearset. H_1 and H_2 are hydraulic machines and the five C are clutches. Regular gears are not shown.

and the torque loss from low pressure is described by:

$$T_{pt} = \begin{cases} k_1(\omega_e) \cdot p_{off}^2 - k_2(\omega_e) \cdot p_{off} & \text{if } p_{off} < 0 \\ 0 & \text{if } p_{off} \geq 0 \end{cases} \quad (7)$$

The fuel per injection is related to the fuel flow according to:

$$\frac{dM_f}{dt} = m_f \frac{n_{cyl}}{2\pi n_r} \omega_e \quad (8)$$

Figure 7 presents the efficiency map of the engine used. The gray lines indicate allowed operating region (minimum speed and maximum torque) and the black line indicates the static optimal operating points for each output power. The figure also shows efficiency levels and output power lines with *kW* markings.

1.4 Transmission Model

The transmission used is the three-mode ($m_T \in \pm[1, 2, 3]$) CVT described in the patent [15], and which has a structure similar to devices used in [5, 16]. The layout is presented in Figure 8. In this figure, the box to the left represents a Ravigneaux planetary gearset and the box to the right represents a regular planetary gearset. The driving direction and the transmission mode are selected by applying the corresponding clutches C_F or C_R, and C_1, C_2 or C_3. The CVT functionality is provided by the two hydraulic machines H_1 & H_2, which together

form a 'variator'. Changing the gear ratio within a mode is done by altering the displacement ratio between the hydraulic machines. The engine-side connection is marked with 'IN' and the wheel-side connection is marked with 'OUT'. The transmission torque at the engine-side is designated T_T and the torque at the wheel-side is designated T_W.

The main source of losses in this concept is the variator, which is modeled according to Equations (9) and (10). This model is based on a model used in [6]:

$$\psi_1 D_v \omega_1 \pm p_v (C_a + (\omega_1 + \omega_2) C_b) - \psi_2 D_v \omega_2 = C_{vT} \dot{p}_v \quad (9)$$

$$\psi_n D_v p_v - T_n \pm (C_c \omega_n + C_d p_v) = 0 \quad (10)$$

The index $n = 1, 2$ denotes the two machines, D_v is maximum displacement, $\psi_n \in (0, 1)$ is relative displacement, ω_n is axle speed, p_v is variator hydraulic pressure, T_n is torque and C_a, C_b, C_c and C_d are efficiency parameters. The signs in the equations depend on the power flow direction. Equation (11) describes hydraulic fluid flow and Equation (12) describes torque in each machine. The variator is constructed so that $\psi_1 + \psi_2 = 1$. The variator pressure dynamics is assumed to be fast compared with other dynamics of the system, *i.e.* it is assumed that the time constant C_{vT} can be set to zero. Mode shifts are performed at the extremals of the variator displacement, and mode shifts at these points do not change the overall gear ratio for a lossless transmission. At mode shifts the speed differences over the involved clutches are close to zero, and the clutch losses are therefore small. This model can be summarized by the two functions:

$$T_T(m_T, \psi_1, \omega_e, v_w) \tag{11}$$

$$T_W(m_T, \psi_1, \omega_e, v_w) \tag{12}$$

1.5 Hydraulics Model

The bucket and boom are hydraulically driven. Pressure and flow of the hydraulic fluid are supplied by a hydraulic pump connected to the engine axle. This pump has variable displacement, so that the same pressure and flow can be provided at different engine speeds. Equations (13) and (14) describe the hydraulic pump:

$$Q_H = \psi_H D_H \omega_e \tag{13}$$

$$Q_H p_H = \eta_H T_H \omega_e \tag{14}$$

D_H is maximum displacement, $\psi_H \in [0, 1]$ is relative displacement and $\eta_H(p_H, \psi_H)$ is pump efficiency. Lowering of the bucket does not require flow from the hydraulic pump.

2 METHODS

2.1 Basic Dynamic Programming Algorithm

Both of the control concepts to be presented are based on the dynamic programming recursion. This method description therefore starts with a recapitulation of this recursion, as used in the following methods. Denote the discretized flow variable $s \in s_k$ with $k = 0, \ldots, N - 1$, states $x \in X$ and controls $u \in U$. The notation $x_k = x(s_k)$ is used. The optimization problem can be formulated, with E referring to the expected value if w_k is stochastic, as:

$$\min_{u_k \in U} E\{J_N(x_N) + \sum_{k=0}^{N-1} g_k(u_k, x_k, w_k)\}$$

$$x_{k+1} = f(x_k, u_k, t), \quad k = 0, \ldots, N - 1 \tag{15}$$

along with equality and inequality constraints. According to [17, 18] the dynamic programming recursion can, for this problem, be stated as:

$$J_k(x_k) = \min_{u \in U} E\{g(x_k, u_k, w_k) + J_{k+1}(x_{k+1}(x_k, u_k, w_k))\} \tag{16}$$

$$J_N(x_N) = g_N(x_N) \tag{17}$$

This recursion is solved according to the following algorithm, expressed for a deterministic load w_k, as previously presented in [19]:
1. For each $x_N \in X_N$, declare $J_N(x) = J_N$
2. **for** $k = N - 1, \ldots, 1$ **do**
3.   For each $x_k \in X_k$, simulate $\frac{dx}{dt}$ for s_k to s_{k+1} for all $u_k \in U$ to find $x_{k+1}(x_k, u_k, w_k)$
4.   For each $x_k \in X_k$

$$J_k(x_k) = \min_{u_k \in U}(g(x_k, u_k, w_k) + \tilde{J}_{k+1}(x_{k+1}(x_k, u_k, w_k))) \tag{18}$$

with $\tilde{J}_{k+1}(x_{k+1})$ interpolated from $J_{k+1}(x_{k+1} \in X)$
5. **end for**

If the load is stochastic, step 3 is performed for each possible load combination $w_l \in W_k$, and Equation (18) is altered to:

$$J_k(x_k) = \min_{u_k \in U} \sum_{w_l \in W_k} p(w_l)(g(x_k, u_k, w_l) + \tilde{J}_{k+1}(x_{k+1}(x_k, u_k, w_l))) \tag{19}$$

in which $p(w_l)$ is the probability of the load being w_l. This first part is used to establish a Cost-To-Go (CTG) map $J(x \in X, s)$. In the following part, this map is used for calculating the optimal trajectory $x^*(s), u^*(s)$:
1. Select an initial state $x_0^* = x_0$
2. **for** $m = 1, \ldots, N$ **do**
3.   For x_{m-1}^*, simulate $\frac{dx}{dt}$ for s_{m-1} to s_m for all $u \in U$ to find $x_m(x_{m-1}^*, u)$
4.   Select

$$u_{m-1}^* = \operatorname*{argmin}_{u \in U}(g(x_{m-1}^*, u, w_{m-1}) + \tilde{J}_m(x_m(x_{m-1}^*, u, w_{m-1}))) \tag{20}$$

in which $\tilde{J}_m(x_m)$ is interpolated from $J_m(x_m \in X)$
5. $x_m^* = x_m(x_{m-1}^*, u_{m-1}^*, w_{m-1})$
6. **end for**

2.2 Dynamic Programming as a One-Step Look-Ahead Controller

The second part of the algorithm presented in the previous section can be seen as a one-step look-ahead simulation. In this case, the load w_k, $k = 0, \ldots, N - 1$ used in the second part is the actual load, which will differ from the load used in the CTG map calculation, unless there is a perfect prediction of future loads. This type of control, assuming a perfect but limited horizon prediction, is used [10 and 20]. If there are differences in the loads in the two parts of the dynamic programming algorithm, the resulting state and control trajectories will in general not be optimal for the second load trajectory. It can, however, be expected that a well-designed CTG map will result in state and control trajectories with a low associated cost for a range of actual loads. In some controllers, such as the one presented in [21], a distance-independent CTG map can be created through assuming distance-independent load probabilities. In the problem treated in this paper, a position-dependent load prediction is available, but there are considerable uncertainties in this prediction. The problem is therefore translated to a problem of selecting states and control signals and constructing a load case for the CTG map calculation, so that the look-ahead control in the second part gives low cost even when the load is altered. Two different concepts have been developed and these are presented in Sections 2.3 and 2.4.

This section discusses the implication of uncertainties, and the impact of disturbances, in each load component as compared with the values predicted in the CTG map calculation. It is assumed that in the one-step look-ahead simulation, the load components represent the desired trajectories derived from driver inputs and the resulting forces experienced by the machine. The components are vehicle speed v_w, longitudinal force F_w, hydraulic flow Q_H and hydraulic pressure p_H.

Component v_w: the load component v_w is part of Equations (3) (vehicle speed dynamics) and (11) and (12) (transmission input and output torque). Note that the vehicle speed dynamics limit the derivative of the possible disturbance. In Equation (3), the impact of changing v_w can be treated as an additional disturbance in F_w. In Equations (11) and (12), the CVT mode and variator displacement ratio can be changed fast. Changes in v_w can therefore be transferred through T_w and T_T to the engine speed dynamics (4).

Component F_w: the load component F_w is part of Equation (3) (vehicle speed dynamics). This component includes longitudinal forces on the bucket, which can change rapidly, *e.g.* if the bucket hits a rock. According to the reasoning for the component v_w, a change in F_w can be transferred through T_w and T_T to the engine speed dynamics (4).

Component Q_H: the load component Q_H is part of Equations (13) (hydraulic flow) and (14) (hydraulic power). In the actual vehicle, the hydraulic flow is related to the bucket lifting speed, so the bucket inertia should limit $\frac{dQ_H}{dt}$. This limitation, however, is lessened by the possibility of forces created through the vehicle pitch dynamics. Therefore, it is assumed that Q_H can change rapidly. Further, the desired hydraulic flow along with the maximum pump displacement $\psi_H D$ causes a lower

limit for the engine speed $\omega_{e,H}$, according to Equation (13). It is not uncommon that $\omega_{e,H}(t_k) > \omega_{e,min}$, and during these instances the limit is often active.

Component p_H: the load component p_H is part of Equation (14) (hydraulic power). This component is related to vertical forces on the bucket, which can change rapidly, *e.g.* if the bucket hits a rock. Changes in this component are transferred through T_H to the engine speed dynamics (4).

The engine torque can be altered instantaneously, though the turbo speed may restrict the magnitude of the change. The component Q_H causes a limitation that is often active, and uncertainty in this load component is therefore the primary obstacle to using dynamic programming as a look-ahead controller. To recapitulate, the limit comes from the relation:

$$Q_H = \frac{dV_H}{dt} = \frac{dV_H}{ds} v_s = \psi_H(t) D_H \omega_e(t) \qquad (21)$$

which if $\frac{Q_H}{\omega_e}$ becomes high enough requires $\psi_H > 1$. Since this is not allowed, other solutions must be found. Since ψ_H is limited, the alternatives identified are to introduce margins through ω_e and v_w, allow for deviation from $V_H(s)$ or introduce a short horizon prediction. These three alternatives are discussed in the following part.

The inertias of the states ω_e and v_s can be seen as the cause of the problem. An instantaneous increase in Q_H would require an instant increase in ω_e or decrease in v_s, both of which are prevented by their inertias. The first alternative is therefore to keep ω_e, as a function of v_s, at such a level that ψ_H will never have to go above 1. Since the actual Q_H is not available a worst-case scenario must be used in the CTG map calculation. The drawback is that both the engine and the hydraulic pump are most efficient at low speeds, so using a preventive increase in the engine speed can be expected to increase fuel consumption. This approach is the motivation and foundation of the 'stochastic dynamic programming' method presented in Section 2.3.

In an actual vehicle, deviating from the desired bucket trajectory is a natural response to an unachievable desired trajectory. In the simulation, however, this approach becomes complicated by several factors. First, each of the measured cycles consists of a bucket trajectory along with corresponding forces. Deviating from the bucket trajectory would produce new forces, and calculating these would require a gravel pile model, which is not readily available. Second, allowing deviations in bucket height corresponds to introducing a freedom in $V_H(s)$, which would require at least one additional state in the system. This is highly undesirable in dynamic programming. For these reasons, this approach is not studied further in this paper.

The availability of a short horizon prediction of future hydraulic flow might seem implausible. In an implementation though, the desired hydraulic flow would be an input from the driver. If a small delay is introduced between driver input and actual flow, this would be equivalent to a short horizon prediction of future hydraulic flow. If a constant time delay is used, no additional state is needed. This approach is the motivation and foundation of the 'free-time dynamic programming' method presented in Section 2.4.

A measurement sequence with 34 short driving cycles is available for the evaluation of the methods. In each evaluation, one cycle is used as the actual cycle in a simulation. In each case it is assumed that the other 33 cycles are available for the CTG map creation. Further, in the second stage the present load is assumed to be known, so that in the simulation, at $s = s_k$, the load w_k is available.

2.3 Stochastic Dynamic Programming

The method presented here is an extension of an algorithm previously presented in [22, 23].

2.3.1 Concept Description

This concept includes the prediction uncertainties in the load cases used in the CTG map creation, by describing the load w_k as a Markov process. In this description, there are at each stage some different alternatives for the load, along with a probability distribution. By assigning an infinite cost to states from which the vehicle cannot complete the cycle, and including a worst-case scenario with a low probability, the CTG map will correspond to a minimization of the cost under the condition that the vehicle must always be able to handle the worst case future load. This method is designated the Stochastic Dynamic Programming (SDP), method.

2.3.2 Implementation

The problem is formulated as a minimization of expected total amount of fuel M_f required for performing a short loading cycle. This can be expressed as:

$$\min E\{M_f(T)\} \tag{22}$$

and the cost function therefore becomes:

$$g(x_k, u_k, w_k) = \sum_{w_l \in W_k} \left(p(w_l) \frac{dM_f}{dt} \right) \tag{23}$$

TABLE 1

States and control signals in the SDP method

Flow	States	Controls
t	ω_e, p_t	m_f

in which W_k is the set of possible loads w_k at $t = t_k$ and p is the probability of that load being w_l. The terminal cost is set to be $J_N = 0$ for all states x_N.

Since $\omega_e(\psi_1)$ is always invertible for this concept either ω_e or ψ, along with m_T, can be used as state. Since the speed will increase for one of the hydraulic machines when ψ_1 gets close to 0 or 1, the losses increase in these regions. Therefore it is desirable to have high state grid density near the extremes of ψ_1, which implies using ψ_1 as state. The possibility of restrictions on $\frac{d\psi_1}{dt}$, especially during mode shifts, also points toward using ψ_1 as state. Since the dynamics are described in terms of ω_e this would imply the following computational scheme:

$$\psi_{1,k} \xrightarrow{W_k} \omega_{e,k} \xrightarrow{\frac{d\omega_e}{dt}} \omega_{e,k+1} \xrightarrow{W_k} \psi_{1,k+1}$$

In the first and last steps, the load is required, since $\omega_e(\psi_1)$ depends on the load. At the last step, a choice has to be made whether to use $\kappa = k$ or $\kappa = k + 1$. Using $\kappa = k$ is equivalent to making a change of variables in Equation (4) from $\frac{d\omega_e}{dt}$ to $\frac{d\psi_1}{dt}$. This choice of κ does not guarantee continuity in ω_e, which makes it possible for the optimizer to draw a net power from the engine inertia. $\kappa = k + 1$, on the other hand, guarantees continuous ω_e and works well for a deterministic load, but in the stochastic case this causes a quadratic increase in load combinations, since $\psi_{1,k+1}$ would have to be calculated for all combinations of W_k, W_{k+1}. This would cause an unacceptable increase in calculation time. This means that for SDP it is not practical to use ψ_1 as a state, and instead ω_e is used. $\omega_e(\psi, m_T)$ may only be non-invertible in small regions near $\psi_1 = \{0, 1\}$, so instead of using m_T as a state, the m_T which gives the highest efficiency is used in ambiguous cases.

The independent, or flow, variable in this calculation is the time t, the states are the engine speed ω_e and the turbo pressure p_t, and the sole control signal is the fuel mass per injection m_f, as summarized in Table 1. The same state and control signals are used in both the CTG map calculation and the look-ahead control simulation.

2.3.3 Load Case Creation for the SDP Method

Using SDP in look-ahead control applications has been studied [21, 24]. In these papers, the load has the Markov property and the probability distribution of the load is also independent of time. In the application at hand,

TABLE 2

Load case components and corresponding probabilities for the SDP Cost-To-Go map calculation

ω_w	T_w	Q_H	p_h
μ (1)	$\mu - \sigma$ (.25)	$\mu - \sigma$ (.25)	$\mu - \sigma$ (.25)
	μ (.5)	μ (.5)	μ (.5)
	$\mu + \sigma$ (.25)	$\mu + \sigma$ (.2)	$\mu + \sigma$ (.25)
		$\mu + 2\sigma$ (.05)	

the load is modeled as a Markov process, but since the intention is to utilize the fact that the vehicle operates in a well-known cycle, the probability distribution of the load does depend on the time, forming a probabilistic short loading cycle. As described in Section 1.1, the operation of a wheel loader can be described by the load components $\omega_w = v_w r^{-1}$, T_w, Q_H and p_H, which are also the components used here. The torque T_w can be calculated, using the measured vehicle speed, from Equation (3). Describing the vehicle speed ω_w as a Markov process is deemed unrealistic, as discussed in Section 2.2, and this component is therefore regarded as deterministic. The load components are calculated from the set of measured loading cycles. First, the time scales in all cycles are adjusted so that the driving direction changes at the same instances in all cycles. The four driving phases are set to be 10 s for the forward and loading phases, and 5 s each for the other three phases. The vehicle speed v_w is adjusted so that the distances driven between each direction change agree with those specified in the FTDP method, for a fair comparison in the subsequent evaluation. All four load components are calculated for each cycle. The mean μ and standard deviation σ of each component over a set of cycles, as functions of time, are calculated. The load W_k for the CTG map calculation consists of all load component combinations, according to Table 2, making a total of 36 possible loads at each instant t_k. This is repeated for all cycles in the measured sequence, producing 34 CTG map load cases, each time excluding one of the basic loading cycles from the set of cycles used in the calculation of μ and σ.

The load case that was excluded in each CTG map load case creation is later used as the load applied in the corresponding simulation, allowing for 34 method evaluations.

2.4 Free-Time Dynamic Programming

The CTG map calculation in the method presented here is partly based on an algorithm previously presented in [25].

2.4.1 Concept

This method reduces the sensitivity to disturbances in Q_H by introducing a short horizon prediction of this load component, and to uncertainties in the prediction of F_w and p_H by introducing a freedom in time. The prediction of Q_H should prevent the vehicle from entering a situation in which the engine speed is too low to allow for the desired hydraulic flow. The freedom in time is introduced through a freedom in vehicle speed. This freedom allows for using the energy stored in the vehicle speed to compensate for temporary high F_w or p_H and for reducing the tractive and hydraulic power by slowing down the flow of time through reducing the vehicle speed. Since a freedom in time is introduced, the components of the load w are redefined as functions of the distances calculated from the vehicle speeds in the measured cycles. This method is designated the Free-Time Dynamic Programming (FTDP) method.

2.4.2 Implementation

CTG Map Calculation

Since a freedom in time is introduced, the problem is reformulated as a combined minimization of the total amount of fuel M_f and time T for performing a short loading cycle. The factor β is introduced to weigh time to fuel in the minimization. This can be expressed as:

$$\min \left\{ M_f(T) + \beta T \right\} \tag{24}$$

and the cost function therefore becomes:

$$g(x_k, u_k, w_k) = \frac{dM_f}{dt} + \beta \tag{25}$$

in which, introducing the vehicle speed $v_s = |v_w|$ and distance driven $s = \int v_s dt$, the time steps $\Delta t = \Delta s / v_{s,k}$ or, if $v_{s,k} \approx 0$, $\Delta t = 2\Delta s / (v_{s,k} + v_{s,k+1})$, are used. The terminal cost is set to be $J_N = 0$ for all states x_N.

By reformulating the cost function and system dynamics to depend on position rather than time, a freedom in time can be introduced without the need to have time as a state of the system. The dynamics for a state x is rewritten, using the chain rule, according to:

$$\frac{dx}{dt} = \frac{dx}{ds}\frac{ds}{dt} = \frac{dx}{ds} v_s = f(x(s), u(s), w(s)) \Rightarrow \tag{26}$$

$$\frac{dx}{ds} = \frac{1}{v_s} f(x(s), u(s), w(s)) \tag{27}$$

During the general driving cycle, the vehicle changes driving direction several times. In these instances the vehicle speed v_s has to go to zero. The state derivatives will then, according to Equation (27), not be well defined. For the vehicle speed dynamics this can be solved by changing the state from speed to kinetic energy according to the description in Section 2.5. Similar state changes would not solve the problem for the engine speed and turbo pressure dynamics, though. Hence the approximation:

$$\Delta s = \bar{v}_s \Delta t, \; \bar{v}_s = \frac{v_{s,k} + v_{s,k+1}}{2} \qquad (28)$$

is used instead when the initial vehicle speed is close to zero, just as in the cost function. In the engine dynamics, this approximation is supplemented with a correction of T_T to ensure that this approximation does not push the transmission efficiency to above 100%. When the approximation is active a constant transmission efficiency of $\alpha = 0.8$ is used. The reformulated minimization criterion becomes:

$$\min \int_0^{S_f} \left(\frac{dM_f}{dt} + \beta \right) \frac{ds}{v_s} \qquad (29)$$

The independent variable in the CTG map calculation is the distance driven s and the states are the vehicle speed $v_s (= |v_w|)$, the engine speed ω_e and the turbo pressure p_t. The control signals are the fuel mass per injection m_f, the CVT mode M_T, the variator displacement ratio ψ_1 and the brake torque T_b. The vehicle speed is forced to zero at the positions of the driving direction changes $s = s_m$ by assigning infinite cost to non-zero vehicle speeds in these instances $J(s_m, v_s > 0) = \infty$. For calculation effort reasons, zero speed is not allowed in any other instance. For the same reason, and since braking is a waste of energy and should be avoided, using non-zero brake torque is only considered if $T_b = 0$ gives infinite cost for all m_f, ψ_1. The gain from the variator ratio ψ_1 to the torques T_T and T_W, according to the functions (11) and (12), is very high and a high density ψ_1 control signal grid must therefore be used. This would, however, have a severe effect on the calculation effort. For this reason, a ψ_1 with high grid density but a narrow range centered around $\psi_1(m_T, \omega_e, v_s)$, such that $T_T(m_T, \psi_1, \omega_e, v_s) = 0$, is used.

Look-Ahead Control Simulation

In the one-step look-ahead simulation, the time t is used as the flow variable, and the time step corresponds to the hydraulic flow delay/short horizon prediction.

In the evaluation, a 0.1 s time step, and hence delay/prediction, is used. This way, an infinite cost can be assigned to controls which give $\psi_H > 1$ at t_{k+1}, and thus state-load combinations which would require $\psi_H > 1$ are avoided. This change of flow variable from that used in the CTG map calculation, means that the positions t_k will not correspond to positions in the grid s. The interpolations in the simulations must therefore also be done over the flow variable, which increases the dimension in the interpolation. This increases the computational load, but the most severe effect occurs in the driving direction changes.

The driving direction changes are included in the CTG maps as infinite cost for all vehicle speeds $v_s > 0$ at the corresponding positions. Say that the vehicle speed must be zero at $s = s_m$. Interpolation will then render $\tilde{J}(s) = \infty$ for all $s_{m-1} < s < s_{m+1}$ except $s = s_m$, $v_s = 0$. The direction changes therefore need special treatment, both in approaching and in leaving these positions. The complete procedure of approaching and leaving a direction change position is illustrated by Figure 9.

Approaching a direction change is detected when $s_k < s_{m-1} - \kappa$ and $s_{m-1} - \kappa < s_{k+1}(u)$, with κ being a small value which acts as a minimum Δt for the next simulation step. When this detection occurs, Δt is adjusted for those u so that $s_{k+1}(u) = s_{m-1}$ and $\tilde{J}$ is interpolated among $J(s_{m-1})$. In the next step those $\Delta t, u$ that give $s_{k+1} = s_m, v_{s,k+1} = 0$ are used and the $\tilde{J}$ interpolation is performed among $J(s_m, v_s = 0)$. The vehicle has now reached the direction change position.

When the vehicle leaves the direction change position, that is, as long as $s_m < s(t_{k+1}) < s_{m+1}$, $\tilde{J}$ is interpolated

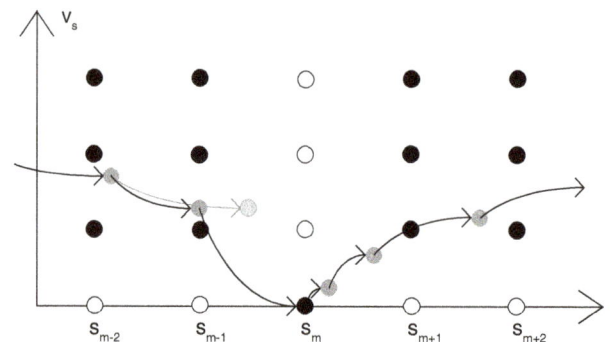

Figure 9

Illustration of the FTDP method simulation at a driving direction change. White nodes represent states with $J(x, s) = \infty$ and black represent states with $J(x, s) < \infty$. The arrows and gray nodes represent the simulated trajectory through the region. The light gray node is one for which Δt is reduced until $s(t_k + \Delta t) = s_{m-1}$.

TABLE 3
States and control signals in the FTDP method

TABLE 3
States and control signals in the FTDP method

	Flow	States	Controls
CTG calculation	s	v_s, ω_e, p_t	m_f, m_T, ψ_1, T_b
Simulation	t	v_s, ω_e, p_t	m_f, m_T, ψ_1, T_b

from the temporary CTG map $[J(s_m, v_{s,1}), J(s_{m+1}, v_{s,n})]$, $n = 2, \ldots, N$, in which N is the size of the v_s grid.

Apart from the change of flow variable, the states and controls are the same in the look-ahead control simulation as in the CTG map calculation. Also, the same calculation effort-saving measures are taken for the control signals in the simulation. The flow, state and control signals in the two parts of the algorithm are summarized in Table 3.

2.4.3 Load Case Creation for the FTDP Method

The basic load components are functions of time, and these need to be reformulated into functions of distance driven. The distance driven must be monotonically increasing for these functions to be well defined. Since the machine drives in the reverse direction part of the time, the velocity v_w is divided into speed $v_s = |v_w|$ and direction $d_s = \text{sign} v_w$, which enables the definition of the distance driven as $s = \int v_s dt$.

The positions at which the vehicle changes the driving direction are specified by the driving cycle, which makes $d_s(s)$ a load component, while v_s is a state of the system. The driving direction changes must occur at the same positions in both the CTG map calculation and the simulation. Therefore, the distance scales are adjusted in all cycles so that each driving phase is 10 m. The tractive force $F_w(t)$ and the hydraulic pressure $p_H(t)$ can be directly shifted to depend on position rather than time $F_w(s), p_H(s)$. The hydraulic flow Q_H, that is the hydraulic fluid volume per time, is transformed to a hydraulic volume per distance, or a hydraulic volume as a function of the distance driven $V_H(s)$:

$$V_H = \int_0^{s_N} \frac{Q_H}{v_s} ds \qquad (30)$$

This hydraulic volume is the integrated flow of hydraulic fluid to the lift and tilt cylinders as a function of the distance driven, while the force F_w and pressure p_H specify the wheel and bucket forces caused by this trajectory. This is repeated for each of the basic loading cycles, producing a total of 34 FTDP load cases. Each load case consists of the components direction of driving d_s,

longitudinal force F_w, hydraulic volume V_H and hydraulic pressure p_H.

2.5 Simulations and Energy Balance

The choice of dynamic programming for the optimization method, combined with the complexity of the system, makes efficient simulation of the functions $x_{k+1}(x_k, u_k, w_k)$ decisive. The Euler forward method is the simplest method for this simulation, and using this method is therefore desirable. Direct application of this method on the aforementioned states, however, does not preserve energy. In fact, using the engine speed dynamics as an example, the Euler step is:

$$\omega_{e,k+1} = \omega_{e,k} + \Delta t \frac{T}{I_e} \qquad (31)$$

and during this step the work performed by the torque is:

$$W_1 = T \omega_{e,k} \Delta t \qquad (32)$$

while the change in kinetic energy is:

$$W_2 = \frac{I_e}{2} \left(\omega_{e,k+1}^2 - \omega_{e,k}^2 \right) = T \omega_{e,k} \Delta t + \frac{(T \Delta t)^2}{2 I_e} \qquad (33)$$

and correspondingly for the vehicle speed dynamics, and also if formulated as functions of the distance driven. There is obviously a discrepancy between the input and output energy. The optimization algorithm has been observed to exploit this discrepancy by fast switching between high positive and negative forces. Similar behavior has also been seen in, *e.g.,* [10] as oscillating controls in the solution. In the system at hand, the gain from the control signal ψ_1 to the torques T_T and T_W is very strong, and the optimizer will therefore be highly inclined to use this shortcut by fast switching between high and low ψ_1, especially in the FTDP method, since the discrepancy can be exploited by moving kinetic energy between the engine and vehicle speeds with higher than 100% efficiency. In some cycles, the magnitude of the discrepancy became large enough for the vehicle to be propelled by this false input alone, requiring no fuel to complete an entire driving cycle. This problem can be prevented by using energy formulations for both vehicle and engine speed dynamics, according to:

$$\frac{d}{dt} \frac{m v_s^2}{2} = v_s (r^{-1} T_W - r^{-1} T_b - F_w) \qquad (34)$$

$$\frac{d}{dt} \frac{I_e \omega_e^2}{2} = \omega_e (T_e - T_T - T_H) \qquad (35)$$

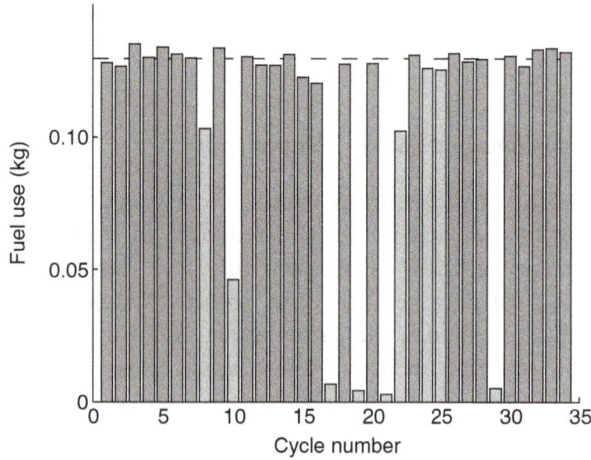

Figure 10

Fuel use in the SDP solutions. The light gray bars indicate
cycles not completed due to infinite cost.

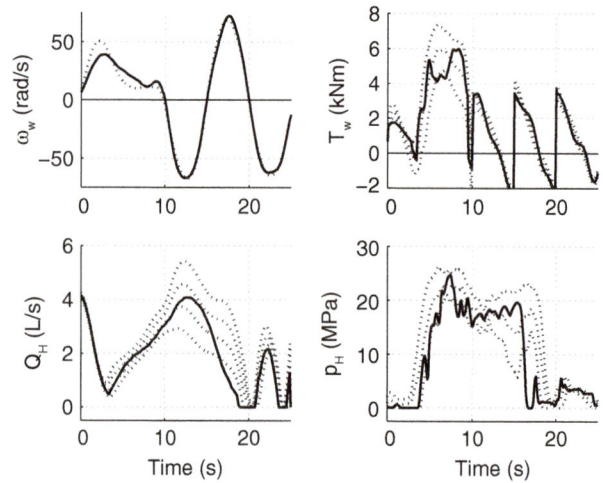

Figure 11

Illustration of the loads used for the 4th evaluation. The dotted
lines are the loads used in the CTG map calculation (see also
Tab. 2) and the solid lines are the cycle used in the simulation.

The Euler method simulation steps can be formulated as:

$$v_{s,k+1} = \sqrt{v_{s,k}^2 + \frac{2v_s \Delta t}{m} \Sigma F} \qquad (36)$$

$$\omega_{e,k+1} = \sqrt{\omega_{e,k}^2 + \frac{2\omega_e \Delta t}{I_e} \Sigma T} \qquad (37)$$

and correspondingly when the distance driven, s, is used as the independent variable, which guarantees the balance of energy. This energy formulation is used in all simulation steps in both the CTG map calculation and the look-ahead control simulation, in both of the methods.

3 EVALUATION

Section 2 describes the one-step look-ahead controller concepts and the corresponding load case creations. In this section, the controllers are evaluated by performing CTG map calculations and subsequent simulations.

3.1 Stochastic Dynamic Programming

In each evaluation of the Stochastic Dynamic Programming (SDP) method, one loading cycle is used in the simulation and all other cycles from the measurement sequence are used in the CTG map calculation. The measurement consists of 34 basic load cases, and

the SDP method is therefore evaluated using 34 simulation loading cycles, each with a corresponding CTG map calculated from the other 33 cycles, according to the description in Section 2.3.3.

Out of the 34 evaluations, 25 rendered a finite cost, which corresponds to 74% of the evaluations being successful. In three of the nine cases of infinite cost, this was caused by low engine speed compared with the minimum required by the hydraulic flow requirement. Most of the other six cases were caused by relatively high vehicle speed, related to the distance driven adjustment as described in Section 2.3.3. Figure 10 illustrates the fuel needed for performing each of the 34 cycles. The light gray bars represent the infinite cost cycles, as the fuel used up until the encountering of the infinite cost, and the dashed line shows the average fuel use of 130 g, in the cycles with a finite cost. The average optimal fuel use over the 34 cycles, that is the fuel required if the simulated cycle is also used for the corresponding CTG-map calculation, is 119 g. The 4th evaluation from the left is used as an example to illustrate the simulation results. Completing this particular cycle required 130 g of fuel.

The CTG and simulation loading cycles for evaluation 4 are shown in Figure 11. The dotted lines are the load alternatives for the CTG map calculation, as specified in Section 2.3.3, and the solid line is the load used in the simulation. This shows that there are significant differences in all components between the simulated cycle and the CTG cycle. In the vehicle speed, the positions of the driving direction changes coincide because of the design of the cycles. The load components in the

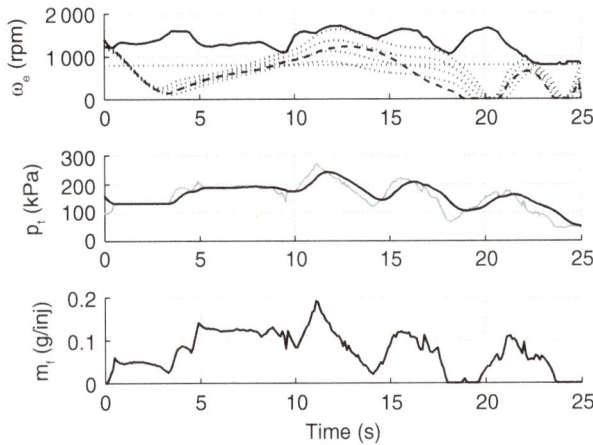

Figure 12

State and control signal trajectories in evaluation 4 (solid). The engine speed ω_e is plotted along with the minimum speeds given by the hydraulic flows in the CTG (dotted, see also Tab. 2) and simulation (dashed) cycles.

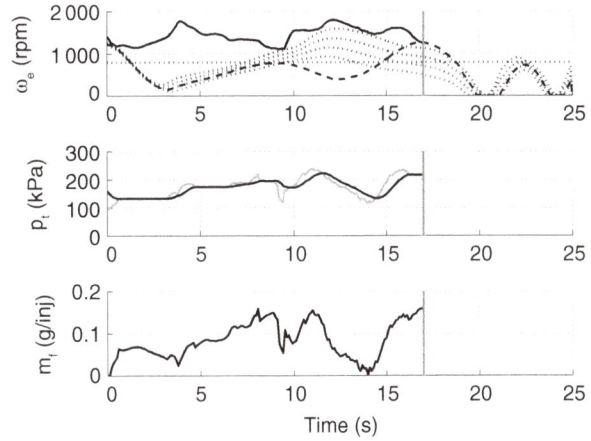

Figure 13

State and control signals in evaluation 8 (solid). The engine speed is plotted along with the minimum speeds given by the hydraulic flow in the CTG (dotted) and simulation (dashed) loads. Infinite cost is encountered at $t = 17$ s.

simulated cycle are more transient than those in the CTG alternatives, since the CTG cycle has been constructed as an average over several cycles. Note that in this example, the hydraulic flow in the simulation is always lower than the highest alternative in the CTG cycle.

Figure 12 shows the state, ω_e and p_t, and control signal, m_f, trajectories from the simulation in evaluation 4. The ω_e state figure also shows the minimum engine speeds specified by the static limit (dotted line), and the hydraulic flow in the CTG load alternatives (four dotted curves) and in the simulation load (dashed curve). This shows that the engine speed is always higher than needed for the highest possible hydraulic flow in the CTG load. This keeps the engine speed higher than required by the actual desired hydraulic flow, which prevents infinite cost. In three of the simulation cases, this was not achieved, but the hydraulic flow in the simulated cycle was higher than the highest alternative in the CTG cycle at a time when the engine speed was close to this limit. This is illustrated by Figure 13, which shows the same signals as in Figure 12 but for the 8th evaluation, referring to Figure 10, in which infinite cost is encountered at $t = 17$ s, as indicated by a vertical gray line. In both figures, the intake pressure p_t is plotted along with the static pressure setpoint p_{set} (gray).

One of the main issues in using dynamic programming is the computational effort, especially when the number of states or control signals increases. Table 4 shows the experienced times needed for calculating the CTG maps and for the look-ahead control simulations.

TABLE 4

Experienced times for CTG map calculation and look-ahead simulation, using the SDP method

	$t_{min}(s)$	$t_{mean}(s)$	$t_{max}(s)$
CTG	877	974	1055
Sim	0.30	0.30	0.32

The simulation times only include the cycles for which the cost is finite. All calculation times are highly dependent on the method implementation and state and control signal grid densities, and should therefore only be considered an indication and are only intended for comparison with the FTDP method. The discretizations have been made as sparse as possible without significantly affecting the optimization results.

3.2 Free-Time Dynamic Programming

In the Free-Time Dynamic Programming (FTDP) method, the creation of a load case for the CTG map calculation only requires a single basic load case. An FTDP load case is therefore created from each of the 34 basic load cases, according to the description in Section 2.4.3. In the evaluation, the CTG map is calculated using one FTDP load case and in the look-ahead control simulation any other FTDP load case can be used. The dataset contains a total of 34 cycles, making a total of 1 122 combinations evaluated. The time to fuel weighting parameter is selected as $\beta = 0.5$ g/s, since this gives cycle times similar to the 25 s specified for the SDP method.

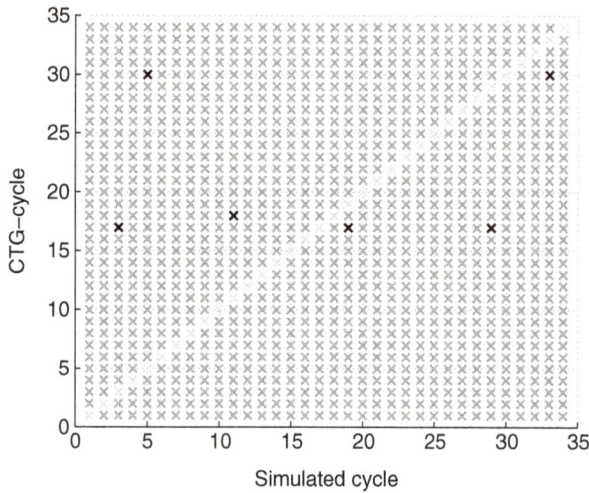

Figure 14

Illustration of success of the simulations in FTDP with $\beta = 0.5\,\text{g/s}$. The light gray markers indicate cycle combinations with perfect prediction, dark gray markers indicate successful simulations and black markers indicate cycle combinations which render infinite cost.

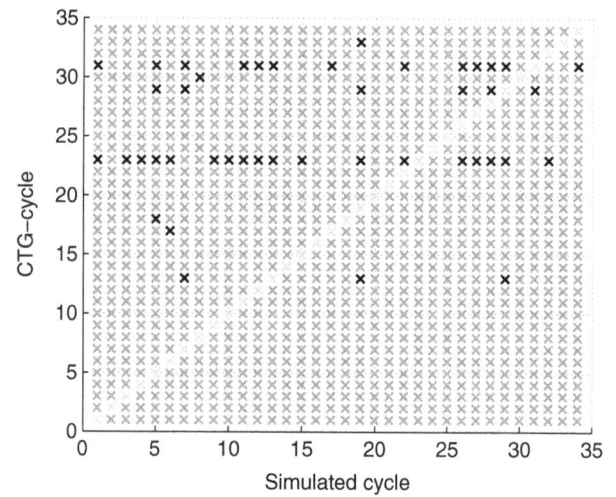

Figure 15

Illustration of success of the simulations in FTDP when time is prioritized by using $\beta = 10\,\text{g/s}$. See Figure 14 for explanation.

Figure 14 summarizes the result of these simulations. The gray markings indicate combinations that rendered a finite cost, while the black markings indicate combinations that rendered an infinite cost. In total, 1 116 combinations were successful, while 6 rendered infinite cost, which translates to success in 99.5% of the combinations. Figure 15 shows the same result, but for $\beta = 10\,\text{g/s}$. In this case, 96.1% of the combinations were successful. It is clear, though, that some cycles were less suited for use in the CTG map calculation. The most prominent of these are cycles 23 and 31. Disregarding these gives a total of 98.8% successful combinations.

Figure 16 shows the fuel and time required for completing each of the 1 116 successfully simulated cycles. The average fuel use is 116 g and the average time use is 24.7 s. The average optimal fuel use over the 34 cycles, which corresponds to the diagonal of Figure 14, is 115 g, and the corresponding average time use is 24.3 s. In the following part, the combination of the 4th cycle for the CTG-map calculation and the 12th cycle for the simulation, referring to Figure 14, is used as an example to illustrate the simulation results. This combination will be referred to as evaluation 4-12. Completing this particular combination requires 24.6 s and 115 g of fuel.

The CTG and simulation load cases for this cycle are shown in Figure 17. The dotted lines are the CTG load, as specified in Section 2.4.3, and the solid lines are the load used in the subsequent simulation. The positions

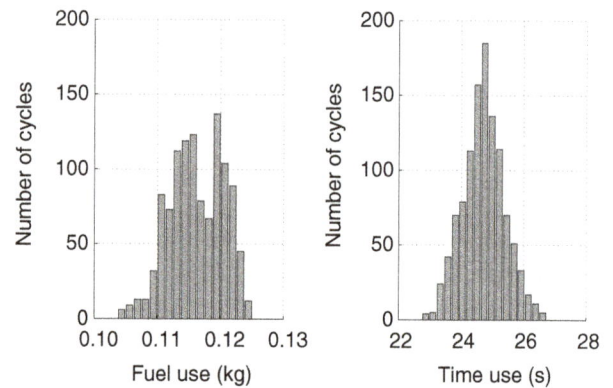

Figure 16

Summary of fuel and time use in the cycle combinations evaluated. The combinations with perfect prediction and those which rendered infinite cost are excluded.

of the driving direction changes are the same in the two cycles, as specified in the design of the cycles. The other components are similar in appearance, though there are significant differences in amplitudes, durations and timing.

Figure 18 shows the state, v_s, ω_e and p_t, and time step Δt trajectories from evaluation 4-12. The ω_e figure also shows the minimum engine speed specified by the static limit (dotted line), and the hydraulic flow in the simulation load (dashed curve). This shows that the vehicle is able to keep the engine speed higher than required by the simulation hydraulic flow, which prevents infinite cost without maintaining a large ω_e margin. The intake

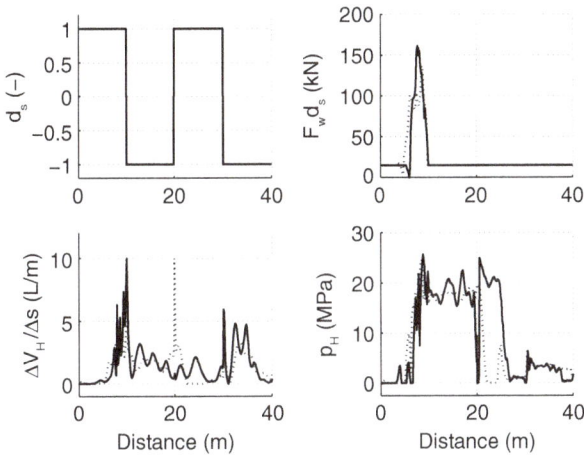

Figure 17

Illustration of the loads used in evaluation 4-12. The dotted lines are cycle 4, which is used for the CTG map calculation, and the solid lines are cycle 12, which is used in the look-ahead control simulation.

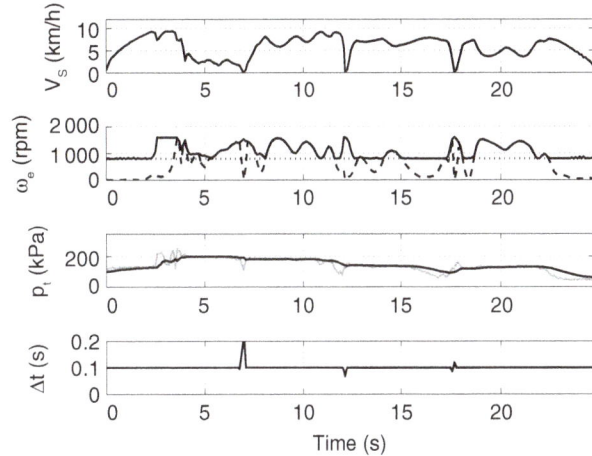

Figure 18

State trajectories in evaluation 4-12, along with the time steps used. The engine speed ω_e is plotted along with the minimum speed given by the hydraulic flow in the simulated cycle (dashed). The intake pressure p_t is plotted along with the set pressure p_{set} (gray).

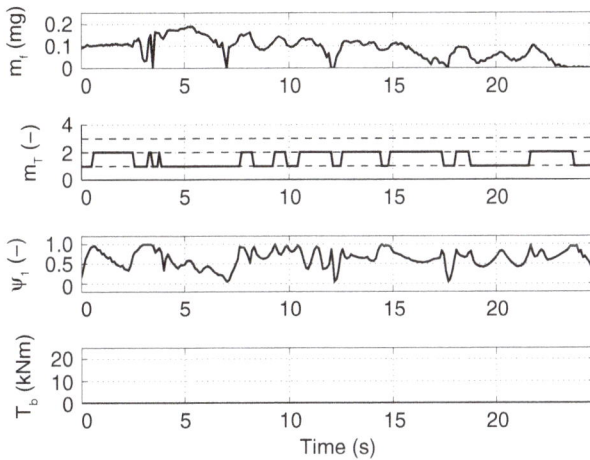

Figure 19

Control signal trajectories in evaluation 4-12.

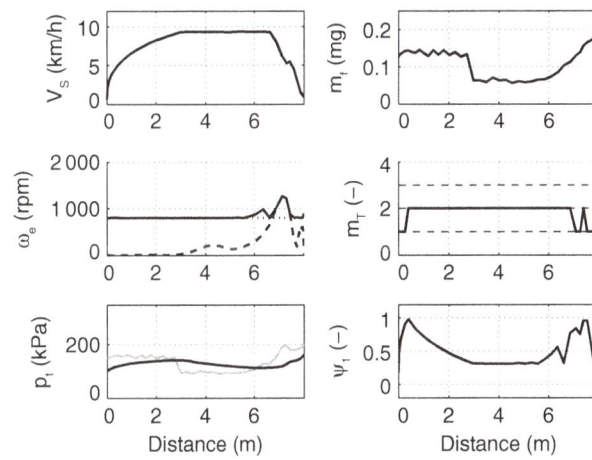

Figure 20

State and control signals before encountering infinite cost in evaluation 23-12 with $\beta = 10\,\mathrm{g/s}$. See Figures 18 and 19 for explanations.

pressure p_t is plotted along with the static pressure setpoint p_{set} (gray). The Δt figure clearly shows the adjustments of Δt made near the driving direction changes.

Figure 19 shows the control, m_f, m_T, ψ_1 and T_b, signal trajectories from evaluation 4-12. The m_f figure shows that the highest fuel flow is experienced during bucket filling. The m_T figure shows that around half the time is spent in CVT mode 1 and half of the time in CVT mode 2, while the ψ_1 figure shows that at mode changes, the variator displacement ratio is near its maximum or minimum, as

required by the transmission model. The T_b figure shows that in this cycle, the vehicle never uses the brakes.

Figure 20 shows the state and control signals from an unsuccessful $\beta = 10\,\mathrm{g/s}$ example. The example is the result from the combination of using the 23rd cycle for the CTG map calculation and the 12th cycle in the subsequent simulation, referring to Figure 15. Figure 21 shows the first part of the load case combination from this example, with the instant of the infinite cost marked. Typical for the cycles which produced CTG maps which

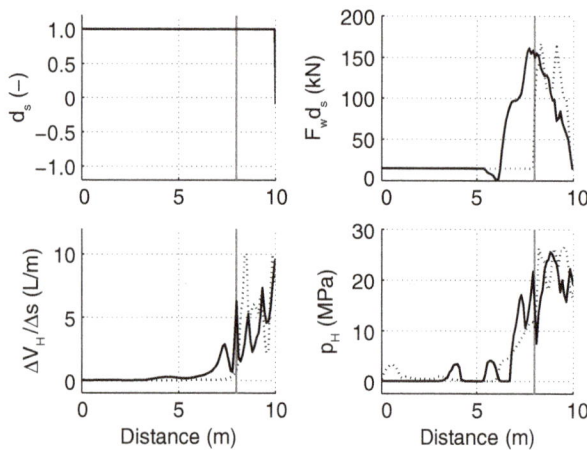

Figure 21

Illustration of the loads in the unsuccessful evaluation 23-12 with $\beta = 10\,\text{g/s}$. The dashed lines are the load used in the CTG map calculation, the solid lines are the load used in the simulation and the vertical gray line shows the instant of infinite cost.

TABLE 5
Impact of different β values in evaluation 4-12

β (g/s)	T (s)	M_f (g)
5	20.9	124.7
2	22.1	119.6
1	23.2	116.9
0.5	24.6	115.0
0.2	25.9	114.7
0.1	26.3	114.4
0.05	26.4	114.2

TABLE 6
Experienced times for CTG map calculation and look-ahead simulation, using the FTDP method

	t_{min} (s)	t_{mean} (s)	t_{max} (s)
CTG	13005	13365	14035
Sim	8.28	9.09	10.6

commonly rendered infinite cost in the subsequent simulations is that the increase in tractive force F_w related to the filling of the bucket is comparatively late and steep.

Table 5 shows seven examples of time and fuel use in evaluation 4-12, performed with different values of the time to fuel weighting parameter β. In the evaluations above,

$\beta = 5 \cdot 10^{-4}$ was used, since this gives an average cycle time similar to the cycle time specified for the SDP method.

Table 6 shows the experienced times needed for calculating the CTG maps and for the look-ahead control simulations. The simulation times only include the cycles for which the cost is finite. All calculation times are highly dependent on the method implementation and state and control signal grid densities, and should therefore only be considered an indication and are only intended for comparison with the SDP method. The discretizations have been made as sparse as possible without significantly affecting the optimization results.

3.3 Method Discussion and Comparison

Two methods are created for using dynamic programming as a one-step look-ahead controller in a wheel loader application. Each of these uses a different approach for increasing the robustness of the look-ahead controller to deviations from the predicted load. This section discusses and compares the two methods, not only with regards to the performance as described in the previous section, but also properties that affect the possibilities of implementing any of the methods as an actual online controller.

3.3.1 CTG Map Creation

In both of the methods, the first part of the algorithm is the creation of a Cost-To-Go (CTG) map $J(x, k)$. The appearance of this map will depend on the load used in the calculation.

In the SDP method, the creation of a load case for use in the CTG calculation can be automated. A dataset containing previously driven cycles, which might, e.g., be from the previous working day, is screened for cycles. A Markov probabilistic cycle is created from the detected cycles, using average and standard deviations of the load in each instance, along with assigned probabilities. The combination of possible loads and the corresponding probabilities are design parameters. The combination used in this evaluation is presented in Table 2. The impact of the combination and probabilities is quite transparent, especially in the impact on robustness to changes in actual hydraulic flow Q_H, while Section 2.2 states that uncertainties in this load component are the most important for the ability to complete the simulation. In the CTG map calculation, engine speeds lower than required by the maximum possible Q_H, according to (21), render infinite cost. In the simulation, the result is that the engine speed never drops below that which corresponds to the highest predicted

Q_H and maximum hydraulic pump displacement $\psi_H = 1$, as shown in Figure 12. Adding even higher possible alternatives for Q_H in the CTG map calculation will therefore increase robustness to high hydraulic flow in the simulation, but will also increase fuel consumption through maintaining a high engine speed even if this is not required by the actual hydraulic flow.

In the FTDP method, a single driving cycle is used as the load in the CTG map calculation. In the evaluation, each of the cycles used in the simulations were also used in the CTG map calculation, but in a real application a particular, and perhaps designed, cycle would be used. When low cycle-time was prioritized, some cycles were less suited for use in the CTG map calculation, which shows that the cycle used has a real impact on the performance in the subsequent simulation. Some care should therefore be put into the selection or creation of the CTG driving cycle. Unfortunately, the impact of the appearance of the CTG cycle is less transparent in the FTDP method than in the SDP method, since the hydraulic volume V_H gives a limit to the combination of the vehicle and engine speeds, rather than to the engine speed alone. The FTDP uses one state more than the SDP method, the vehicle speed, which is directly related to the time needed for completing a driving cycle. Completing a cycle faster generally requires more fuel, and the weighting of cycle time to fuel use is governed by the weighting parameter β. Increasing β increases the v_s dependency in $J(x, k)$, which pushes the vehicle toward higher speeds in the simulation. Predicting the impact of different β in the CTG map calculation, in the simulation of a specific cycle, is not trivial and deciding upon a suitable value may require iteration of CTG map calculations and look-ahead simulations.

3.3.2 Performance

In evaluating the performance of each of the methods, the first requirement is that the vehicle should be able to perform the specified driving cycles. This is fulfilled if the simulation is completed without the system violating any bound, deviating from the desired trajectory or going into an infinite cost region in the CTG map. This requirement was not fulfilled in all cycles for any of the two methods, but the ratio of successful to unsuccessful simulations differs for the two, and can therefore be considered a first performance measurement. In the SDP method, 73.5% of the simulations were successful, compared with 99.5% in the FTDP method. These numbers depend on the design parameters used in the CTG map calculations, though. Some of the failed SDP simulations did hit the minimum engine speed needed for

fulfilling the hydraulic flow requirement, while most fails were related to the vehicle speed adjustment made for obtaining the same distances driven in both method evaluations. The robustness to hydraulic flow uncertainties could have been increased through adding an even higher hydraulic flow component in the CTG load case. However, as long as the highest hydraulic flow component in the CTG load case does not require maximum engine speed at all times, there will always be a possibility of a higher requirement in the simulation. In the FTDP method, the cause of each unsuccessful simulation was less clear. The number of these simulations increases when speed is prioritized, and in this case, the success of a simulation was more related to the cycle that was used in the CTG map calculation than that used in the simulation. Many of the unsuccessful simulations occurred when cycles in which the longitudinal force increase related to bucket filling occurs late were used in the CTG map calculation.

Another requirement is low fuel consumption, and the fuel use is therefore the second performance measurement. In the SDP method evaluations the average fuel use was 130 g and the pre-specified cycle time was 25 s. If the robustness to hydraulic flow is increased through increasing the maximum predicted flow, the average fuel use can be expected to increase. In the example of Figure 10, the fuel use is already noticeably higher than the optimal (119 g), because of the implemented margin towards high hydraulic flow. In the FTDP method evaluations, with $\beta = 0.5$ g/s, the average fuel use was 116 g and the average time use was 24.7 s. Changing β so that the FTDP average time increases to closer to 25 s might reduce the average fuel consumption somewhat, but Table 5 indicates that this reduction would be small. The fuel use in the FTDP method is close to that achieved with perfect prediction (115 g), though there is a difference in time use (24.3 s). The optimal fuel use is lower for the FTDP method because of the addition of another state, or degree of freedom. The actual fuel use is also closer to the optimum for the FTDP method since this method does not need to maintain a power margin through the engine speed, causing a reduction of the efficiency.

3.3.3 Implementation

If either of the two proposed methods is to be used in an online application, there are a few issues still to be addressed.

Both methods rely on the actual position, as referred to the flow variable, being known and the length of the cycle, including the points of driving direction change, being fixed. In a real application, this will not be the case, as illustrated by Figure 5. In both methods, the position can be reset when a driving direction change occurs, but after these, disturbances in time or position, depending on the method, must be handled.

In the simulations, it is, at each instant t_k, assumed that the present load w_k is known and constant during each interval. Neither of these assumptions can be expected to hold in a real implementation, and an expected, and possibly probabilistic, load must therefore be used in the one-step look-ahead choice of control signals performed in the simulations. The need for preventing the states of moving into regions with infinite CTG will require overestimating the load, which can be expected to cause higher fuel consumption.

The time required for calculating a new CTG map restricts the adaptability of the controllers, regardless of the method used. In the implementations evaluated here, the CTG map calculation time corresponds to around 50-500 times the length of each loading cycle, depending on the method used. It will therefore not be possible to quickly create a new CTG map if the general driving cycle changes, but the CTG map, or maps, must be created beforehand. This must be addressed if the working site changes from day to day or if different drivers operate the machine. The most critical calculation effort, however, is in the look-ahead control simulations. This part must be completed online and using a much less powerful computer. The simulations of the 25 s cycles required around 0.3 s using the SDP method and around 10 s using the FTDP method. Despite requiring a shorter time than the length of the cycle, the calculation effort is too high for an implementation. Improvements might be possible through approximation of the true CTG map and improvement of the method for searching for the optimal control action in the one-step look-ahead, or even by calculation of an $u^*(x, t, P)$ map in advance.

SUMMARY AND CONCLUSIONS

Wheel loader operation is often highly repetitive. This repeating of similar motions may be used as the basis of a prediction of future operation. If a prediction of the future load trajectory is available, this can be used in an optimization of engine and transmission operation. In this paper a wheel loader with a three-mode CVT is studied. Predictions of future loads have been used in actual control systems before, e.g. in [26], but only for on-road vehicles. A prediction based only on repetition, however, will become approximate and

contain uncertainties. The complexity of the wheel loader system and its operation, along with the introduction of considerable uncertainties in the load prediction, makes it necessary to expand previously presented methods. Two conceptual methods, based on dynamic programming, for one-step look-ahead control of a wheel loader transmission are developed and presented in this paper.

A wheel loader driving cycle can be represented by a bucket trajectory and the corresponding vertical and longitudinal forces. A measurement sequence which contains 34 short loading cycles, described by vehicle speed, hydraulic flow (change of bucket height), hydraulic pressure (vertical force) and tractive (longitudinal) force, is used throughout the paper. The most important prediction uncertainties are in the hydraulic flow. The two controller concepts are evaluated through their performance in each of these 34 cycles, in each case having the other 33 cycles available for use as a load prediction. Deviating from the desired trajectory is not allowed, since this would require introducing another state in the optimization and a gravel pile model for calculating new forces, a model which is not readily available.

The first method presented is based on stochastic dynamic programming and is designated SDP. In this method, the 33 cycles available for the prediction are condensed into a statistical cycle with several possible loads at each instant in time, and an estimated CTG map is calculated from this cycle. The second method, designated FTDP, has vehicle speed as a state of the system, and introduces a fixed 0.1 s delay from driver input to bucket movement, a delay equivalent to a prediction of bucket movement. Again, a CTG map is calculated. The CTG maps are in both methods used in a one-step look-ahead controller for, in each instant, selecting the control action that can be expected to minimize the cost for completing the driving cycle.

The SDP method implementation turns out to require about 1/10th of the computational time of the FTDP method, both in the CTG map calculations and in the subsequent simulations. The lower time is because the SDP method has two states while the FTDP method has three, and this persists even though the SDP method has several load alternatives in each instant. The most important performance measurement is the ratio of cycles for which the look-ahead simulation could be completed without violating any bound, deviating from the desired trajectory or going into an infinite cost region in the CTG map. These simulations are regarded as successful. In the SDP method evaluation 74% of the 34 simulations were successful. In the FTDP method evaluation, the ratio of successful simulations depends on the value of the time to fuel weighting parameter β. Using a β which gives cycle times similar to the one specified in the SDP solving rendered 99.5% of 1 122 evaluations successful. Increasing the weight on time in the CTG map calculation increases the importance of the choice of cycle to use in the CTG map calculation and reduces the ratio of successful simulations. The second performance measurement is the fuel use. In the SDP method evaluations, the average fuel use was 130 g and the pre-specified cycle time was 25 s. In the FTDP method evaluations, with $\beta = 5 \cdot 10^{-4}$, the average fuel use was 116 g and the average time use was 24.7 s.

The driving cycle used in the CTG map creation affects the result of the one-step look-ahead simulation. In the SDP method, the impact is relatively transparent, especially with respect to robustness to different hydraulic flows. The CTG load can be used to trade increased robustness to hydraulic flow for higher fuel consumption. The FTDP method seems to be less sensitive to the load used in the CTG map calculation, unless cycle time is prioritized. In any case, the impact of the CTG cycle is less transparent in FTDP than in SDP.

In all, this evaluation shows that both methods may have a potential for use in a one-step look-ahead controller for a wheel loader transmission, but that there are still issues to be addressed before implementation, especially the treatment of uncertainties in the prediction of distance driven. In the evaluation, the SDP method required about 1/10th of the computational effort of the FTDP method and has better transparency of the impact of the CTG load. On the other hand, the vehicle was unable to complete the cycle in 26% of the evaluations when using the SDP method, as compared with a fail rate of less than 1% for the FTDP method, while the FTDP method also showed a 10% lower fuel consumption.

REFERENCES

1 Filla R. (2008) Alternative systems solutions for wheel loaders and other construction equipment. In 1st International CTI Forum Alternative and Hybrid Drive Trains, CTI

2 Rydberg K-E. (1998) Hydrostatic drives in heavy mobile machinery new concepts and development trends, In *International Off- Highway & Powerplant Congress & Exposition*, SAE, 981989.

3 Carl B, IIvantysynova M, Williams K. (2006) Comparison of operational characteristics in power split continuously variable transmissions, *Comercial Vehicle Engineering Congress and Exhibition*, SAE 2006-01-3468.

4 Grammatico S, Balluchi A, Cosoli E. (2010) A series-parallel hybrid electric powertrain for industrial vehicles, In *2010 IEEE Vehicle Power and Propulsion Conference*, IEEE, pp. 1-6.

5 Savaresi S, Taroni F, Previdi F, Bittanti S. (2004) Control system design on a power-split cvt for high-power agricultural tractors, *IEEE/ASME Transactions on Mechatronics* **9**, 3, 569-579.

6 Lennevi J. (1994) *Hydrostatic Transmission Control, Design Methodology for Vehicular Drivetrain Applications*, Dissertation, Linköping University.

7 Zhang R. (2002) *Multivariable Robust Control of Nonlinear Systems with Application to an Electro-Hydraulic Powertrain*, Dissertation, University of Illinois.

8 Filla R. (2005) An event driven operator model for dynamic simulation of contruction machinery, In *Proceedings from the Ninth Scandinavian International Conference on Fluid Power*, Linköping University.

9 Wang F, Zhang J, Sun R, Yu F. (2012) Analysis on the performance of wheel loaders in typical work cycle, *Applied Mechanics and Materials* **148**, 526-529.

10 Hellström E, Åslund J, Nielsen L. (2010) Design of an efficient algorithm for fuel-optimal look-ahead control, *Control Engineering Practice* **18**, 11, 1318-1327.

11 Asadi B, Vahidi A. (2011) Predictive cruise control: utilizing upcoming traffic signal information for improved fuel economy and reduced trip time, *IEEE Transactions on Control Systems Technology* **19**, 707-714.

12 Khayyam H, Nahavandi S, Davis S. (2012) Adaptive cruise control look-ahead system for energy management of vehicles, *Expert Systems with Applications* **39**, 3, 3874-3885.

13 Frank B, Pohl J, Palmberg J-O. (2009) Estimation of the potential in predictive control in a hybrid wheel loader, In *SICFP'09 proceedings, The 11th Scandinavian International Conference on Fluid Power*.

14 Rizzoni G, Guzzella L, Baumann B.M. (1999) Unified modeling of hybrid electric vehicle drivetrains, *IEEE/ASME Transactions on Mechatronics* **4**, 246-257.

15 Mattsson P, Åkerblom M. (2012) Continuously variable transmission and a working maching including a continuously variable transmission, Patent, WO 2012/008884 A1.

16 Lauinger C, Englisch A, Gotz A, Teubert A, Muller E, Baumgartner A. (2007) Cvt components for powersplit commercial vehicle transmissions, In *Proceedings of the 6th International CTI Symposium*, CTI.

17 Bellman R. (1957) *Dynamic Programming*, Princeton University Press.

18 Bertsekas D.P. (2005) *Dynamic Programming and Optimal Control*, volume 1, Athena Scientific, 3 ed.

19 Nilsson T, Fröberg A, Åslund J.(2012) Optimal operation of a turbocharged diesel engine during transients, *SAE International Journal of Engines* **5**, 2, 571-578.

20 Silva J.E, Sousa J.B. (2011) Dynamic programming techniques for feedback control, In *Preprints of the 18th IFAC World Congress*, IFAC, pp. 6857-6862.

21 McDonough K, Kolmanovsky I, Filev D, Yanakiev D, Szwabowski S, Michelini J. (2012) Stochastic dynamic programming control policies for fuel efficient in-traffic driving, In *American Control Conference (ACC)*, IEEE, pp. 3986-3991.

22 Nilsson T, Fröberg A, Åslund J. (2012) Fuel potential and prediction sensitivity of a power-split cvt in a wheel loader, In *IFAC Workshop on Engine and Powertrain Control, Simulation and Modeling*, IFAC, pp. 49-56.

23 Nilsson T, Fröberg A, Åslund J. (2012) On the use of stochastic dynamic programming for evaluating a power-split cvt in a wheel loader, In *8th IEEE Vehicle Power and Propulsion Conference*, IEEE, pp. 840-845.

24 Leroy T, Malaize J, Corde G. (2012) Towards real-time optimal energy management of hev powertrains using stochastic dynamic programming, In *8th IEEE Vehicle Power and Propulsion Conference*, IEEE, pp. 383-388.

25 Nilsson T, Fröberg A, Åslund J. (2013) Fuel and time minimization in a cvt wheel loader application, In *7th IFAC Symposium on Advances in Automotive Control*, IFAC.

26 Hellström E. (2010) *Look-ahead Control of Heavy Vehicles*, PhD Thesis, Linköping University.

9

Shale Gas Pseudo Two-Dimensional Unsteady Seepage Pressure Simulation Analysis in Capillary Model

Lai Fengpeng[1]*, Li Zhiping[1], Li Zhifeng[2], Yang Zhihao[1] and Fu Yingkun[1]

[1] School of Energy Resources, China University of Geosciences, Beijing 100083 - China
[2] China Huadian Engineering Co., Ltd, Beijing 100035 - China
e-mail: laifengpeng@cugb.edu.cn

* Corresponding author

Abstract — *Shale gas is rapidly growing as a source of natural gas in China. Compared with the conventional gas reservoir, the shale gas reservoir is characterized by low porosity, low permeability, and adsorbed gas, making the flow mechanism of shale gas reservoir more complex. In this study, we investigated six factors influencing the gas flow: the Darcy flow, the slippage effect, the Knudsen diffusion effect, the desorption of gas on pore walls, the diffusion effect of gas in organic matter, and the matrix deformation effect. We simplified gas flow in the development process to only include gas flow in the capillaries and then considered the six influence factors. This study establishes a shale gas pseudo two-dimensional unsteady capillary seepage mathematical model based on the continuity equation, using the implicit difference method to solve the mathematical model. Certain capillary parameters were added to the calculation, and the study analyzed the effect of the different factors on both the pressure distribution and the cumulative gas production. Results show that the Knudsen diffusion effect and the desorption of gas from pore walls have lower impact on the pressure than the others factors. The diffusion effect of gas in organic matter, the slippage effect, and the matrix deformation effect have a stronger impact on the pressure. The gas in organic matter continuously diffuses into the capillary with the increasing of the production time, and the pressure drop becomes slow because of the gas diffusion.*

Résumé — **Analyse des instabilités d'une simulation pseudo-bidimensionnelle de la pression de filtration de gaz de schiste dans un modèle capillaire** — Le gaz de schiste connaît une croissance rapide en tant que source de gaz naturel en Chine. En comparaison du réservoir de gaz conventionnel, le réservoir de gaz de schiste se caractérise par une faible porosité, une faible perméabilité, et par le gaz adsorbé qui rend le mécanisme d'écoulement du réservoir de gaz de schiste plus complexe. Dans cette étude, nous avons étudié six facteurs qui influencent le débit de gaz : l'écoulement selon le modèle de Darcy, l'effet de glissement, l'effet de diffusion Knudsen, la désorption du gaz sur les parois poreuses, l'effet de diffusion du gaz dans la matière organique et l'effet de déformation de la matrice. Nous avons simplifié le débit de gaz dans un modèle pour inclure uniquement le débit de gaz dans les capillaires et ensuite étudier les six facteurs d'influence. Cette étude développe un modèle mathématique de pénétration capillaire instable et pseudo-bidimensionnelle du gaz de schiste en se fondant sur l'équation de continuité et en utilisant la méthode de différence implicite pour résoudre le modèle mathématique.

Certains paramètres des capillaires ont été ajoutés au calcul et l'étude a analysé l'effet des différents facteurs sur la distribution de pression et sur la production cumulée de gaz. Les résultats montrent que l'effet Knudsen et la désorption du gaz à partir de parois poreuses ont un impact moins important sur la pression que les autres facteurs. L'effet de diffusion du gaz dans une matière organique, l'effet de glissement et la déformation de la matrice ont une plus grande influence sur la pression. Le gaz dans la matière organique diffuse en continu dans le capillaire au fur et à mesure de la production, et la baisse de pression est ralentie à cause de la diffusion du gaz.

INTRODUCTION

In shale-gas systems, nanometer- to micrometer-size pores, along with natural fractures, form the flow-path network that allows flow of gas from the mud rock to induced fractures during production (Loucks et al., 2012). Shale gas production involves many different stages, such as reservoir and well bore. The production of a new drilling well comes first from a fracture and macro pore, and then from a small pore. The thermodynamic equilibrium between the gas and organic/clay changes as the gas is desorbed from the surface of the organic/clay during the development process. Driven by the changing thermodynamic equilibrium, the gas molecules diffuse from within the organic matter to its outer surface and then into the pore network. The gas flowing model that considers pore size has been applied to shale gas reservoir; it considers the change in the state of gas molecules according to Darcy's law. The Klinkenberg slippage effect contributes to the analysis of the production increase of tight gas and shale gas reservoirs (Ertekin et al., 1986). A comprehensive slippage concept is used to modify Darcy's law (Clarkson et al., 2012). In addition to Darcy's law, the concentration difference diffusion also influences gas flow (Ozkan et al., 2010). The contributions of slip flow and Knudsen diffusion increase the apparent permeability of the reservoir while gas production takes place. The effects of both mechanisms explain the higher-than-expected gas production rates commonly observed in these formations (Shabro et al., 2011).

Shale gas reserves have been exploited in Sichuan and Bohai bay basins in China. However, the study of the percolation mechanism and the capacity evaluation of shale gas is still in its infancy. A very high proportion of shale gas research has been focused towards understanding and improving the hydraulic fracturing in shale gas and comparatively less work has been done towards understanding the flow mechanism in the shale matrix (Swami and Settari, 2012). The CoalBed Methane (CBM) model is typically used to simulate the production characteristics of shale gas since shale undergoes organic carbon adsorption. Conventional Darcy seepage models are also used to describe the production characteristics of shale gas. Darcy's law is modified by

introducing apparent permeability to obtain the constitutive equation for gas seepage (Yao et al., 2012). However, these models are not based on shale's nanometer pore structure, which can cause some deviations in the description of its percolation mechanism, leading to incorrect reservoir evaluations and prediction.

In this paper, we first study the impact of the factors influencing gas flow and establish a capillary model to simplify the shale gas flow development process. The factors include Darcy flow, Knudsen diffusion, slippage effect, the desorption effect of adsorbed gas on pore wall surface, the diffusion effect of organic matter gas and matrix deformation effect. Based on the continuity equation, we deduce a shale gas pseudo two-dimensional unsteady capillary seepage mathematical model and use the implicit difference method to solve the mathematical model. Finally, we suggest some parameter values for the capillary model and analyze the impact of different factors influencing the pressure distribution.

1 SHALE GAS FLOW MECHANISM IN CAPILLARY

Several studies have proposed flow models capturing pore scale flow mechanism in shale reservoirs. Probably the simplest models from which Darcy's law may be derived are those made of capillary tubes in one arrangement or another. The starting point in all these models is Hagen-Poisseuille's law governing the steady flow through a single, straight circular capillary tube. Other models based on capillary tubes are described (Bear, 1972; Kanellopoulos, 1985). Ertekin et al. (1986) incorporated Klinkenberg's slippage effect to account for the higher than predicted value of production in shale reservoirs. Javadpour (2009) stated that both slippage and Knudsen diffusion become important at the shale nano pore scale.

1.1 Influence Factor Analysis of Gas Flow

1.1.1 Darcy Flow

While shale gas reservoir is different from conventional natural gas reservoir, the Darcy seepage law is used in both to describe the relationship between the gas flow and pressure in microscopic pores.

1.1.2 Slippage Effect

Gas slippage effect was the first time found by Kundt and Warburg (Scheidegger, 1957). Muskat (1946) applied gas flow in porous media theory to the development of oil and gas fields, making the gas slippage effect theoretical research possible for low permeability gas reservoirs industrialized mining. Compared to the rocks in conventional reservoirs, gas shales generally have much smaller pores and pore throats. Therefore, in gas shale, collisions between molecules and pore walls are more frequent than collisions between the molecules themselves. Under these circumstances, the process of gas slippage becomes important, resulting in matrix permeability that is dependent upon the type of gas flowing through the shale as well as the gas pressure (Letham, 2011).

1.1.3 Knudsen Diffusion Effect

Gas diffusion can be described by superposition of viscous and Knudsen flow (Markovic et al., 2009), and the phenomena of gas diffusion through porous materials has been discussed extensively by Kerkhof and Geboers (2005). Knudsen diffusion occurs when the scale length of a system is comparable to or smaller than the mean free path of the particles involved, e.g. in a long pore with a narrow diameter (2-50 nm), because molecules frequently collide with the pore wall (Malek and Coppens, 2003). In the Knudsen diffusion regime, the molecules do not interact with one another, but move in straight lines between points on the pore channel surface.

1.1.4 Desorption Effect of Adsorbed Gas on Pore Wall Surface

The impact of the adsorption layer on gas flowing in the pores is worth considering. Hartman et al. (2011) extend their discussion to the adsorption layer effect on multicomponent natural gases. Their approach is based on the thermodynamically consistent ideal adsorbed solution model to accurately predict adsorbed gas storage capacity for gas mixtures. Swami (2012) postulates that a significant amount of gas is also stored in the bulk of organic matter or kerogen. In his conceptual model of one shale pore, he models flow behavior taking into account the free gas, adsorbed gas, and gas dissolved in kerogen, and found that after the reservoir pressure becomes less than the critical pressure, gas adsorption capacity decreases as reservoir pressure decreases.

1.1.5 Diffusion Effect of Organic Matter Gas

There is a lot of adsorbed gas in shale, with an adsorbed gas portion ranging anywhere from 20% to 85% (Curtis, 2002). Much of the shale gas in organic matter is in the adsorbed state. When reservoir pressure decreases, shale gas spreads from organic matter into nanoscale pores and then diffuses by percolation. The diffusion rate of gas in organic matter is slow and the diffusion amount is small, so the production period of a shale gas well can reach up to several decades. Fick's laws can be used to describe the flow caused by a change in concentration of adsorbed gas due to pressure changes.

1.1.6 Matrix Deformation Effect

The effective stress in a shale reservoir increases with reservoir pressure, thus reducing the rate of the development process. When a reservoir is compacted, its physical properties such as porosity and permeability decrease. However, when reservoir pressure becomes lower than the critical desorption pressure, shale gas desorbs from the adsorption layer, which causes matrix shrinkage, and an improvement in the reservoir's physical properties.

1.2 Shale Gas Flow Mechanism

In the shale gas development process, gas mainly comes from within and outside of the reservoir's organic matter, diffuses into the large pores and fractures and eventually flows into the well bore. In this paper, we use the capillary flow model to simplify the whole flow process and consider the factors influencing gas flow identified above.

During the simulation of gas flow in the capillaries, the reduction of pressure causes free gas to flow, according to Darcy's flow and also results in the gas molecules slippage effect. Gas molecules in the capillary wall are in motion, so Knudsen diffusion should be considered because capillary size is nanoscale. The adsorbed gas in the capillary wall is desorbed as pressure continues to decline. The adsorbed gas in organic matter that is outside the capillary walls begins to diffuse into the capillaries, and this diffusion causes an increase of the pressure. The gas flowing into the capillaries is affected by the six factors listed above.

Shale gas in reservoirs has three kinds of storage forms according to flow law. The first is free gas in pores, the second is adsorbed gas on the surface of pore walls, and the last is adsorbed gas in organic matter. The flow process of these three forms of gas is controlled by six mechanisms as described above.

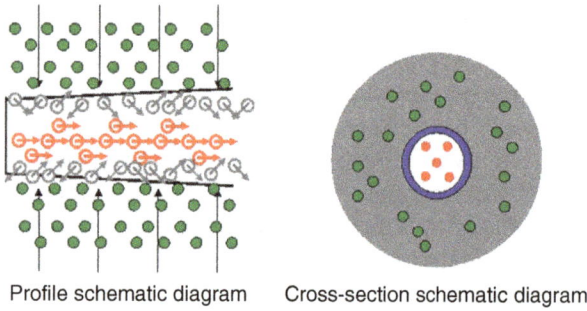

Profile schematic diagram   Cross-section schematic diagram

Figure 1

Capillary gas flow diagram.

2 SHALE GAS FLOW MODEL IN CAPILLARY

2.1 Model Assumes

2.1.1 Assumptions

1) The capillary model is composed of a matrix pore system and a large number of organic inclusions;
2) Flow in the reservoir is isothermal, free gas is the real gas;
3) The diffusion of gas from organic inclusions flows radially, the diffusion of gas from the capillaries flows linearly;
4) In the initial state, a certain amount of free gas stays in the capillaries, and most of the gas stays on the inner surface of the matrix and in the adsorbed state in organic matter;
5) The flow in the capillaries reflects a multi-flow mechanism, without regard to gravity or the pressure effects;
6) Radial flow reflects a process of diffusion, and this gas diffusion process is in a non-equilibrium pseudo-steady state, according to Fick's first law.

2.1.2 Schematic Diagram

The shale nanoscale pore flow channel is assumed to be a two-dimensional cylindrical flow pipe, with a volume of organic matter outside the cylindrical pipe (*Fig. 1*). A large volume of gas absorbed on the organic matter desorbs with the change of reservoir pressure, then diffusing from organic matter block into the pore through pore wall (green point shown in *Fig. 1*). Meanwhile, gas adsorbed on the surface of pore begins its desorption, then directly flowing into the pore channels (grey point shown in *Fig. 1*). The free gas now flows in the pore channel under pressure difference (red point shown in *Fig. 1*).

There is a closed boundary at one end of the cylindrical pipe, the initial pressure is P_i, the pressure at the drain tunnel is P_w, and there is a layer of gas molecules adsorbed on the

surface of pore wall. During the reduction of pore pressure, the adsorbed gas in the organic matter is desorbed, and then diffuses into the pores through surface contact between the organic matter and the pore.

2.2 Capillary Gas Flow Models

2.2.1 Darcy Seepage Equation

Considering Darcy seepage, both the motion equation and the ideal gas state equation are considered in the continuity equation, and the quality basic differential equation can be described as follows:

$$\frac{M}{2RT}\frac{K_m}{z\mu_g}\frac{\partial}{\partial x}\left(\frac{dp^2}{dx}\right) = \frac{M}{RTz}C_g\frac{\partial p^2}{\partial t} \quad (1)$$

where M is the molecular mass in g/mol, K_m is the permeability in mD, R is the universal gas constant, T is the temperature in K, z is the gas compressibility factor, μ_g is the gas viscosity in mPa.s, x is the capillary position in cm, C_g is the gas compressibility in MPa^{-1}, p is the pressure in MPa, and t is the time in s.

2.2.2 Desorption Effect of the Pore Wall Adsorbed Gas

The Langmuir equation can be used to describe the phenomenon of the CBM adsorbed on the coal seam surface, and is also used in the study of shale gas reservoirs because shale gas is similar to CBM. In the Langmuir isotherm equation, the relationship between gas pressure and gas adsorption can be expressed as follows:

$$V = V_L\frac{bp}{1 + bp} \quad (2)$$

where V is the gas adsorption in m^3/t, V_L is the gas Langmuir volume in m^3/t, and b is the gas Langmuir adsorption constant in MPa^{-1}.

As the pressure decreases, gas flows into the well bore from the pore walls, and the gas flow equation at the time of resolution on the wall can be described as in Equation (3):

$$J_d = \rho_r\rho_g V\frac{b}{(1 + bp)^2}\frac{\partial p}{\partial t} \quad (3)$$

where ρ_r is the shale density in kg/m^3, and ρ_g is the methane gas density in kg/m^3.

2.2.3 Diffusion Effect of the Gas in Organic Matter

A large amount of shale gas is adsorbed given large amounts of organic matter outside of the capillaries.

Gas desorbs from the organic matter when the pressure drops below the critical desorption pressure, and gas flows into the capillary by concentration diffusion.

Given the outer radius of the enfolded peripheral cylinder r, the flow equation of gas diffusion from the organic system to the inner capillaries can be described as in Equation (4):

$$J_k = \frac{\rho_r \pi r^2 L \rho_g (V_L - \overline{V}_i) D_k}{\pi \left(\frac{d}{2}\right)^2} \tag{4}$$

where L is the core length in m, $\overline{V}_i$ is the equilibrium adsorption quantity in m^3/t, d is the nano pore current diameter in m and D_k is the diffusion coefficient of organic matter in m^2/s.

2.2.4 Matrix Deformation Effect

The nano pore diameter changing equation, which considers matrix shrinkage and stress sensitivity can be described as in Equation (5):

$$d = d_i \left(1 + \left(1 + \frac{2}{\phi_i}\right)\right.$$
$$\left.\left\{\frac{V_L \rho_r RT}{EV_0}[\ln(1 + bp_i) - \ln(1 + bp)] + C_p(p_i - p)\right\}\right)^{0.5} \tag{5}$$

where d_i is the original diameter of nano pore in m, ϕ_i is the original porosity in fractional, p_i is the initial pressure in MPa, p is the currently reservoir pressure in MPa, C_p is the elastic compression factor in 10^{-4} MPa^{-1}, E is the Young's modulus in MPa, and V_0 is the gas molar volume in 10^{-3} m^3/mol.

3 UNSTEADY FLOW MODELS OF SHALE GAS CAPILLARY MODEL

Since the gas flow in the capillary is influenced by the six factors identified above, the continuity equation can be

described as follows:

$$\frac{M}{2RT}\frac{C_g D\mu_g + FK_m}{z\mu_g}\frac{\partial}{\partial x}\left(\frac{dp^2}{dx}\right) -$$
$$\rho_r \rho_g V_L \frac{b}{(1+bp)^2}\frac{\partial p}{\partial t} + \frac{\rho_r \rho_g \left(\frac{bp_L}{1+bp_L} - \frac{bp}{1+bp}\right)D_k}{\pi\left(\frac{d}{2}\right)^2} = \frac{M}{RTz}C_g\frac{\partial p^2}{\partial t} \tag{6}$$

where D is Knudsen diffusion constant in m^2/s, and p_L is the Langmuir pressure in MPa.

$$D = \frac{d}{3}\sqrt{\frac{8RT}{\pi M}} \tag{7}$$

The gas molecule near the capillary wall flows more easily because of the slippage effect. The dimensionless slippage factor F is used to modify the slip velocity in capillary tubes (Brown et al., 1946; Swami and Settari, 2012). The factor F may be looked upon as the ratio of μ_g to the effective viscosity applicable to Poiseuille's equation at low pressures (Brown et al., 1946):

$$F = 1 + 1000\left(\frac{8\pi RT}{M}\right)^{0.5}\frac{\mu_g}{\overline{p}r}\left(\frac{2}{\alpha} - 1\right) \tag{8}$$

where $\overline{p}$ is the average pressure in MPa, r is the nano pore radius in m, and α is the dimensionless distribution coefficient of diffusion flow.

K_m is the Darcy permeability and is given by Poiseuille equation for circular capillaries as:

$$k_m = \frac{r^2}{8} = \frac{d^2}{32} \tag{9}$$

Equation (5), (7), (8) and Equation (9) are put into Equation (6), and the resulting equation is as follows: see Equation (10) below

$$\frac{M}{2RT}\frac{C_g d_i\left(1 + \left(1 + \frac{2}{\phi_i}\right)\left\{\frac{V_L\rho_r RT}{EV_0}[\ln(1+bp_i) - \ln(1+bp)] + C_p(p_i - p)\right\}\right)^{0.5}\sqrt{\frac{8RT}{9\pi M}}\mu_g}{z\mu_g}\frac{\partial}{\partial x}\left(\frac{dp^2}{dx}\right)$$
$$+ \frac{M}{2RT}\frac{Fd_i^2\left(1 + \left(1 + \frac{2}{\phi_i}\right)\left\{\frac{V_L\rho_r RT}{EV_0}[\ln(1+bp_i) - \ln(1+bp)] + C_p(p_i - p)\right\}\right)}{32z\mu_g}\frac{\partial}{\partial x}\left(\frac{dp^2}{dx}\right) - \rho_r\rho_g V_L\frac{b}{(1+bp)^2}\frac{\partial p}{\partial t}$$
$$+ \frac{\rho_r\rho_g\left(\frac{bp_L}{1+bp_L} - \frac{bp}{1+bp}\right)D_k}{\pi\left(d_i^2\left(1 + \left(1 + \frac{2}{\phi_i}\right)\left\{\frac{V_L\rho_r RT}{EV_0}[\ln(1+bp_i) - \ln(1+bp)] + C_p(p_i - p)\right\}\right)\right)} = \frac{M}{RTz}C_g\frac{\partial p^2}{\partial t} \tag{10}$$

The initial condition for Equation (10) can be described as follows:

$$t = 0,\ P = P_i,\ 0 \le x < L$$

The inner boundary condition for Equation (9) can be described as follows:

$$x = 0,\ P = P_w,\ t > 0$$

The outer boundary condition for Equation (9) can be represented as:

$$x = L,\ \frac{\partial p}{\partial x} = 0,\ t > 0$$

4 SHALE GAS UNSTEADY FLOW SIMULATION ANALYSIS

To emulate gas flow, we represent a nanopore by a capillary tube. Figure 1 shows the physical model. The entire system is at initial reservoir pressure in the beginning. Right boundary of the pore is open to a constant pressure, and the left boundary is a no flow boundary. Shabro et al. (2011) and Swami and Settari (2012) listed some data in their papers. We referred to these data, but did not copy the data. For example, the pore radius is 2 nm in their papers, we assume the radius is 1 nm. We compare some data in Table 1.

Table 2 shows the simulation-related parameters and values for this simulation model. Results from this model are used to analyze the impact of the six factors influencing pressure and gas production.

The radial flow of gas from the organic matter to the capillaries also flows along the capillary strings at the same time. We use the implicit method to solve the flow process. Following the gas seepage model for the capillaries, we first study the pressure distribution in the capillaries. Figure 2 shows the nonlinear pressure distribution from the closed boundary to the constant

pressure drainage boundary, which is calculated with consideration to the six factors identified above.

The pressure drops slowly near the closed boundary, indicating that the drop in pressure is smaller at the closed boundary than at the drainage boundary. As the pressure reduces, there is greater influence from the six factors.

Figure 3 shows the pressure distribution under the influence of different factors. Figure 4 shows the simulation pressure when considering these six factors, and the pressure result when not considering the matrix deformation effect. In Figure 4, the pressure at the closed

TABLE 2

Simulation related parameters and value

Parameter	Value	Units
Nano pore radius	1	nm
Nano pore length	1	m
Reservoir temperature	60	°C
Gas viscosity	0.015	mPa·s
Knudsen diffusion coefficient	2E-11	m^2/s
Outer boundary pressure	25	MPa
Inner boundary pressure	5	MPa
Langmuir constant	0.4	MPa^{-1}
Langmuir volume	0.035	m^3/kg
Inclusions radius	0.0000001	m
Gas density	0.7174	g/L
Rock density	2 560	kg/m^3

TABLE 1

Comparison of some data

Parameter	Value in this paper	Value in the reference	Units
Nano pore radius	1	2	nm
Nano pore length	1	1×10^{-7} to 1	m
Langmuir volume	0.035	0.02	m^3/kg
Inner boundary pressure	5	8.6	MPa
Outer boundary pressure	25	17.2	MPa
Rock density	2 560	2 500	kg/m^3

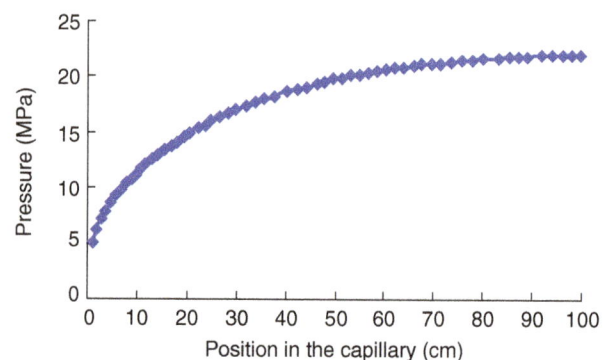

Figure 2

Pressure distribution at different position in the capillary.

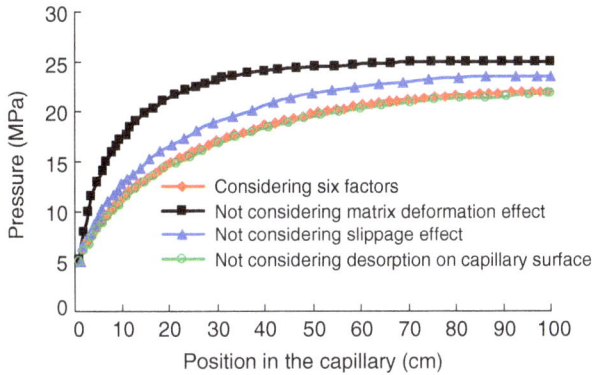

Figure 3

Capillary pressure distribution under different influence factors.

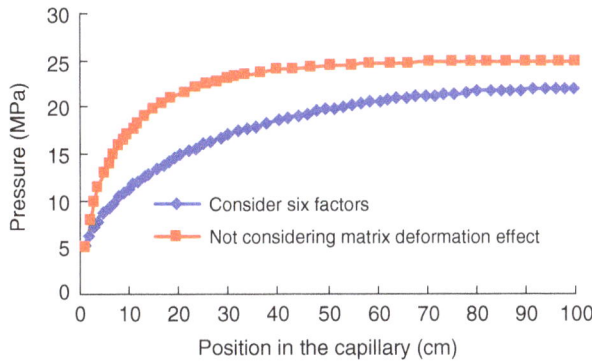

Figure 5

Capillary pressure distribution under not considering slippage effect.

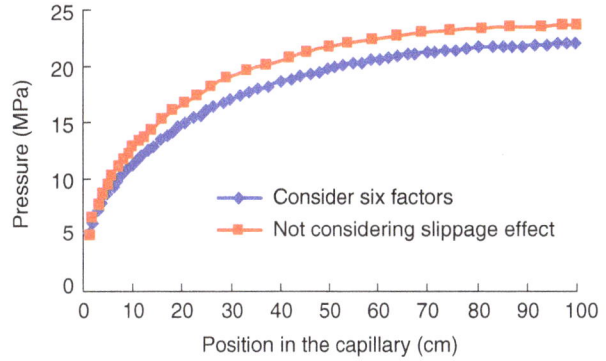

Figure 4

Capillary pressure distribution under not considering matrix deformation effect.

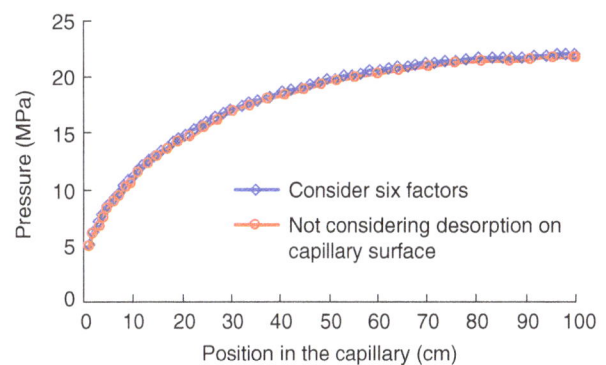

Figure 6

Capillary pressure distribution under not considering desorption on capillary surface.

boundary is 25 MPa when not considering the matrix deformation effect. On the same curve, the pressure is 24.5 MPa at the 50 cm point, and the pressure is 20 MPa at 16 cm point. The nano pore diameter is fixed in the whole simulation process if the matrix deformation effect is not considered. Compared with considering overall factors, the nano pore diameter can be assumed to become narrower when ignoring the matrix deformation effect. The pressure transmission is slow, the pressure consumption increases near the well bore.

Figure 5 shows the pressure contrast with one curve considering all factors, and the other one not considering the slippage effect. At the initial state, the pressure that does not consider slippage effect is higher than the pressure considering all factors, with pressure values of 23.64 and 21.93 MPa, respectively. This means that the slippage effect is an important factor in shale gas production, and should not be ignored. When slippage effect

is not considered, the pressure is 21.6 MPa at the 50 cm point, and 16.35 MPa at the 16 cm point. When calculating the slippage effect, the value of the gas slippage correction factor becomes larger as pressure decreases. Compared with the pressure that considers all factors, the flow ability in the capillaries reduces if the slippage effect is not considered. The pressure decrease mainly occurs near the well bore.

Figure 6 shows the pressure variation with one curve considering all factors, and the other one not considering desorption on the capillary surface. At the initial state, the pressure not considering desorption on capillary surface is slightly less than the pressure considering all factors, with pressure values at 21.93 and 21.91 MPa, respectively. When desorption on capillary surface is not considered, the pressure is 19.7 MPa at the 50 cm point, and 13.5 MPa at the 16 cm point. The adsorbed gas desorbs from the capillary surface, flowing into the

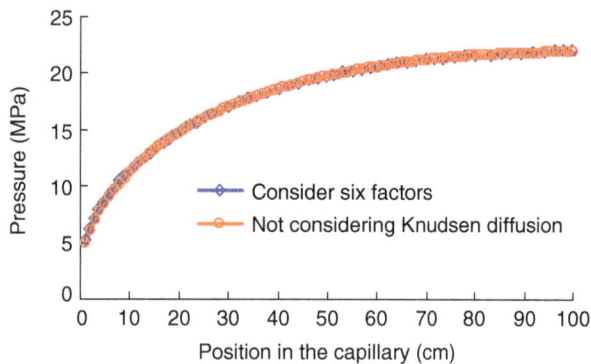

Figure 7

Capillary pressure distribution under not considering desorption on capillary surface.

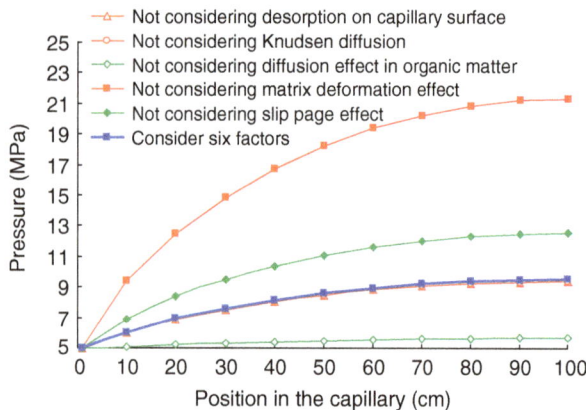

Figure 8

Capillary pressure distribution under different influence factors after a certain time.

Figure 8 shows the pressure distribution in the capillary under the action of various factors with the simulation time increasing by 10 times. The pressure has changed dramatically after a certain production time.

With the production time increasing by a factor 10, the desorption on the capillary surface and the Knudsen diffusion have a more obvious effect on the pressure than the original condition. However, the other three factors have more influence on the pressure. The closed boundary pressure is 21.26 MPa without considering the matrix deformation effect. The pressure is 12.52 MPa without considering slippage effect. If we do not consider the diffusion effect in organic matter, the pressure is 5.69 MPa. The pressure is 9.55 MPa when considering six factors. The gas in the organic matter continuously diffuses into the capillary with the increasing of production time. The gas diffusion from organic matter helps to maintain the pressure, and the pressure drop becomes slow because of the gas diffusion.

CONCLUSIONS

In this study, we establish a capillary model to simplify the shale gas flow development process. Six factors of influence are studied to analyze the mechanism of shale gas flow in capillaries. Our shale gas pseudo two-dimensional unsteady capillary seepage mathematical model establishes a mechanism for the analysis of shale gas flow, solved by the implicit difference method. Our simulation results show that the Knudsen diffusion effect and desorption of gas on capillary surfaces has less impact on pressure than the other factors. The diffusion effect of gas in organic matter, the slippage effect and the matrix deformation effect have greater impact on the pressure. The gas diffusion from organic matter helps to maintain the pressure.

ACKNOWLEDGMENTS

This research is supported by the Fundamental Research Funds of the Central Universities, and the National Special Fund (No. 2011ZX05009-006).

REFERENCES

Bear J. (1972) *Dynamics of Fluids in Porous Media*, Dover Publications, Inc., New York.

Brown G.P., Dinardo A., Cheng G.K., Sherwood T.K. (1946) The Flow of Gases in Pipes at Low Pressures, *Journal of Applied Physics* **17**, 10, 802-813.

capillary string, and the desorption gas interferes with the capillary gas, thus influencing the gas flowing although the influence is not obvious because the quantity of gas adsorbed is low.

Figure 7 shows the pressure variation with one curve considering all factors, and the other one not considering Knudsen diffusion. Knudsen diffusion increases the mobility of gas molecules, but the influence is not obvious because the velocity of the Knudsen diffusion is very slow at the beginning of production. At the same time, the velocity of gas diffusion is also very slow, so the amount of gas diffusing from the organic matter to the capillary strings is tiny and it has no obvious effect on the pressure.

Clarkson C.R., Nobakht M., Kaviani D., Ertekin T. (2012) Production Analysis of Tight Gas and Shale Gas Reservoirs Using the Dynamic-Slippage Concept, *SPE Journal* **17**, 1, 230-242.

Curtis J.B. (2002) Fractured shale-gas system, *AAPG Bulletin* **86**, 11, 1921-1938.

Ertekin T., King G.A., Schwerer F.C. (1986) Dynamic Gas Slippage: A Unique Dual-Mechanism Approach to the Flow of Gas in Tight Formations, *SPE Formation Evaluation* **1**, 1, 43-52.

Hartman R.C., Ambrose.R.J., Akkutlu I.Y., Clarkson C.R. (2011) Shale Gas-in-Place Calculations Part II - Multicomponent Gas Adsorption Effects, *North American Unconventional Gas Conference and Exhibition*, Texas, USA, 14-16 June.

Javadpour F. (2009) Nanopores and Apparent Permeability of Gas Flow in Mudrocks (Shales and Siltstone), *Journal of Canadian Petroleum Technology* **48**, 8, 16-21.

Kanellopoulos N.K. (1985) Capillary models for porous media: Newtonian and non-Newtonian flow, *Journal of Colloid and Interface Science* **108**, 1, 11-17.

Kerkhof P.J.A.M., Geboers M.A.M. (2005) Analysis and extension of the theory of multi-component fluid diffusion, *Chemical Engineering Science* **60**, 12, 3129-3167.

Letham E.A. (2011) *Matrix Permeability Measurements of Gas Shales: Gas Slippage and Adsorption as Sources of Systematic Error, Bachelor Thesis*, The University of British Columbia.

Loucks R.G., Reed R.M., Ruppel S.C., Hammes U. (2012) Spectrum of pore types and networks in mudrocks and a descriptive classification for matrix-related mudrock pores, *AAPG Bulletin* **96**, 6, 1071-1098.

Malek K., Coppens M.O. (2003) Knudsen self- and Fickian Diffusion in Rough Nanoporous Media, *Journal of Chemical Physics* **119**, 5, 2801-2811.

Markovic A., Stoltenberg D., Enke D., Schlnder E.U., Seidel-Morgenstern A. (2009) Gas permeation through porous glass membranes Part I. Mesoporous glasses-Effect of pore diameter and surface properties, *Journal of Membrane Science* **336**, 1-2, 17-31.

Muskat M., (1946) *The Flow of Homogeneous Fluids through Porous Media*, Edwards J.W., Inc. Ann Arbor, Michigan.

Ozkan E., Raghavan R., Apaydin O.G. (2010) Modeling of Fluid Transfer From Shale Matrix to Fracture Network, *SPE Annual Technical Conference and Exhibition*, Florence, Italy, 19-22 Sept.

Scheidegger A.E. (1957) *The physics of flow through porous media*, University of Toronto Press, Toronto.

Shabro V., Torres-Verdin C., Javadpour F. (2011) Numerical Simulation of Shale-Gas Production: from Pore-Scale Modeling of Slip-Flow, Knudsen Diffusion, and Langmuir Desorption to Reservoir Modeling of Compressible Fluid, *North American Unconventional Gas Conference and Exhibition*, Texas, USA, 14-16 June.

Swami V. (2012) Shale Gas Reservoir Modeling: From Nanopores to Laboratory, *SPE Annual Technical Conference and Exhibition*, Texas, USA, 8-10 Oct.

Swami V., Settari A.T. (2012) A Pore Scale Gas Flow Model for Shale Gas Reservoir, *SPE Americas Unconventional Resources Conference*, Pennsylvania, USA, 5-7 June.

Yao T.Y., Huang Y.Z., Li J.S. (2012) Flow Regim for Shale Gas in Extra Low Permeability Porous Media, *Chinese Journal of Theoretical and Applied Mechanics* **44**, 6, 990-995.

Integrated Energy and Emission Management for Diesel Engines with Waste Heat Recovery Using Dynamic Models

Frank Willems[1,2]*, Frank Kupper[2], George Rascanu[1] and Emanuel Feru[1]

[1] Eindhoven University of Technology, Faculty of Mechanical Engineering, P.O. Box 513, 5600 MB Eindhoven - The Netherlands
[2] TNO Automotive, Steenovenweg 1, 5708 HN Helmond - The Netherlands
e-mail: f.p.t.willems@tue.nl - frank.kupper@tno.nl - g.c.rascanu@student.tue.nl - e.feru@tue.nl

* Corresponding author

Abstract — *Rankine-cycle Waste Heat Recovery (WHR) systems are promising solutions to reduce fuel consumption for trucks. Due to coupling between engine and WHR system, control of these complex systems is challenging. This study presents an integrated energy and emission management strategy for an Euro-VI Diesel engine with WHR system. This Integrated Powertrain Control (IPC) strategy optimizes the CO_2-NO_x trade-off by minimizing online the operational costs associated with fuel and AdBlue consumption. Contrary to other control studies, the proposed control strategy optimizes overall engine-aftertreatment-WHR system performance and deals with emission constraints. From simulations, the potential of this IPC strategy is demonstrated over a World Harmonized Transient Cycle (WHTC) using a high-fidelity simulation model. These results are compared with a state-of-the-art baseline engine control strategy. By applying the IPC strategy, an additional 2.6% CO_2 reduction is achieved compare to the baseline strategy, while meeting the tailpipe NO_x emission limit. In addition, the proposed low-level WHR controller is shown to deal with the cold start challenges.*

Résumé — Une stratégie intégrée de gestion des émissions et de l'énergie pour un moteur Diesel avec un système WHR (*Waste Heat Recovery*) — Les systèmes WHR basés sur le cycle de Rankine sont des solutions prometteuses pour réduire la consommation de carburant pour les camions. En raison du couplage physique entre le moteur et le système WHR, l'asservissement de ces systèmes est particulièrement difficile. Cette étude présente une stratégie intégrée de gestion des émissions et de l'énergie pour un moteur Diesel EURO-VI avec un système WHR. Cette stratégie IPC (*Integrated Powertrain Control*) optimise le compromis entre CO_2 et NO_x en minimisant les coûts d'exploitation liés à la consommation de carburant et d'AdBlue. Contrairement à d'autres algorithmes d'asservissement, la stratégie proposée ici optimise les performances globales du système moteur – post-traitement des gaz d'échappement – WHR tout en respectant les contraintes d'émissions. À partir de simulations, le potentiel de cette stratégie IPC est montré sur un WHTC (*World Harmonized Transient Cycle*) en utilisant un modèle de simulation haute-fidélité. Ces résultats sont comparés à une stratégie d'asservissement de référence. En appliquant la stratégie IPC, une réduction supplémentaire de 2,6 % CO_2 est obtenue, tout en respectant la limite légale de NO_x. En plus, la loi de commande bas niveau pour le WHR arrive à gérer les problèmes de démarrage à froid.

NOMENCLATURE

DOC   Diesel Oxidation Catalyst
DPF   Diesel Particulate Filter
EGR   Exhaust Gas Recirculation
IPC   Integrated Powertrain Control
PM   Particule Matter
SCR   Selective Catalytic Reduction
VTG   Variable Turbine Geometry
WHR   Waste Heat Recovery
WHTC   Word Harmonized Transient Cycle

INTRODUCTION

With the implementation of Euro-VI emission legislation, tailpipe emissions are forced towards near zero impact levels. During the last two decades, nitrogen oxides (NO_x) and Particulate Matter (PM) emissions are reduced by 86% and 95%, respectively, for trucks. To meet these targets, a combination of engine measures (common rail fuel injection equipment, advanced turbocharging, Exhaust Gas Recirculation (EGR)) and aftertreatment systems (soot filters, catalysts) are applied.

As illustrated in Figure 1 [1], it has been increasingly challenging to keep the fuel consumption (and thus CO_2 emission) around the current level for each emission phase. However, driven by concerns about global warming and energy security, attention for heavy-duty applications currently also moves towards CO_2 emission reduction. On top of the current targets for pollutants,

up to 20% CO_2 reduction has to be achieved in 2020 compared to the 2010 standards in the US [2]. Similar measures are discussed now in Europe [3].

For distribution trucks, garbage trucks and city buses, hybrid-electric drivetrains attract much attention to reduce CO_2 emissions. These drivetrains are less effective for long haul truck applications. In these cases, Waste Heat Recovery (WHR) seems a very promising technology [4-6]; in WHR systems, energy is recovered from heat flows, as illustrated in Figure 2 [7].

Up to 6% fuel consumption reduction has been demonstrated [4, 6]. However, control studies mainly focus on low level WHR system control [8-12]. Only a very few studies concentrate on energy management strategies for the complete engine [13]. However, these studies do not deal with the impact of the WHR system on emissions.

In this study, a cost-based optimization strategy is presented that explicitly deals with the requirements for CO_2 and pollutant emissions [7]. This strategy integrates energy and emission management and exploits the interaction between engine, aftertreatment and WHR system: Integrated Powertrain Control (IPC). Contrary to earlier work [14], a high fidelity WHR model is applied, which includes WHR dynamics and a low-level WHR controller. As a result, the simulation model combines a detailed aftertreatment and WHR system model and differs from the simplified control model that is embedded in the IPC strategy.

This work is organized as follows. First, the studied powertrain and applied simulation and control models are presented in Section 1. Section 2 discusses the

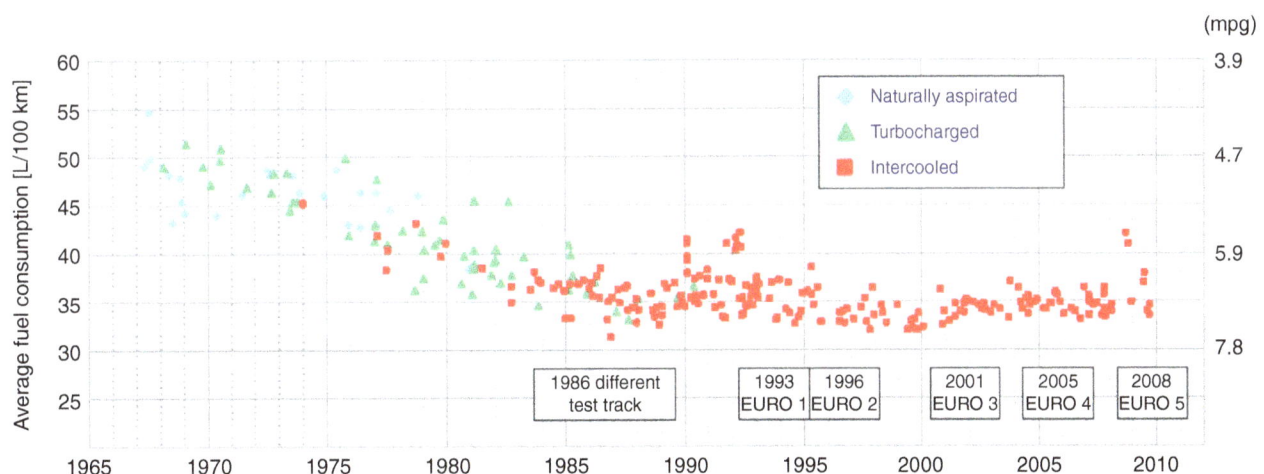

Figure 1

Historic fuel consumption for 40 ton trucks [1].

developed IPC strategy, whereas the control design is described in Section 3. For a World Harmonized Transient Cycle (WHTC), the results of this IPC strategy are compared with the results of a baseline engine control strategy in Section 4. Finally, conclusions are drawn and directions for future research are sketched.

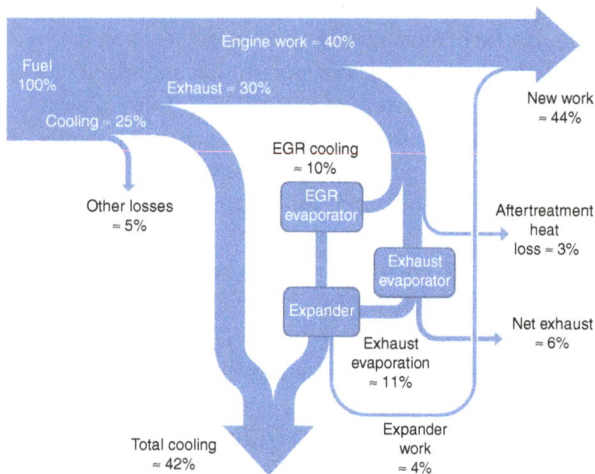

Figure 2

Sankey diagram of engine with WHR system [7].

1 SYSTEM DESCRIPTION

Figure 3 shows a scheme of the examined engine platform. It is based on a 6 cylinder, 13 liter, 375 kW Euro-VI Diesel engine, which is equipped with a cooled Exhaust Gas Recirculation (EGR) system and a turbocharger with Variable Turbine Geometry (VTG). Furthermore, an exhaust gas aftertreatment system is installed. This system consists of a Diesel Oxidation Catalyst (DOC), a Diesel Particulate Filter (DPF) and an urea-based Selective Catalytic Reduction (SCR) system.

The DPF system removes the particulates from the exhaust flow. To avoid clogging of the filter, fuel is periodically injected upstream of the DOC. As a result, the exhaust gas temperature is raised, such that the trapped particulates are oxidized. The remaining NO_x emissions downstream of the DPF system are converted into harmless products (nitrogen and water) over the 32.6 liter Cu-Zeolite SCR catalyst. For this catalytic process, ammonia (NH_3) is required. This is partly formed upstream of the catalyst by decomposition of the injected aqueous urea solution (trade name: AdBlue) in the hot exhaust gases. Further decomposition takes place in the SCR catalyst. To avoid unacceptable NH_3 slip, an ammonia oxidation catalyst (AMOX) is installed.

In this study, the Euro-VI engine is extended with a Waste Heat Recovery (WHR) system. This system is

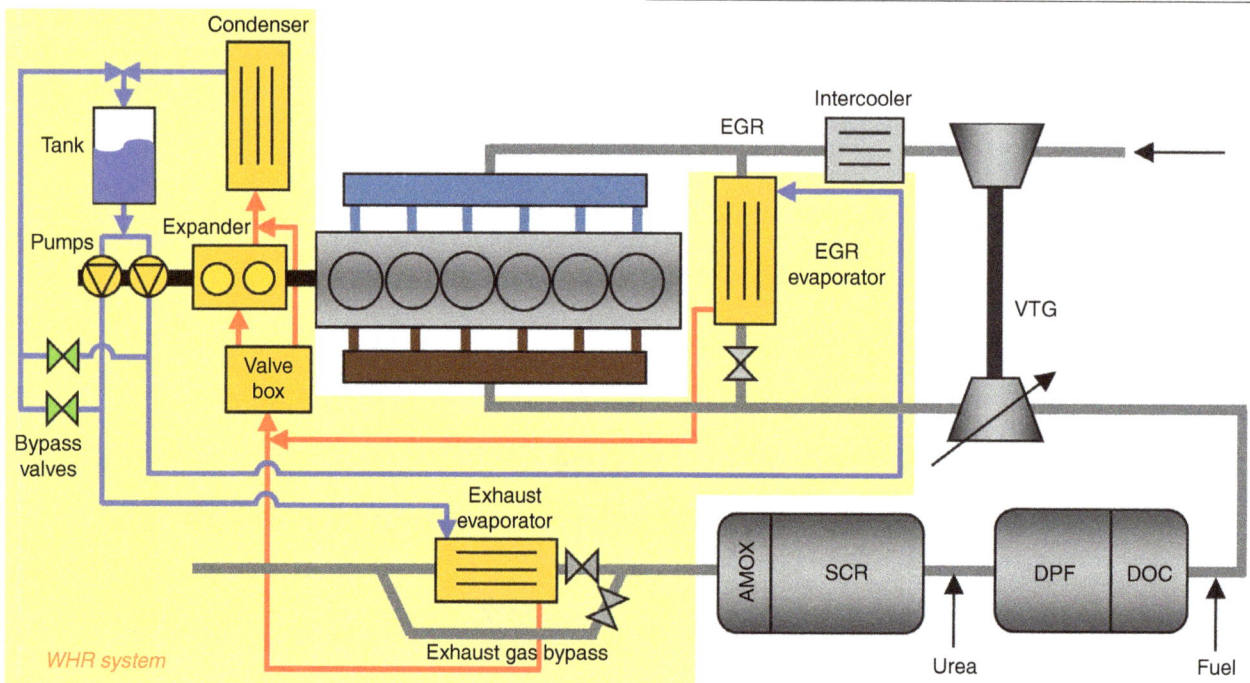

Figure 3

Scheme of the studied Euro-VI engine with WHR system [16].

based on a Rankine cycle, in which thermal energy is recovered from both the EGR line and exhaust line using two individual evaporators. The recovered exhaust heat is converted to mechanical power using a piston expander. The piston expander and pumps are mechanically coupled with the engine crankshaft. Consequently, the recovered power is directly transmitted to the engine. The working fluid mass flow through both evaporators is controlled by two bypass valves. As this is a closed system, the working fluid is cooled over the condenser, such that it is in liquid phase at the entrance of the pumps. It is noted that the heat input from the exhaust gas can be limited using the exhaust gas bypass. However, this operation mode is not considered in this study.

The following sections give a description of the applied simulation and control model.

1.1 Simulation Model

1.1.1 Engine

To describe the engine behavior, engine maps of the exhaust gas mass flow $\dot{m}_{exh}$, exhaust gas temperature T_{exh} and engine out NO_x mass flow $\dot{m}_{NO_x}$ are applied. These four-dimensional maps $f(N_e, \tau_e, u_{EGR}, u_{VTG})$ are constructed using a validated mean-value engine model. For varying combinations of EGR valve position u_{EGR} and VTG position u_{VTG}, the fuel mass flow $\dot{m}_f$ is varied such that the requested torque $\tau_{d,req} = \tau_e$ is realized (with constant engine speed N_e (rpm)). Note that these maps are determined for the engine without WHR system.

1.1.2 Aftertreatment System

A high fidelity aftertreatment model is implemented to simulate the DOC/DPF and SCR system. This modular model is built up using one-dimensional submodels of a pipe with urea decomposition, pre-oxidation catalyst (DOC), DPF, SCR catalyst, and ammonia oxidation (AMOX) catalyst. All catalyst models are based on first principle modeling and consist of mass and energy balances. By dividing the catalyst in various segments, these validated models describe the spatial distribution of pressure, temperature and chemical components. Further details on the model approach and SCR model can be found in [15].

1.1.3 Waste Heat Recovery System

For the description of the WHR system, we applied the model that is presented in [16]. This model is fitted and validated over a wide operating range. In this model, the pumps and expander are described by stationary

maps, whereas the valves are described by stationary relations. The piping within the WHR system model are represented by volumes. These volumes are modeled as incompressible pressure volumes and compressible pressure volumes for the liquid case and vapor case, respectively.

As the main WHR system dynamics correspond to the evaporators, condenser and pressure volumes, the modeling of these components is briefly reviewed in the sequel of this section.

1.1.3.1 Evaporators and Condenser

The evaporators and condenser are counter flow type of heat exchangers. Their models can be separated in three parts: the working fluid, the heat exchanger wall and the secondary fluid. The working fluid is pure ethanol, while the secondary fluid is exhaust gas for the evaporator case and coolant for the condenser case. The model is given by a set of non-linear partial differential equations which describe the conservation of mass and energy.

Conservation of mass (working fluid):

$$V_{wf}\frac{\partial \rho_{wf}}{\partial t} + L\frac{\partial \dot{m}_{wf}}{\partial z} = 0 \tag{1}$$

Conservation of energy:

$$\rho_{wf}V_{wf}\frac{\partial h_{wf}}{\partial t} = -\dot{m}_{wf}L\frac{\partial h_{wf}}{\partial z} + \alpha_{wf}S_{wf}(T_w - T_{wf}) \tag{2a}$$

$$\rho_g V_g\frac{\partial h_g}{\partial t} = \dot{m}_g L\frac{\partial h_g}{\partial z} - \alpha_g S_g(T_g - T_w) \tag{2b}$$

Conservation of energy at the wall:

$$\rho_w V_w c_{pw}\frac{\partial T_w}{\partial t} = \alpha_g S_g(T_g - T_w) + \alpha_{wf}S_{wf}(T_{wf} - T_w) \tag{3}$$

where h_{wf} is the working fluid enthalpy, h_g is the exhaust gas enthalpy, $\dot{m}_{wf}$ and $\dot{m}_g$ are the mass flow rates of the working fluid and exhaust gas, respectively. The parameters used in the heat exchanger model are listed in Table 1. As a consequence of the evaporation process, the heat transfer coefficient on the working fluid side α_{wf} is characterized by three heat transfer coefficients: α_{wf}^l for liquid, α_{wf}^{tp} for two-phase and α_{wf}^v for vapor. To account for phase change, mathematical equations for the working fluid properties are derived using Figure 4.

The model given by Equations (1, 2) and (3) is discretized with respect to time and space based on a finite difference approximation. The resulting expressions are a set of difference dynamic equations, which are used to

TABLE 1

Heat exchanger and pressure volume model parameters

Symbol	Unit	Description
c_p	J/kg·K	Specific heat capacity at constant pressure
c_{pg}	J/kg·K	Exhaust gas specific heat capacity
c_{pw}	J/kg·K	Wall mass specific heat capacity
c_v	J/kg·K	Specific heat capacity at constant volume
L	m	Heat exchanger tube length
R	J/kg·K	Ideal gas constant
S_{wf}	m^2	Surface area of the working fluid
S_g	m^2	Surface area of the exhaust gas
V	m^3	Volume
V_g	m^3	Volume of the exhaust gas side
V_w	m^3	Wall volume
V_{wf}	m^3	Volume of the working fluid side
α_{wf}	W/m^2·K	Working fluid heat transfer coefficient
α_g	W/m^2·K	Exhaust gas heat transfer coefficient
ρ_{wf}	kg/m^3	Working fluid density
ρ_g	kg/m^3	Exhaust gas density
ρ_w	kg/m^3	Wall density

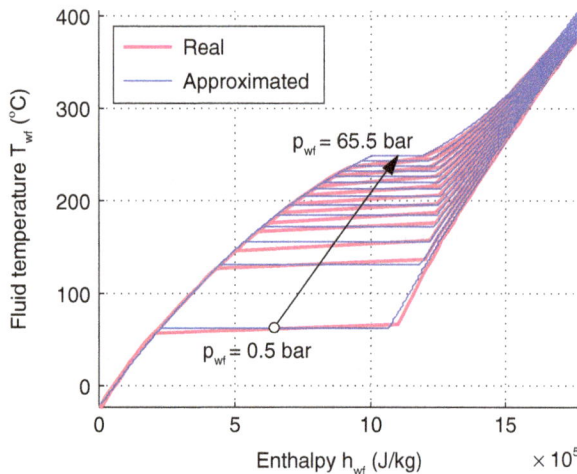

Figure 4

Ethanol temperature as a function of specific enthalpy and pressure [16].

compute the evaporator model output variables. These output variables are the outlet temperatures for the exhaust gas and working fluid side, respectively.

1.1.3.2 Pressure Volumes

The compressible pressure volume model assumes that the working fluid is in superheated vapor state and that it behaves like an ideal gas. Based on the laws for mass and energy conservation, the following equations are derived:

$$\frac{dm}{dt} = \dot{m}_{in} - \dot{m}_{out} \qquad (4a)$$

$$m\frac{dT}{dt} = bT_{in} + aT \qquad (4b)$$

$$mT\frac{dp}{dt} = (bT_{in} + aT)p + mT\frac{R}{V}(\dot{m}_{in} - \dot{m}_{out}) \qquad (4c)$$

where a and b are defined as:

$$
\begin{aligned}
a &= -\dot{m}_{in} - \left(\frac{c_p}{c_v} - 1\right)\dot{m}_{out} \\
b &= \frac{c_p}{c_v}\dot{m}_{in}
\end{aligned}
\qquad (5)
$$

Note that these equations depend on both temperature and mass flow rates. For steady state conditions, Equations (4b, c) reduce to $T = -\frac{b}{a}T_{in} = T_{in}$, respectively.

1.2 Control Model

This section presents the control model that is embedded in the optimal control strategy in Section 2. For real-world implementation, this simplified model has to represent the main system characteristics and has to be evaluated in real-time. Compared to the simulation model, the main difference lies in the description of the aftertreatment and WHR system; identical engine maps are applied.

1.2.1 Waste Heat Recovery System

In contrast with [14], the actual net mechanical WHR power output $P_{WHR} = \tau_{WHR} \cdot \omega_{WHR}$ is assumed to be available in the supervisory controller. Consequently, no explicit WHR model description is used in the controller.

By assuming ideal torque management, the requested engine torque is determined from:

$$\tau_{e,req} = \tau_{d,req} - \tau_{WHR} \qquad (6)$$

with the actual produced WHR torque τ_{WHR} (Nm) available from the simulation model.

1.2.2 Aftertreatment System

The thermal behavior of the total DOC-DPF-SCR system is described by two coupled differential equations, Equation (8). Note that the DOC-DPF system behavior is lumped in one equation. For the SCR conversion efficiency η_{SCR}, a set of three stationary maps is used, which are determined for different pre-SCR concentration ratios $C_{NO_2}/C_{NO_x} = (0, 0.5, 1.0)$ and for a specified ammonia slip level. From Figure 5, it is seen that the individual SCR efficiency map depends on the average SCR catalyst temperature T_{SCR} (°C) and space velocity SV (1/h):

$$SV = 3\,600 \cdot \frac{\dot{m}_{exh}}{\rho_{exh} V_{cat}} \qquad (7)$$

with normal condition exhaust gas density $\rho_{exh}(g/m^3)$ and SCR catalyst volume $V_{cat}(m^3)$. Using the predicted C_{NO_2}/C_{NO_x} ratio from a stationary DOC efficiency map, the NO_x conversion efficiency is computed by interpolation.

In summary, the control model is written in state space form $\dot{x} = f(x)$:

$$\dot{x} = \begin{bmatrix} c_1 \cdot \dot{m}_{exh}(T_{exh} - T_{DOC}) \\ c_2 \cdot \dot{m}_{exh}(T_{DOC} - T_{SCR}) - c_3(T_{SCR} - T_{amb}) \\ \dot{m}_{NO_x}(1 - \eta_{SCR}(T_{SCR}, SV, C_{NO_2}/C_{NO_x})) \end{bmatrix} \qquad (8)$$

with state variables:

$$x = \begin{bmatrix} T_{DOC} \\ T_{SCR} \\ m_{NO_x,tp} \end{bmatrix} = \begin{bmatrix} \text{DOC catalyst temperature} \\ \text{SCR catalyst temperature} \\ \text{Tailpipe } NO_x \text{ mass} \end{bmatrix}$$

The applied model parameters are specified in Table 2.

TABLE 2

Control model parameters

Constant	Unit	Definition	Value
c_1	kg^{-1}	$\frac{c_{p,exh}}{C_{DOC}}$	0.1163
c_2	kg^{-1}	$\frac{c_{p,exh}}{C_{SCR}}$	0.0512
c_3	$s \cdot kg^{-1}$	$\frac{h}{C_{SCR}}$	1.0000

2 CONTROL STRATEGY

The main goal of the engine control system is to determine the settings for the control inputs:

$$u^T = [\dot{m}_f \ \dot{m}_a \ u_{EGR} \ u_{VTG} \ u_{WHR,EGR} \ u_{WHR,exh}]$$

such that fuel consumption $\dot{m}_f$ is minimized within the constraints set by emission legislation. As illustrated in Figure 6, the available manipulated variables are the AdBlue dosing quantity $\dot{m}_a$, EGR valve position u_{EGR}, VTG rack position u_{VTG}, and WHR bypass valve position $u_{WHR,EGR}$ and $u_{WHR,exh}$.

To satisfy these requirements, the Integrated Powertrain Control (IPC) approach, which is introduced in [17], is followed. It is a model-based control design philosophy for combined engine-aftertreatment-energy recovery systems that:

– minimizes fuel consumption, while meeting emission constraints;
– offers a robust emission control solution for both test cycles and real-life operation;

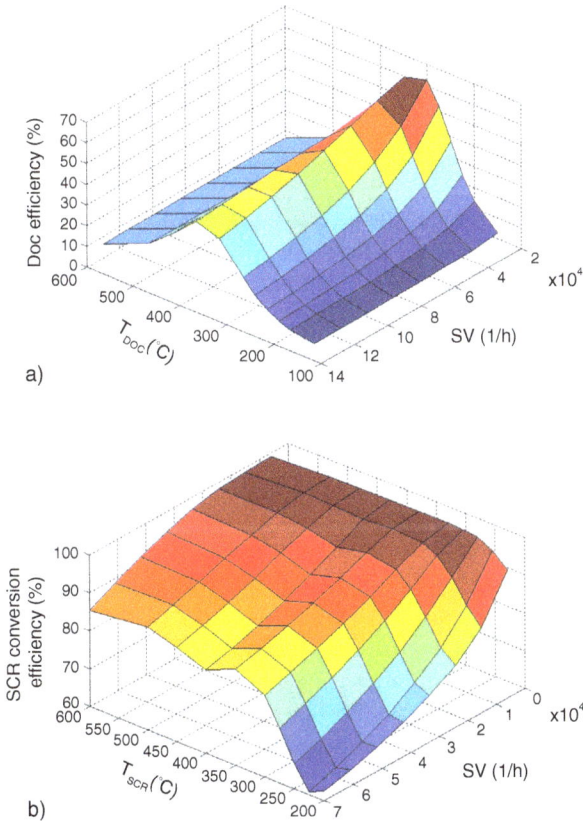

Figure 5

a) DOC efficiency map C_{NO_2}/C_{NO_x} and b) SCR efficiency map for $C_{NO_2}/C_{NO_x} = 0.5$.

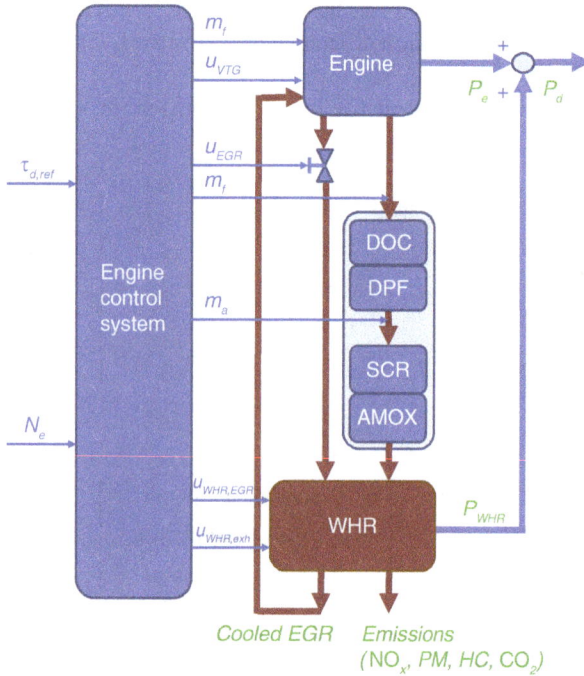

Figure 6

Overview of the engine control problem.

- deals with complex system interactions;
- uses models and optimal control theory to derive optimal control strategies;
- relaxes the calibration complexity.

In this systematic approach, the performance of the separate low-level controllers is coordinated by a supervisory controller, Figure 7. Based on information of the actual status and the driver's torque request, this controller determines the desired control settings using on-line optimization. With the available prediction of NO_x and PM reduction of the DPF-SCR system and WHR power output, the IPC strategy specifies the engine settings that give the required exhaust gas temperature, EGR and air flow, and engine out emissions to minimize operational costs (and thus fuel consumption) within the limits set for tailpipe emissions.

The developed IPC strategy is compared with a baseline engine control strategy, for reference. The examined control strategies are described below.

2.1 Optimization Problem

Following the IPC approach, the studied control problem is formulated in the optimal control framework. We propose to minimize the total operational costs associated with fuel, AdBlue consumption and active

DPF regeneration. Consequently, the following objective function is defined:

$$\min_{u_d} \int_0^{t_e} w(N_e, \tau_d) \cdot \left[\pi_f \dot{m}_f + \pi_a \dot{m}_a + \pi_{PM} \dot{m}_{PM}\right] dt \quad (9)$$

subject to:

$$\frac{\int_0^{t_e} \dot{m}_{NO_{xtp}} dt}{\int_0^{t_e} \frac{P_d}{3.6 \times 10^6} dt} \leq Z_{NOx} \, (\text{tail} - \text{pipe } NO_x \text{ limit}) \quad (10)$$

with Diesel price $\pi_f = 1.34 \times 10^{-3}$ Euro/g, AdBlue price $\pi_a = 0.50 \times 10^{-3}$ Euro/g, and fuel costs associated with active DPF regeneration per gram of accumulated soot $\pi_{PM} = 7.10 \times 10^{-2}$ Euro/g. In this case, the EGR valve position and VTG rack position are the selected decision variables: $u_d^T = [u_{EGR} u_{VTG}]$.

Assuming that all injected urea decomposes in ammonia and is available for NO_x conversion, the desired AdBlue dosage $\dot{m}_a$ (g/s) in Equation (9) is determined from:

$$\dot{m}_a = c_5 \cdot \eta_{SCR}(T_{SCR}, SV, C_{NO_2}/C_{NO_x}) \cdot \dot{m}_{NO_x} \quad (11)$$

where $c_5 = 2.0067$ and $\dot{m}_{NO_x}$ (g/s) is the engine out NO_x emission. With the weighting function $w(N_e, \tau_d)$, it is aimed to capture the desired performance independent of the applied test cycle, see also Section 3.

2.2 IPC Strategy

For this optimization problem, Pontryagin's Minimum Principle is applied to find an optimal solution, see, e.g., [18]. Accordingly, a Hamiltonian is formulated which entails the objective function from Equation (9) augmented with Lagrange multipliers λ and the state dynamics $f(x)$ from Equation (8):

$$H = w(N_e, \tau_d) \cdot \left[\pi_f \dot{m}_f + \pi_a \dot{m}_a + \pi_{PM} \dot{m}_{PM}\right] + \lambda^T f(x) \quad (12)$$

These Lagrange multipliers represent equivalence price parameters and have the following interpretation:
- λ_1 represents a cost-equivalent parameter for a DOC/DPF temperature rise of 1°C within 1 s. A larger value will result in higher T_{DOC};
- λ_2 represents a cost-equivalent parameter for a SCR temperature rise of 1°C within 1 s. By increasing its value, a better heat transfer between DOC/DPF and SCR can be achieved, and so a better SCR conversion efficiency;
- λ_3 takes into account the accumulated tailpipe NO_x emissions. A higher value will more penalize the raw engine out NO_x emissions;

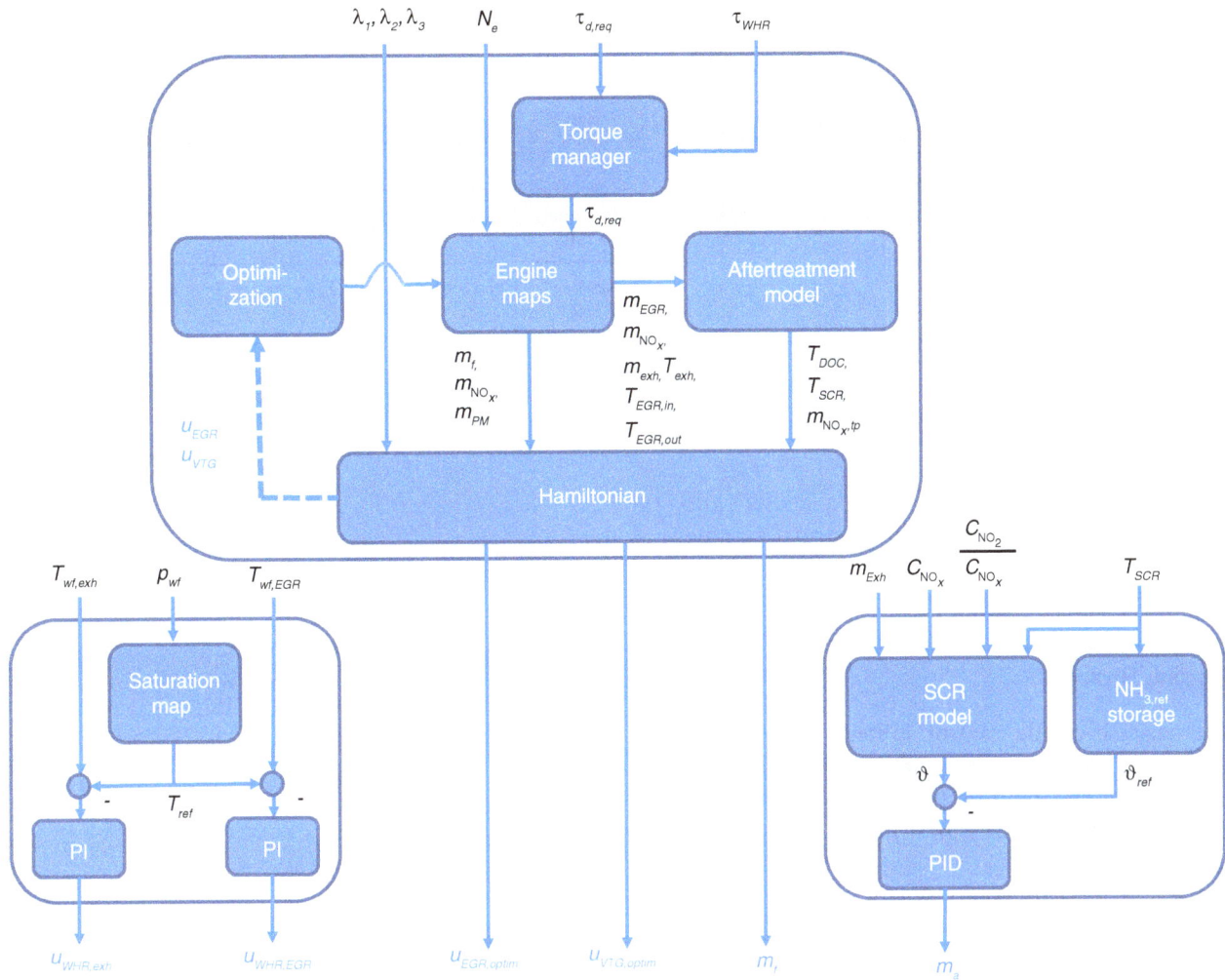

Figure 7

Scheme of the proposed engine control system.

Two necessary conditions for optimality of the solution u_d can be formulated:

$$-\frac{\partial H}{\partial x} = \dot{\lambda} \quad (13)$$

$$\frac{\partial H}{\partial u_d} = 0 \quad (14)$$

From these conditions, it is easily seen that λ_3 remains constant for the optimal solution, and only depends on its initial conditions $\lambda_3(0)$. More important, the dynamics of λ_1 and λ_2 are unstable and have end-point constraints. These two facts make the solution to this optimal control problem difficult to implement in practice, as it requires the entire drive cycle to be known *a priori*.

As we want to use the presented systematic framework, the pragmatic approach that is described in [19] is followed; the course of λ_1 and λ_2 are determined by a heuristic, postulated rule parameterized by λ_T, ΔT_1 and ΔT_2. This approach is illustrated in Figure 8. The effort to heat up the aftertreatment system is assumed to be proportional to the SCR inefficiency $1 - \eta_{SCR}$. When T_{DOC} is lower or marginally higher than T_{SCR}, it seems better to invest in raising the engine-out exhaust temperature rather than promoting heat convection from DOC/DPF to SCR (high λ_1). The converse holds when T_{DOC} T_{SCR} (high λ_2).

This sub-optimal controller is implemented in the presented simulation model. At every time step over the studied test cycle, the Hamiltonian, Equation (12), is

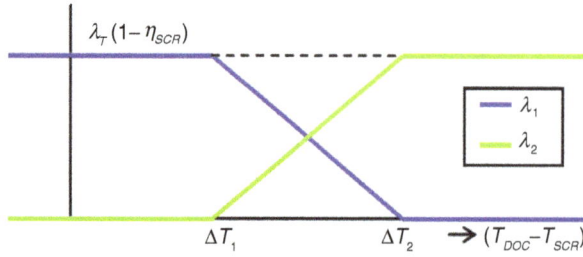

Figure 8

Heuristic rule for λ_1 and λ_2 [19].

TABLE 3
Selected control parameters

Control strategy	Control parameters			
	$-\lambda_{(1,M1)}$	$\lambda_{(1,M2)}$	λ_2	λ_3
Baseline-(WHR)	2.0×10^{-5}	0	0	6.8×10^{-3}
Recal-WHR	1.2×10^{-5}	0	0	4.88×10^{-3}
	$-\lambda_T$	ΔT_1	ΔT_2	λ_3
IPC-WHR	1.6×10^{-5}	67.4	92.5	2.0×10^{-4}

TABLE 4
Emission targets for control design

Cycle	$NO_{x,eo}$ (g/kWh)	η_{SCR} (%)	Weight (%)	$NO_{x,tp}$ (g/kWh)
Cold WHTC	3.5	80	16	0.112
Hot WHTC	3.5	90	84	0.294
Weighted WHTC				0.406

numerically optimized on-line using a bounded 2D gradient descent method for the specified set of Lagrange multipliers λ. In Section 3, the off-line calibration of these multipliers is discussed.

2.3 Baseline Strategy

For the baseline engine control strategy, we mimic a state-of-the-art air management strategy for a standard Euro-VI engine configuration (without WHR system). This strategy is characterized by switching between two control modes:
- thermal management mode (M1) for rapid heat-up of the aftertreatment system ($T_{SCR} < 200°C$);
- low NO_x mode (M2) for normal operation ($T_{SCR} \geq 250°C$).

A fundamental difference with the IPC strategy is that the baseline strategy relies on fixed control settings (u_{EGR}, u_{VTG}) for each engine operating point (N_e, τ_e). For both modes, these settings are pre-determined in an off-line optimization procedure, which is often based on stationary test conditions.

As we want to use the same control structure for both strategies in simulations, two different sets of constant λ are used for the control modes (Tab. 3). The current engine calibration is mainly optimized using steady state measurements. Therefore, anticipated steady-state T_{DOC} and T_{SCR} values from the engine maps are used in the Hamiltonian to evaluate the SCR efficiency maps, instead of the momentary temperatures.

3 CONTROL DESIGN

This section discusses the control design procedure for the applied strategies. An overview of the selected control parameters is given in Table 3.

3.1 Baseline Strategy with WHR System

For the baseline strategy, the control design boils down to the determination of:
- air management: engine maps for EGR valve and VTG settings by specifying the corresponding λ set for the control modes;
- SCR control: θ_{ref} map and PID-control settings;
- WHR control: PI-control settings.

These controllers have to be designed, such that the specified engineering target of 0.41 g/kWh is met. To realize this, a NO_x emission budget and averaged SCR conversion efficiencies η_{SCR} are specified for both cold and hot World Harmonized Transient Cycle (WHTC), Table 4.

3.1.1 Air Management

Following [19], two different sets of constant (λ_1, λ_3) are determined to specify the control modes of the baseline controller. For the low NO_x mode, $\lambda_{1,M2}$ and λ_2 are set to zero (no promotion of aftertreatment heat up), while λ_3 is tuned such that the engine out NO_x emission target is reached over the WHTC. For the thermal mode, λ_3 is kept unchanged, whereas $\lambda_{1,M1}$ is tuned to get maximal T_{exh} increase within the targets set for engine out NO_x emission. This baseline Euro-VI case is the reference for the other studied strategies.

The baseline engine controller is also applied to the engine with WHR system (referred to as Baseline-WHR). In this case, the applied engine maps and controller settings are identical to the baseline strategy. However, the main difference is the implemented torque manager: the requested torque $\tau_{d,req}$ is realized by an ideal torque split, as described by Equation (6). This means that, compared to the baseline case, the engine will run in different operating points depending on the power delivered by the WHR system.

In the baseline-WHR case, the controller does not account for the effect of the WHR system on emissions. This can lead to relatively large deviations from the targets set for emissions. Consequently, this controller is tuned such that powertrain with WHR system is closely meeting the 0.41 g/kWh target again. This case is referred to as Recal-WHR and the corresponding new set (λ_1, λ_3) can be found in Table 3.

3.1.2 Low-Level SCR Control

For AdBlue dosing control, a model-based ammonia storage controller is applied. This low-level controller is based on a SCR catalyst model, which estimates the ammonia storage θ from SCR catalyst temperature T_{SCR} and pre-SCR NO_x emissions $\dot{m}_{NO_x}$ in real-time. This estimated value is compared with a reference value θ_{ref}. The difference is fed to the PID controller. By controlling θ, we aim to achieve high NO_x conversion efficiency and avoid excessive NH_3 slip in case of a sudden temperature increase. More details can be found in [20].

For the standard Euro-VI engine with baseline strategy, the static map $\theta_{ref}(T_{SCR})$ is calibrated, such that tailpipe NO_x emission meets the specified standards over the studied WHTC. Furthermore, cycle-averaged and peak tailpipe NH_3 emissions are kept within 10 and 25 ppm, respectively. The applied θ_{ref} map is shown in Figure 9. This SCR control calibration is used in all simulations.

3.1.3 Low-Level WHR Control

With the introduction of the high fidelity WHR system model, the two working fluid bypass valves $u_{WHR,exh}$ and $u_{WHR,EGR}$ have to be controlled. The main goal of this controller is to maximize power output P_{WHR}, within the

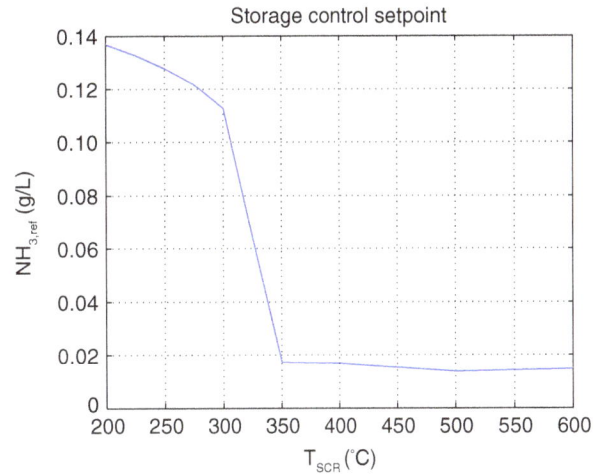

Figure 9

Reference ammonia storage θ_{ref}.

constraints set by safe operation: the WHR system has to produce vapor in order to avoid damaging the expander. In this study, we focus on the power production mode; e.g., start-up and shut down procedures are not considered. Nevertheless, results for the cold WHTC are also presented in Section 4.

Analogue to [12], two parallel PI controllers are implemented to control the post-EGR and post-exhaust evaporator temperature to their desired values (Fig. 7):

$$u_{WHR,i} = K_{P,i} \cdot (T_{ref} - T_{wf,i}) + K_{I,i} \int_0^{t_e} (T_{ref} - T_{wf,i})dt$$

(15)

where $i = \{EGR, exh\}$ and T_{ref} is determined from the saturation vapor curve of the working fluid:

$$T_{ref}(p_{wf}) = T_{sat}(p_{wf}) + \Delta T_{sat}$$

(16)

with safety margin $\Delta T_{sat} = 10°C$.

As in [21], a PI controller with bumpless transfer mode and anti-windup method is implemented. In this study, the WHR control parameters $K_{P,i}$ and $K_{I,i}$ are manually tuned and chosen to be constant over the complete operating envelope, Table 5. For details on the low-level WHR control design, the reader is referred to [22].

3.2 IPC Strategy

For the IPC strategy, the following sets of control parameters have to be specified:

TABLE 5

Selected WHR control parameters

EGR bypass valve		Exhaust bypass valve	
$K_{P,EGR}$	−8	$K_{P,exh}$	−6
$K_{I,EGR}$	−7	$K_{I,exh}$	−6

- weighting function $w(N_e, \tau_d)$;
- Lagrange multipliers and their related variables: $\Delta T_1, \Delta T_2, \lambda_T$ and λ_3.

3.2.1 Weighting Function

Figure 10 shows the applied weighting function $w(N_e, \tau_d)$. For the studied cold and hot WHTC, typical operating points corresponding to high way driving are weighted more heavily, such that more attention is paid to minimize the operational costs during long haul driving conditions.

3.2.2 Lagrange Multipliers

To minimize the objective function over the studied cycle, a numerical minimization method is applied. This method aims to find the control parameters $\Delta T_1, \Delta T_2, \lambda_T$ and λ_3 that minimize the operational costs over the hot WHTC cycle, while the weighted tailpipe NO_x emissions stay within the specified limits. For this purpose, the cumulative cycle costs are evaluated. By applying the Nelder-Mead simplex method, the optimal set of control parameters is found that corresponds to the lowest costs over the studied duty cycle (*Tab. 3*).

4 SIMULATION RESULTS

To evaluate the performance of the four proposed controllers in Table 3, simulations are done over the WHTC, which is shown in Figure 11. This cycle specifies the requested engine speed N_e and torque $\tau_{d,req}$. Three parts can be distinguished: urban driving conditions (0-900 s), rural driving conditions (900-1380 s), and highway driving conditions.

As we focus on Euro-VI emission targets, results are generated for both cold and hot cycle conditions. In case of a cold cycle, the initial temperatures of the aftertreatment components and WHR system are set to 20°C; engine heat up is not modeled yet. According to Euro-VI legislation, results for the cold and hot cycle are combined using weights of 16% and 84%, respectively, to provide the overall cycle-averaged WHTC result. Table 6 summarizes the results of the studied cases.

4.1 Overall Powertrain Results

4.1.1 CO$_2$-NO$_x$ Trade-off

Figure 12 shows the trade-off between the cycle-averaged CO_2 and NO_x emissions. The baseline case, representing a conventional Euro VI engine without WHR system, is used as reference. Results from the other cases are expressed as a percentage of the baseline case. The figure shows that simply adding a WHR system (Baseline-WHR) reduces CO_2 emission with 2.6% and tailpipe NO_x emission with 2.8%. Recalibrating the controller to exploit the tailpipe NO_x margin created by the WHR system (Recal-WHR) yields an extra 0.8% of CO_2 emission reduction (*Tab. 6*).

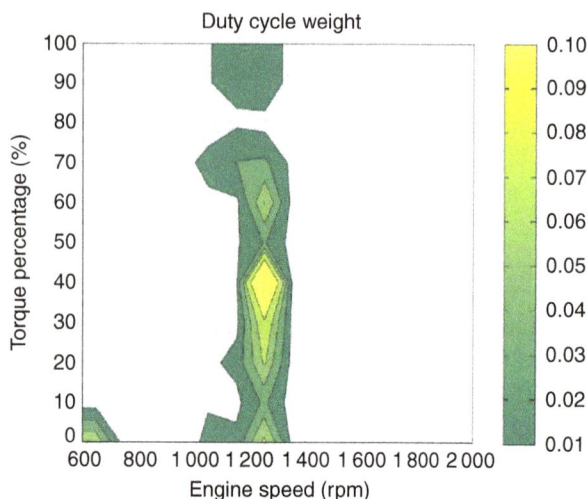

Figure 10

Weighting function $w(N_e, \tau_d)$ [19].

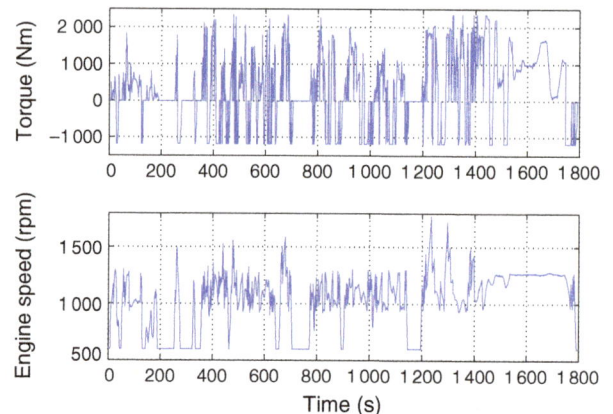

Figure 11

World harmonized transient cycle.

TABLE 6
Overview of WHTC results

Quantity	Control strategy			
	Baseline	Baseline -WHR	Recal -WHR	IPC -WHR
$NO_{x,eo}$ (g/kWh)				
Hot	3.45	3.32	3.38	5.08
Cold	3.61	3.49	3.57	5.06
Weighted	3.48	3.35	3.41	5.08
$NO_{x,tp}$ (g/kWh)				
Hot	0.348	0.338	0.356	0.369
Cold	0.730	0.716	0.689	0.616
Weighted	0.410	0.398	0.410	0.409
$NH_{3,max}$ (ppm)				
Hot	3	3	3	2
Cold	3	3	3	2
CO_2 (%)				
Hot	100	97.3	96.6	94.0
Cold	100	97.6	96.7	93.8
Weighted	100	97.4	96.6	94.0
WHTC Costs (%)				
Fuel	97.2	94.6	93.9	91.3
AdBlue	1.0	1.0	1.0	1.6
PM	1.8	1.8	1.7	1.0
Total	100	97.4	96.6	93.9

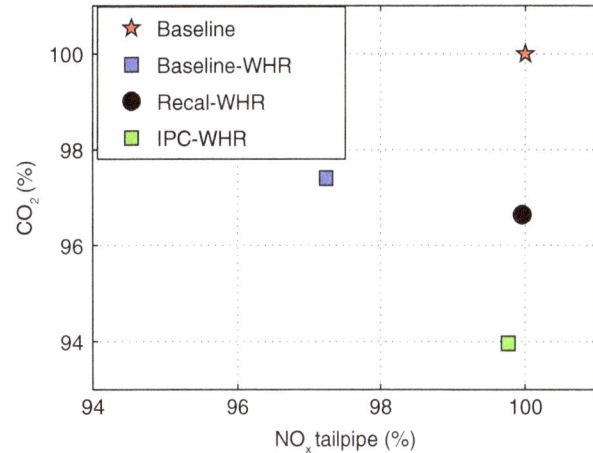

Figure 12

CO_2-NO_x trade-off for WHTC.

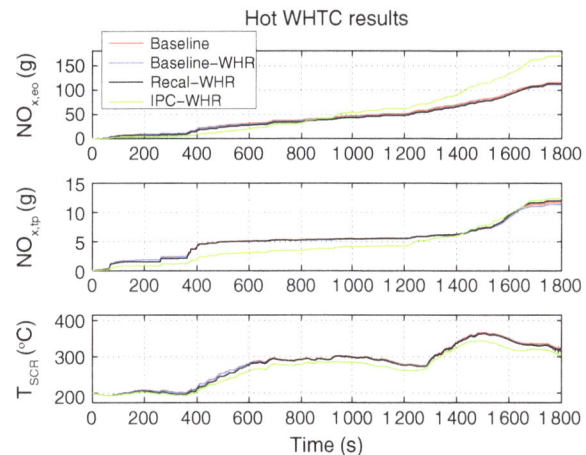

Figure 13

Cumulative emission results and SCR temperature (hot WHTC).

By implementing the IPC strategy on the powertrain equipped with WHR system, a further 2.6% reduction of CO_2 emission is achieved compared to the Recal-WHR case, without increasing tailpipe NO_x emission. This sums up to a total CO_2 reduction of 6% compared to the baseline strategy.

4.1.2 EGR-SCR Balancing

The IPC strategy is able to achieve the additional CO_2 reduction by determining on-line the cost-optimal balance between engine-out NO_x emission and SCR conversion efficiency at every time step: EGR-SCR balancing. From Figure 13, it is seen that the SCR temperature (and conversion efficiency) is relatively low at the start of the WHTC (0-400 s). During this period, the IPC strat-

egy keeps the tailpipe NO_x emission low by applying more EGR and does not promote SCR heat-up, unlike the baseline strategies. Consequently, engine out NO_x emission is significantly lower compared to the baseline strategies. When the SCR temperature increased to a level where NO_x conversion is sufficiently high, the IPC strategy minimizes EGR, such that operational costs are minimized within the emission constraint.

Figure 14 shows the corresponding fuel and AdBlue consumption as well as the resulting operational costs over time. These results are given relative to the baseline. At every time instant t_k, the relative fuel consumption is determined by:

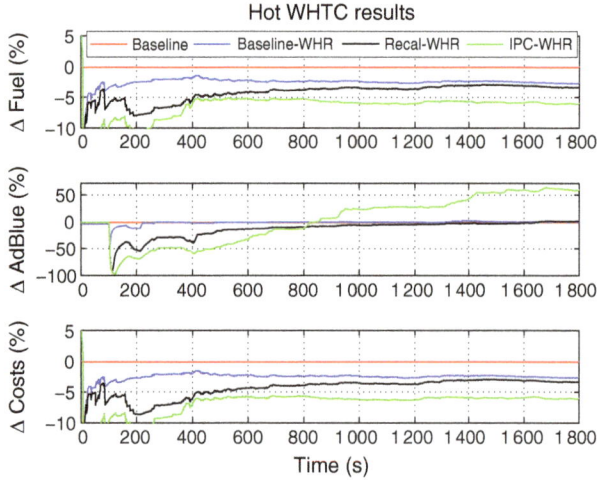

Figure 14

Operational costs and corresponding fuel and AdBlue consumption (hot WHTC). All results relative to the baseline.

$$\Delta Fuel(t_k) = 100 \cdot \frac{\int_0^{t_k} (\dot{m}_f - \dot{m}_{f,Baseline})dt}{\int_0^{t_k} \dot{m}_{f,Baseline}dt} \qquad (17)$$

In a similar way, the relative AdBlue and total operational costs are computed.

In the hot WHTC, the WHR system is active from the onset of the cycle. Due to the typical low load (and thus low waste heat) of the urban part of the WHTC, the WHR output is initially low, Figure 14. At $t = 400$ s, the WHR system alone (Baseline-WHR) has managed to save 1.5% of fuel and operational costs, while the Recal-WHR and IPC-WHR case have both saved around 5%. After this moment, the SCR system is heated up sufficiently; the IPC strategy starts to reduce the amount of EGR in order to reduce fuel consumption. This is done at the cost of increased engine-out NO_x emission and AdBlue consumption.

Over the entire hot WHTC, the IPC strategy managed to save around 6% of fuel, although the AdBlue consumption increased with over 50%. Due to the relatively low AdBlue consumption and costs in combination with reduced DPF regeneration costs, the total operational costs are reduced compared to the baseline strategies. More details on the operational costs can be found in Table 6.

4.1.3 WHR Effect

Figure 15 illustrates the WHR effect on powertrain performance for the hot WHTC. In the upper graph, the ratio of engine power (without WHR) P_e and requested power P_d is shown. It is observed that the WHR contribution is negligible during most of the rural part of the WHTC and during large idling periods (Fig. 11). Momentary, the WHR contribution shoots up to 100%. This occurs when the desired power P_d is very small, such that the WHR system fully produces the desired power. Note that some fuel will still be burnt in the diesel engine, since it still has to overcome (most of) its friction. It is not motoring during these periods as the desired power is still positive. Most WHR power is produced during the highway part of the WHTC, because the supply of waste heat to the evaporators is relatively high and constant. Recall that the IPC strategy is applying less EGR at the highway part, so it actually produces less WHR power compared to the other control strategies.

The two lower graphs of Figure 15 show the engine efficiency and the total powertrain efficiency, respectively:

$$\eta_{e,avg}(t_k) = \frac{\int_0^{t_k} P_e dt}{\int_0^{t_k} P_{fuel} dt}$$

$$\eta_{total.avg}(t_k) = \frac{\int_0^{t_k} (P_e + P_{WHR})dt}{\int_0^{t_k} P_{fuel} dt}$$

with $P_{fuel} = \dot{m}_f \cdot Q_{LHV}$ and the lower heating value Q_{LHV} of Diesel. In these graphs, the moving average of the efficiency is shown relative to the baseline:

$$\Delta\eta_{e,avg}(t_k) = \eta_{e,avg}(t_k) - \eta_{e,avg,Baseline}(t_k) \qquad (18)$$

$$\Delta\eta_{total,avg}(t_k) = \eta_{total,avg}(t_k) - \eta_{total,avg,Baseline}(t_k) \qquad (19)$$

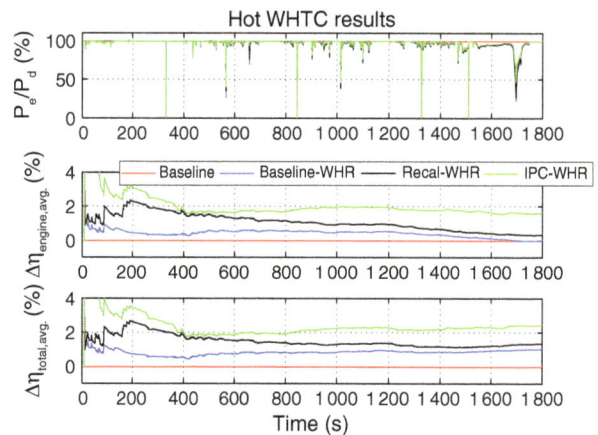

Figure 15

WHR effect on power output and efficiency (hot WHTC).

It is concluded that the IPC strategy is able to achieve a significant improvement in engine efficiency compared to the other strategies. This is the result of EGR-SCR balancing; due to different EGR valve and VTG rack settings, engine efficiency is increased. Comparison of $\eta_{e,avg}$ and $\eta_{total,avg}$ shows that the WHR system only has limited contribution for the IPC case. The Baseline-WHR and Recal-WHR strategies gain most in total efficiency due to the WHR system.

4.2 Low-Level WHR Controller

For the hot WHTC, the functionality of the low-level WHR controllers is illustrated in Figure 16. Based on the difference between $T_{ref}(p_{wf})$ and the actual post-evaporator temperature $T_{wf,i}$, the corresponding bypass valves are controlled. Recall from Equation (16) that the working fluid is in vapor phase, when $T_{wf,i} - T_{ref}(p_{wf}) > -\Delta T_{sat}$. From the two upper graphs, it is seen that both PI controllers are effective over the entire engine operating envelope; $T_{ref}(p_{wf})$ is tracked with a maximum absolute error of 10°C . The exhaust evaporator generates vapor over the entire hot WHTC. However, two phase flow is encountered downstream of the EGR evaporator around $t = 750$-800 s, despite the efforts of the PI controller: the EGR bypass valve is fully opened ($u_{WHR,EGR} = 0\%$). This is due to the low heat input during the preceding engine idling phase.

In general, the heat input during the urban part of the WHTC is relatively low. In this part, the working fluid flow through the evaporators is minimized by fully opening the EGR and exhaust bypass valves. As a result, the WHR power output P_{WHR} is small (Fig. 15). With increasing heat input in the rural and high way part, both bypass valves are gradually closed, especially the exhaust bypass $u_{WHR,exh}$. This leads to an increasing WHR power output: up to 7 kW for the IPC-WHR case. Note that $u_{WHR,EGR}$ is more closed in the baseline-WHR and recal-WHR cases towards the end of the WHTC. In this case, the IPC strategy reduces the EGR mass flow. To keep the working fluid in vapor phase, the controller responds by further opening the bypass valve, such that the EGR evaporator flow is reduced.

4.3 Cold Start WHTC Results

For the cold WHTC, the WHR system first has to heat up the working fluid to a superheated state before it can start to generate power. Although the current low-level WHR controller is not optimized for cold starts, the way it deals with cold starts is more than acceptable.

As shown in Figure 17, the EGR evaporator heats up significantly faster than the exhaust WHR evaporator. This is because the EGR evaporator is exposed to almost instant high temperatures, whereas the exhaust evaporator has to wait for the upstream aftertreatment system to heat up first. The ethanol inside the evaporators starts at 20°C for the cold start. This implies that the initial tracking error of the low level controller is around 70°C.

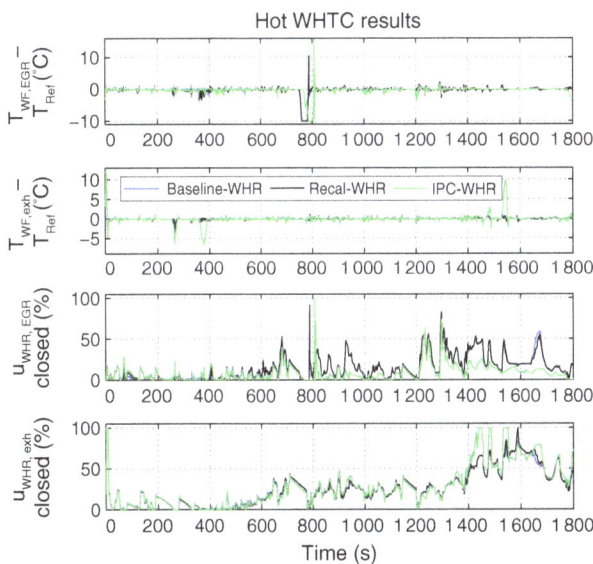

Figure 16

Operation of both PI WHR controllers during hot WHTC.

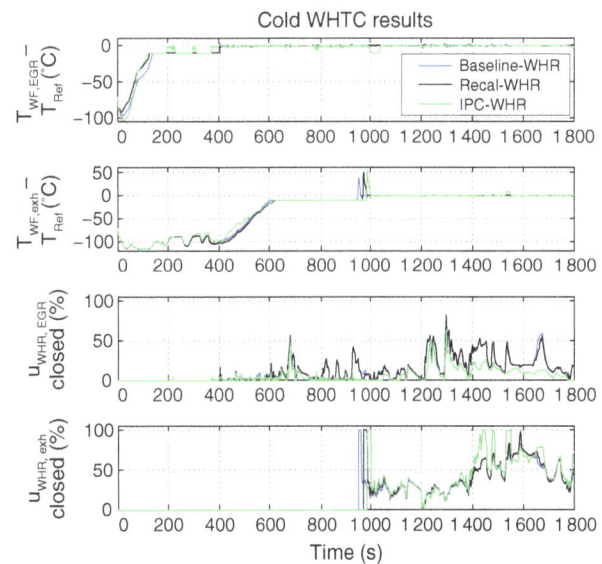

Figure 17

Operation of both PI WHR controllers during cold WHTC.

During heat-up, the low-level WHR controller keeps the bypass valves fully opened, such that a minimum amount of ethanol reaches the evaporators. The temperature of the ethanol in the evaporators rises until the saturation temperature is reached and evaporation starts. After evaporation is complete, the ethanol temperature quickly reaches its setpoint value. Then, the low-level WHR controller starts to close the working fluid bypass valves to increase the flow of ethanol to the evaporators. The second graph of Figure 17 shows that for the exhaust evaporator overshoot occurs at the end of the evaporation phase. Anti-windup is applied in the low-level WHR controllers to prevent excessive overshoot after the end of the evaporation phase. This same graph also shows that the tracking error can increase if the bypass valve saturates: $u_{WHR,exh} = 100\%$ (between 1 400 and 1 600 s).

CONCLUSIONS AND FUTURE WORK

A supervisory controller is presented for an Euro-VI engine with Waste Heat Recovery (WHR) system. This controller is rooted in the IPC approach and integrates energy and emission management. From simulation results over a WHTC, it is concluded that a recalibration of the baseline engine controller is required to use the full CO_2 reduction potential of the WHR system. With the IPC strategy, a systematic approach is introduced, which optimizes the CO_2-NO_x trade-off: additional 2.6% CO_2 reduction compared to the recalibrated baseline strategy (Recal-WHR).

Current research is dedicated to further development of the low-level WHR controller. Furthermore, tests will be performed on an engine dynamometer to demonstrate the potential of the proposed controllers. For the IPC strategy, focus is on the robustness for different duty cycles and on total energy management, which also includes the impact of the WHR system on the cooling system.

REFERENCES

1 ACEA (2011) *Commercial vehicles and CO₂* Report, ACEA.

2 EPA (2011) *EPA and NHTSA adopt first-ever program to reduce greenhouse gas emissions and improve fuel efficiency of medium-and heavy-duty vehicles.* Regulatory Announcement, August.

3 ACEA (2008) Vision 20-20. www.acea.be, September, ACEA Press Conference at the *IAA 2008*, Hanover, Germany.

4 Bredel E., Nickl J., Bartosch S. (2011) Waste heat recovery in drive systems of today and tomorrow, *MTZ Worldwide* **72**, 52-56.

5 Nelson C. (2009) Exhaust energy recovery, *Directions in Engine-Efficiency and Emissions, Research (DEER) Conference*, Dearborn, Michigan, 3-6 Aug.

6 Park T., Teng H., Hunter G.L., van der Velde B., Klaver J. (2011) A Rankine cycle system for recovering waste heat from HD Diesel engines - Experimental results, *SAE Paper* 2011-01-1337.

7 Kupper F. (2012) Integrated Powertrain Control for an Euro-VI heavy-duty Diesel engine with Waste Heat Recovery system, *Master's Thesis*, Eindhoven University of Technology.

8 Abbe Horst T., Rottengruber H.-S., Seifert M., Ringler J. (2013) Dynamic heat exchanger model for performance prediction and control system design of automotive waste heat recovery systems, *Applied Energy* **105**, 293-303.

9 Hou G., Sun R., Hu G., Zhang J. (2011) Supervisory predictive control of evaporator in Organic Rankine Cycle (ORC) system for waste heat recovery, *2011 International Conference on Advanced Mechatronic Systems*, Zhengzhou, China, 11-13 Aug., pp. 306-311.

10 Howell T., Gibble J., Tun C. (2011) Development of an ORC system to improve HD truck fuel efficiency, *Directions in Engine-Efficiency and Emissions, Research (DEER) Conference*, Detroit, Michigen, 3-6 Oct.

11 Quoilin S., Aumann R., Grill A., Schuster A., Lemort V., Spliethoff H. (2011) Dynamic modeling and optimal control strategy of waste heat recovery organic Rankine cycles, *J. Applied Energy* **88**, 6, 2183-2190.

12 Tona P., Peralez J., Sciarretta A. (2012) Supervision and control prototyping for an engine exhaust gas heat recovery system based on a steam Rankine cycle, *2012 IEEE/ASME International Conference on Advanced Intelligent Mechatronics*, Kaohsiung, Taiwan, 11-14 July, pp. 695-701.

13 Hounsham S., Stobart R., Cooke A., Childs P. (2008) Energy recovery systems for engines, *SAE Paper* 2008-01-0309.

14 Willems F., Kupper F., Cloudt R. (2012) Integrated energy & emission management for heavy-duty Diesel engines with waste heat recovery system, Proceedings of the *2012 IFAC Workshop on Engine and Powertrain Control, Simulation and Modeling (ECOSM'12)*, Rueil-Malmaison, France, 23-25 Oct., pp. 203-210.

15 Cloudt R., Saenen J., van den Eijnden E., Rojer C. (2010) Virtual exhaust line for model-based Diesel aftertreatment development, *SAE Paper* 2010-01-0888.

16 Feru E., Kupper F., Rojer C., Seykens X., Scappin F., Willems F., Smits J., de Jager B., Steinbuch M. (2013) Experimental validation of a dynamic waste heat recovery system model for control purposes, *SAE Paper* 2013-01-1647.

17 Willems F., Foster D. (2009) Integrated Powertrain Control to meet future CO_2 and Euro-6 emission targets for a Diesel hybrid with SCR-deNOx system, *IEEE Proc. of 2009 American Control Conference*, St. Louis, MO, USA, 10-12 June, pp. 3944-3949.

18 Geering H. (2007) *Optimal control with engineering applications*, Springer Verlag.

19 Cloudt R., Willems F. (2011) Integrated Emission Management strategy for cost-optimal engine-aftertreatment operation, *SAE International Journal of Engines* **4**, 1, 1784-1797.

20 Willems F., Cloudt R. (2011) Experimental demonstration of a new model-based SCR control strategy for cleaner heavy-duty diesel engines, *IEEE Transactions on Control Systems Technology* **19**, 5, 1305-1313.

21 Pramudya Indrajuana A. (2012) Control development for Waste Heat Recovery system on heavy duty trucks, *Master's Thesis*, Delft University of Technology.

22 Rascanu G. (2013) Integrated Powertrain Control for truck engines with Waste Heat Recovery system, *Master's Thesis*, Eindhoven University of Technology.

11

Design Methodology of Camshaft Driven Charge Valves for Pneumatic Engine Starts

Michael M. Moser*, Christoph Voser, Christopher H. Onder and Lino Guzzella

Institute for Dynamic Systems and Control, ETH Zurich, Sonneggstrasse 3, 8092 Zurich - Switzerland
e-mail: mimoser@ethz.ch - voserc@ethz.ch - onder@ethz.ch - lguzzella@ethz.ch

* Corresponding author

Abstract — *Idling losses constitute a significant amount of the fuel consumption of internal combustion engines. Therefore, shutting down the engine during idling phases can improve its overall efficiency. For driver acceptance a fast restart of the engine must be guaranteed. A fast engine start can be performed using a powerful electric starter and an appropriate battery which are found in hybrid electric vehicles, for example. However, these devices involve additional cost and weight. An alternative method is to use a tank with pressurized air that can be injected directly into the cylinders to start the engine pneumatically. In this paper, pneumatic engine starts using camshaft driven charge valves are discussed. A general methodology for an air-optimal charge valve design is presented which can deal with various requirements. The proposed design methodology is based on a process model representing pneumatic engine operation. A design example for a two-cylinder engine is shown, and the resulting optimized pneumatic start is experimentally verified on a test bench engine. The engine's idling speed of 1200 rpm can be reached within 350 ms for an initial pressure in the air tank of 10 bar. A detailed system analysis highlights the characteristics of the optimal design found.*

Résumé — **Méthodologie pour le design des valves de chargement opérées par arbre à cames** — Les pertes à vide représentent une partie essentielle de la consommation des moteurs à combustion interne. La mise à l'arrêt du moteur pendant la marche à vide peut, par conséquent, en améliorer son efficacité générale. Pour être accepté par le conducteur, le redémarrage du moteur doit être rapide. On peut réaliser ce démarrage rapide du moteur, moyennant un démarreur électrique puissant conjointement avec un accumulateur approprié, solution retenue par exemple, pour les véhicules à système hybride électrique. Cependant, ces derniers augmentent le coût et le poids. Une alternative consiste dans le démarrage pneumatique du moteur en utilisant de l'air comprimé stocké dans un réservoir sous pression et injecté directement dans les cylindres. Cette étude présente le démarrage pneumatique du moteur en utilisant des valves de chargement commandées par arbre à cames. On présente une méthodologie visant à une consommation de l'air optimale en mesure de respecter des exigences différentes. La démarche proposée s'appuie sur le modèle d'un processus représentant l'opération pneumatique du moteur. La vérification expérimentale du démarrage pneumatique est réalisée et optimisée sur un moteur 2 cylindres sur banc d'essai. Avec une pression initiale de 10 bar dans le reservoir d'air, la vitesse de rotation à vide de 1 200 tr/min peut être atteinte en 350 ms. Une analyse détaillée confirme les caractéristiques du système optimisé.

INTRODUCTION

An inherent property of Internal Combustion Engines (ICE) is their limited operability below a minimum rotational speed. Starting an engine involves accelerating it up to a specific minimal speed of operation. Due to its inertia and the low starter power, a conventional start of an ICE takes up to 1 second. An engine shutdown during idling phases implying such long start times is not accepted by the driver. Therefore, conventional ICE are typically not shut down during idling phases.

However, idling losses constitute a significant amount of the total fuel consumption. During the New European Driving Cycle (NEDC), they amount to 4-8% depending on the engine type and size as shown in [1-4]. In order to exploit this fuel saving potential by eliminating the idling phases while still satisfying driver demands, the duration of an engine start needs to be reduced.

A common approach to reducing the engine start time is the installation of a powerful Electric Starter (ES) and an appropriate battery. This strategy can be pursued with hybrid electric vehicles since they are equipped with more powerful electric motors and batteries than conventional engines. However, they induce additional weight and cost. In [5], it is shown that for this setup start times as low as 300 ms are achievable.

Several authors, *e.g.* those of [3, 6-8], investigated a method for fast starts of gasoline engines with direct fuel injection. The idea behind that approach is to inject fuel into the stopped engine. If the engine has stopped at an appropriate position, the fuel can be ignited at standstill to restart the engine without using the ES. However, a controlled engine shutdown is essential to enforce the engine to stop at an appropriate position. Robustness under all operating conditions and emissions due to incomplete combustion at low engine speeds are critical issues with this approach. The authors of [5] extended the investigation to ES-assisted engine starts to reduce the start time and the emissions.

A cost and weight effective alternative is to use compressed air that can be injected directly into the cylinders. The compressed air is stored in a tank. The tank can be recharged by running the engine as a piston compressor (as *e.g.* done in hybrid pneumatic engines [1]) or by an additional compressor which can be engaged to the crank shaft or which is driven electrically. For the rest of the paper, it is assumed that the compressed air is available and the recharging method is not discussed further. Figure 1 shows a schematic representation of the engine setup considered. Figure 2 shows the setup of the engine on the testbench on which the measurements presented later in this paper are performed.

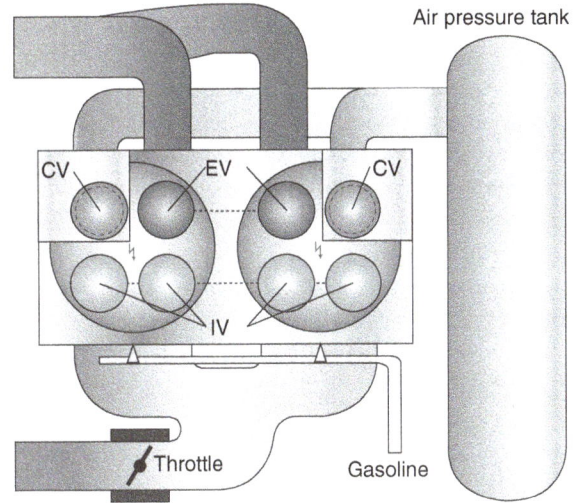

Figure 1

Schematic illustration of the engine setup. A two-cylinder engine with one Exhaust Valve (EV) and two Intake Valves (IV) per cylinder is shown. One EV per cylinder is replaced by a Charge Valve (CV).

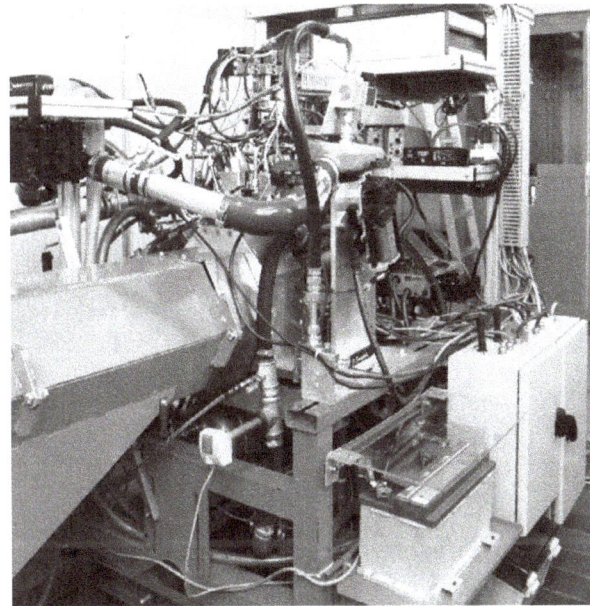

Figure 2

Photograph of the engine including the testbench equipment on which the measurements presented in this paper are performed.

To start the engine pneumatically, air is injected during the expansion stroke where it produces a positive torque that accelerates the engine. The engine is then driven purely pneumatically without burning any fuel. In [9], pneumatic engine starts using a fully variable valve system for the actuation of the charge valves (CV) are investigated. Using this setup, the CV lift profile can be adapted on a cycle-to-cycle basis. However, a fully variable valve system increases the complexity and cost of the system.

In this paper, pneumatic engine starts using camshaft driven CV are investigated. Camshaft driven CV are characterized by a valve lift profile which cannot be adapted during the operation of the engine. The CV lift profile must be defined during the design process. Furthermore, the camshaft driven valve system has an on/off capability, i.e. the valves can either be activated or deactivated. The level of complexity can be significantly reduced by using camshaft driven CV instead of fully variable CV. The goal of this work is to provide a methodology for the optimization of the CV design for minimum air consumption with the constraint of a maximally allowable engine start time. Parts of this work were already presented in [10], the content of which was revised and extended with a detailed system analysis for various operating conditions.

The paper is structured as follows: in Section 1, the model and the setup of the engine as well as the boundary conditions for the optimization of the CV design are introduced. Additionally, an engine stop strategy is proposed. In Section 2, the design problem is formulated as a constrained minimization problem. A design methodology to solve this optimization problem is presented. Section 3 shows the application of this methodology in a design example whose results are verified experimentally on a test bench engine. The CV design found is analyzed with respect to varying initial engine positions, tank pressures and tank temperatures. The paper finishes with a conclusion summarizing the contribution of this work and an outlook on future research.

1 MODEL AND SETUP

For the determination of the optimal CV parameters a process model of the engine is used to simulate the start. A detailed description of the model can be found in Appendix A. This section deals with the electric starter, the parameters of the valve lift profile and the role of the initial engine position. The influence of the pneumatic start on the catalyst temperature is a practical issue which is briefly discussed in Appendix B.

Note that for the crank angle $\phi = 0°$, the piston is located at the Top Dead Center (TDC) after the compression stroke. Crank angles with values of $\phi > 0°$ are located after TDC.

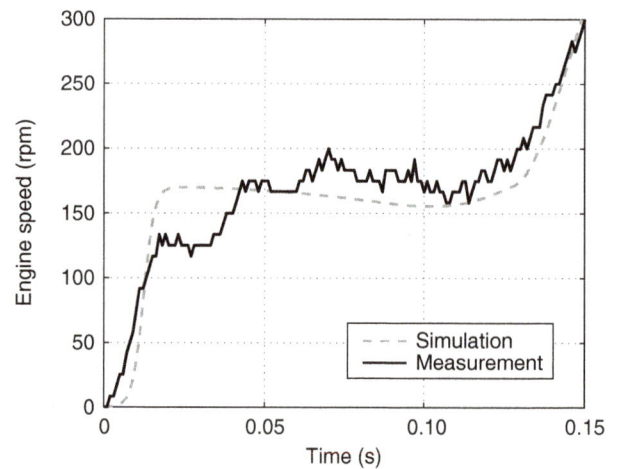

Figure 3

Graph of the measured and simulated engine speed during an engine start (first half of the first engine revolution, where the ES is active). The measurements are taken on the engine described in Section 3.

1.1 Electric Starter

In contrast to [9], this paper focuses on camshaft driven CV. In such a realization, the valve lift profile is fixed. Furthermore, the valve can only be actuated by a rotating engine. For the initial actuation of the CV and for other reasons as described in [7] an ES is essential. Furthermore, the ES helps to reduce the start time since it serves as an additional torque supplier during the pneumatic engine start.

In order to precisely predict the behavior of the engine during the start phase, a good model of the ES is crucial. Since the operation of the ES takes place under highly dynamic conditions, the model of the ES needs to be identified during transient operation. The behavior of the ES is modeled as a static torque-speed relationship $T_{ES}(\omega_e)$ for engine speeds below 200 rpm. For higher engine speeds the engine and the ES are assumed to be disconnected. A least squares approach is used to find a polynomial fit for the torque-speed relationship. Figure 3 shows the experimental validation of the relationship found by contrasting the simulated and the measured engine speed trajectories. Only the first half of the first engine revolution is shown because in this phase the dynamics of the ES are dominant.

1.2 Valve Lift Profile

The valve lift profile is assumed to have a simplified valve acceleration profile in the Crank Angle (CA) domain, i.e.

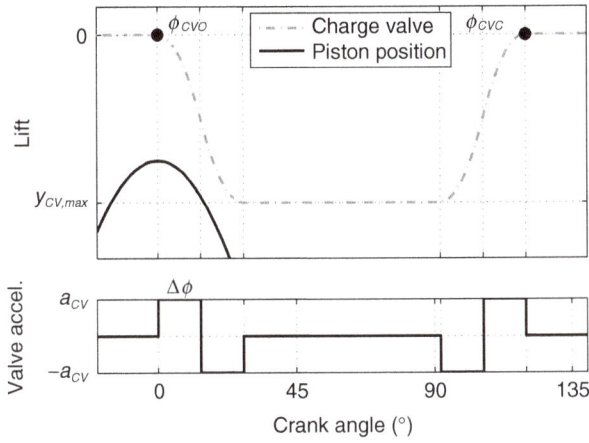

Figure 4

Sample CV lift and acceleration profile.

the magnitudes of the acceleration and the deceleration are equal and piecewise constant. Accordingly, the valve lift profile $y_{CV}(\phi)$ becomes a function of the acceleration in the CA domain a_{CV}, the CV opening angle ϕ_{CVO}, the CV closing angle ϕ_{CVC} and the maximum valve lift $y_{CV,max}$ that can be performed with the actuator chosen. The angular durations of the acceleration crenels are:

$$\Delta\phi = \sqrt{\frac{y_{CV,max}}{a_{CV}}} \qquad (1)$$

To avoid the collision of the valve with the piston the following condition must be fulfilled:

$$y_{CV}(\phi) \leq \min\left\{\frac{V_{cyl}(\phi)}{A_{cyl}}, y_{CV,max}\right\} \qquad (2)$$

where $V_{cyl}(\phi)$ is the cylinder volume at the CA position ϕ and A_{cyl} is the cross-sectional area of the cylinder. Figure 4 shows a sample CV lift profile.

To minimize throttling losses across the CV, the valve should be opened and closed as fast as possible. The valve acceleration thus has to be maximal. In contrast to the Intake Valves (IV) and the Exhaust Valves (EV), the CV is only actuated at low engine speeds. Since mechanical stresses apply in the time domain, the maximum CV acceleration in the CA domain can be chosen higher than that of the IV and EV. A reasonable value for the CV acceleration a_{CV} can be found with:

$$a_{CV} = a_{IV} \cdot \left(\frac{\omega_{e,max}}{\omega_{e,s,max}}\right)^2 \qquad (3)$$

where a_{IV} denotes the IV acceleration in the CA domain, $\omega_{e,max}$ is the maximum engine speed and $\omega_{e,s,max}$ is the maximally allowable engine speed during the pneumatic start operation.

Given the maximum lift and the valve acceleration, the remaining parameters of the CV lift profile are ϕ_{CVO} and ϕ_{CVC}. A methodology to find appropriate values is presented in Section 2.

1.3 Initial Engine Position

The duration of the pneumatic engine start strongly depends on the time that elapses until pressurized air is injected for the first time. The sooner pressurized air is injected, the faster the engine starts. The duration until the CV opens at ϕ_{CVO} is mainly influenced by the size of the ES and the initial engine position ϕ_0. Hence, the engine should be shut down in a way such that the piston which is in the compression stroke comes to rest closely to the CV opening angle.

The rest position can be influenced by the amount of air that is sucked into the cylinder during the intake stroke. A higher air mass inside the cylinder results in more compression work to be done during the compression stroke. Hence, the piston comes to rest more closely to the Bottom Dead Center (BDC). For smaller air masses inside the cylinder the gas spring force induced by the compression is smaller resulting in piston rest positions closer to TDC.

The actuation of the throttle during the engine shutdown can be used to limit the pressure in the intake manifold and thus to adjust the amount of air that enters the cylinders. For engines equipped with a variable valve timing system, the variability of the opening duration of the IV can also be used to adjust the amount of air inside the cylinders.

Since the engine considered in this paper (for details, see Sect. 3) is not equipped with a variable valve timing system for the IV, the throttle is used to adjust the engine rest position. To find an appropriate value for the initial engine position in the model, several experiments are conducted. Several engine shutdowns with constant throttle opening are performed on the warmed-up engine and the rest position is recorded. The shutdowns are initialized at the engine's idling speed of 1 200 rpm. Figure 5 shows the results for the two-cylinder parallel twin engine considered. The best performance is achieved with a throttle opening of 2% because the variance is smallest and the mean position is closest to the TDC. Higher values yield a rest position which is further away from the TDC. Smaller throttle opening values yield a poor repeatability of the rest position, which is not

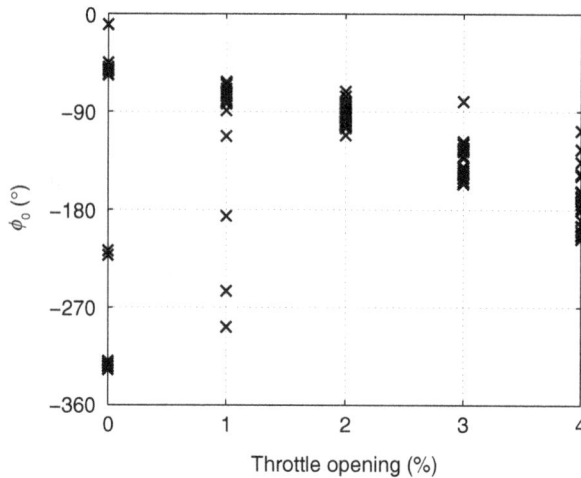

Figure 5

Measured engine rest positions for various throttle openings during engine shutdown. For details concerning the engine see Section 3.

desirable. Thus, an initial position of $\phi_0 = -95°$ CA is chosen.

1.4 Throttle Control during Engine Start

The throttle position setting found above is only relevant during engine shutdown. As shown in [9], completely closing the throttle during the engine start phase is air- and time-optimal. A closed throttle implies that the intake manifold pressure decreases during the start. Hence, the amount of air inside the cylinder is reduced, which causes less compression work to be required. Minimizing the compression work results in minimal negative torque during the engine start, which leads to shorter start times. Therefore, for the design methodology described below the throttle is always assumed to be completely closed during the engine start.

If a variable valve timing system for the IV is available, the compression work is minimized by setting the opening duration of the IV to the minimum.

2 DESIGN METHODOLOGY

In this section, the underlying optimization problem of the CV design is stated. The optimization problem is analyzed and discussed. A stepwise procedure for its solution is presented. The feasible set satisfying all the constraints is determined and specific design choices are introduced.

The objective of the proposed design methodology is to find values for the design variables which enable the engine to reach a prescribed start speed $\omega_{e,s}$ within less than a prescribed start time $t_{s,max}$ and which at the same time minimize the amount of pressurized air used. The relevant design variables are: ϕ_{CVO}, CV opening angle, ϕ_{CVC}, CV closing angle, d_{CV}, CV diameter.

Let $\Delta = \{\phi_{CVO}, \phi_{CVC}, d_{CV}\} \in \mathbb{R} \times \mathbb{R} \times \mathbb{R}^+$ be the parameter space considered for the relevant design variables. The variable δ denotes a single design in the set Δ.

The start time and the air consumption also depend on the initial tank pressure p_t. During the operation of the engine on a drive cycle, the tank pressure varies. The variation of the tank pressure is induced by the emptying and recharging of the tank caused by operating modes that use or provide pressurized air, respectively. Fully variable valves can adjust their valve timing to account for the changing tank pressure. However, the valve timing of camshaft driven CV is fixed. Hence, the design procedure has to consider the entire operating range of the tank pressure. To that end, the weighted sum of the amounts of air consumption resulting for n different tank pressures $\vec{p}_t \in \mathbb{R}^{n,+}$ is minimized.

The tank temperature impacts the start time and the air consumption as well, due to its influence on the mass transfer into the cylinders. However, for the proposed design methodology the tank temperature is assumed to be constant due to the following reasons. According to Equation (A2) the mass flow only depends on the square root of the tank temperature in contrast to the tank pressure on which the mass flow depends linearly. Furthermore, the relative temperature change of the air tank during the operation of the engine is significantly smaller than the relative tank pressure change. Due to the uninsulated tank, the temperature varies according to measurements between 10°C and 50°C or 283 K and 323 K, respectively, i.e. ±7%, whereas the tank pressure varies between 6 bar and 14 bar, i.e. ±40%. The influence of the air temperature in the tank on the resulting optimal CV design is analyzed in Appendix C. It turns out that the assumption of a constant air temperature in the tank is justified.

Considering only the variation in the tank pressure, the optimization problem can be written as:

$$\min_{\delta \in \Delta} \sum_{i=1}^{n} w(\vec{p}_t(i)) \cdot m_a(\delta, \vec{p}_t(i)) \text{ s.t. } t_s(\delta, \vec{p}_t(i)) \leq t_{s,max}$$

$$(4)$$

where $w(\vec{p}_t(i))$ are the pressure dependent weighting factors, t_s is the time needed to reach the desired engine speed $\omega_{e,s}$ and m_a denotes the amount of air used for the start. The values of t_s and m_a are calculated with the nonlinear process model f_{PM}:

$$[m_a(\delta, p_t), t_s(\delta, p_t)] = f_{PM}(\delta, p_t, \omega_{e,s}) \qquad (5)$$

The air consumption m_a is calculated by taking the integral over the mass flow through the CV $\dot{m}_{CV}$:

$$m_a = \int_0^{\hat{t}_s} \dot{m}_{CV} \, dt \qquad (6)$$

where $\hat{t}_s$ implies an extension of the integration interval to $\hat{t}_s \geq t_s$. It is defined as the smallest value of $\hat{t}_s$ satisfying the following constraints:

$$\omega_e(\hat{t}_s) \geq \omega_{e,s} \qquad (7)$$

$$\dot{m}_{CV}(\hat{t}_s) = 0 \qquad (8)$$

Hence, the variable $\hat{t}_s$ defines the instant when the CV are closed and the pneumatic start is completed. This definition accounts for the fact that the CV cannot be closed and deactivated immediately when $\omega_e = \omega_{e,s}$ is reached. In contrast to the fully variable valves used in [9], the camshaft driven CV can only be deactivated if $\phi \notin [\phi_{CVO}, \phi_{CVC}]$. Figure 6 visualizes the definitions of t_s and $\hat{t}_s$ for a pneumatic engine start.

The scalar weighting function $w(p_t) \geq 0$ only depends on the initial tank pressure. It is used to select and penalize specific initial tank pressures in the optimization. Promoting high initial tank pressures results in an

inferior design for low initial tank pressures and *vice versa*. A reasonable choice of $w(p_t)$ is to make it large at low initial tank pressures. For low initial tank pressures, the air consumption can become a critical issue during a drive cycle. On the other hand, if the initial tank pressure is high, an increased air consumption can be accepted. Special attention should be paid to frequently occurring initial tank pressures during the drive cycle.

2.1 Properties of the Optimization Problem

Due to the reciprocating behavior of the engine and the highly nonlinear process model f_{PM} the optimization problem has special features. These are discussed below.

The feasible set is defined as:

$$\Omega(p_t) := \{\Delta | t_s(\Delta, p_t) \leq t_{s,max}\} \qquad (9)$$

i.e., it contains all designs which fulfill the performance constraint for the initial tank pressure p_t. The smallest initial tank pressure for which $\Omega(p_t)$ is not empty is defined as the minimum tank pressure $\tilde{p}_{t,\Omega}$ of the feasible set. The tank pressures considered have to fulfill the inequality $\vec{p}_t(i) \geq \tilde{p}_{t,\Omega} \, \forall i$. However, since the minimum tank pressure is not known *a priori*, the choice of $\vec{p}_t$ is difficult. This fact makes it also difficult to formulate a reasonable weighting function $w(p_t)$, which ensures that the correct tank pressure range is promoted or penalized.

Furthermore, the reciprocating behavior of the engine implies a further difficulty for the solution of the optimization problem. The function $[m_a, t_s] = f_{PM}$ is piecewise continuous in both the air mass and the start time. This fact is shown by introducing the following consideration. Let δ_1 and δ_2 be two designs with $||\delta_1 - \delta_2||_\infty < \varepsilon$ where $||\cdot||_\infty$ denotes the infinity norm and $\varepsilon > 0$ is a very small number. Hence, the two designs differ just slightly. However, for specific choices of δ_1 and δ_2 the evaluation of $f_{PM}(\delta_i, p_t)$ for the same initial tank pressure yields completely different results. Such a discrepancy occurs when $\omega_{e,s}$ is reached in different numbers of power strokes: $N_{ps}(\delta_1) \neq N_{ps}(\delta_2)$. Thus, there is a discontinuity in m_a and t_s between δ_1 and δ_2. Discontinuities in the objective function can cause problems with numerical optimization algorithms. Further details on this discontinuity property are shown in Section 3.3 for a design example.

2.2 Solution to the Optimization Problem

Depending on the objective of the optimization and the prior knowledge two solving methods are proposed, namely numerical optimization and a brute-force approach. For the numerical optimization, the Particle

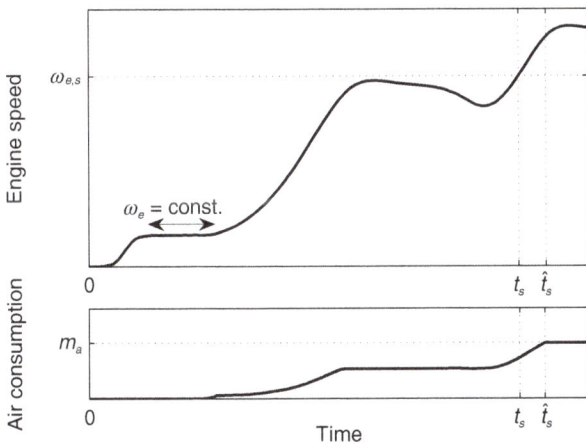

Figure 6

Engine speed and cumulative air consumption for a pneumatic engine start. During the constant speed phase, the engine is only driven by the ES. The variables t_s and $\hat{t}_s$ are labeled to clarify their definitions.

TABLE 1

Comparison of the solving methods

Method	Numerical optimization	Brute-force
Min. tank pressure	Must be known	Is found
Calculation time	Fast	Slow
Precision of solution	High	Lower
Sensitivity analysis	Not suited	Suited

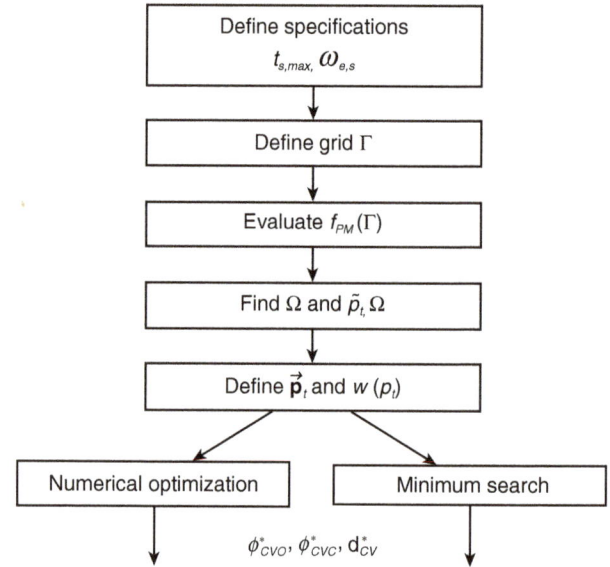

Figure 7

CV design procedure for pneumatic engine start.

Swarm Optimization (PSO) is proposed since it copes well with the properties of the problem described above. Table 1 lists the properties of the proposed methods. The numerical optimization is suited if the minimum tank pressure is known, such that $\vec{p}_t$ can reasonably be selected. The derivation of the optimal solution is rather fast, and a precise solution is found.

In the brute-force approach the process model is simulated for a multitude of CV design variables and tank pressures. It is computationally more demanding. Of course, the computational demand depends on the coarseness of the considered grid. The advantage of this method lays in the fact that the minimum tank pressure does not need to be known in advance. It is found within the approach. Furthermore, the simulation results can be used for sensitivity analyses. The precision of the solution is limited by the coarseness of the grid.

In the following, the two solution methods are explained. Regardless of the solving method, the specifications, i.e. $\omega_{e,s}$ and $t_{s,max}$, have to be formulated first.

2.2.1 Numerical Optimization

If the minimum tank pressure $\tilde{p}_{t,\Omega}$ is known, a reasonable tank pressure range $\vec{p}_t$ and a reasonable weighting function can be defined. For the solution the PSO algorithm presented in [11] is proposed.

2.2.2 Brute-Force

The brute-force approach visualized in Figure 7 starts with the definition of a grid $\Gamma = \{\Delta, p_t\}$ on which the process model is evaluated. Then, the evolution of the process model on Γ follows. The computational effort for the evaluation depends on the discretization chosen. Given $t_s(\Gamma) = f_{PM}(\Gamma, \omega_{e,s})$, the feasible set Ω and the minimum tank pressure $\tilde{p}_{t,\Omega}$ can be derived. A reasonable tank pressure range $\vec{p}_t$ and a reasonable weighting function thus can be defined.

The optimal solution $\delta^* = \{\phi^*_{CVO}, \phi^*_{CVC}, d^*_{CV}\}$ is found by evaluating Equation (4) on the feasible set Ω. This

minimum search is computationally not demanding if $\vec{p}_t \in \Gamma$, i.e., if $\vec{p}_t$ consists only of tank pressures that are also part of Γ. Then, no further model evaluations are necessary. Alternatively, a numerical optimization can be conducted with the defined tank pressure range and weighting function. It is computationally more demanding than the minimum search on an already existing grid. However, the solution is more precise.

3 DESIGN EXAMPLE

The following design example shows the application of the CV design methodology presented. The engine under consideration is a two-cylinder parallel twin engine. For more details on the engine, see [9] where the same engine was used. The specifications of the engine are given in Table 2. The maximum CV lift is $y_{CV,max} = 4\,mm$, which corresponds to the value that can be performed on the test bench engine. The IV acceleration a_{IV} that guarantees safe operation up to the maximum engine speed $\omega_{e,max}$ is given in Table 2. By applying Equation (3), the maximally allowable CV acceleration in the CA domain is found to be $a_{CV} = 3.04 \times 10^{-5}\,m/^\circ CA^2$.

Specifications

The desired start engine speed is set to $\omega_{e,s} = \omega_{e,idle} = 1\,200\,rpm$. The maximally allowable start

TABLE 2

Design example parameterization

Parameter	Variable	Value	Unit
Number of cylinders	N_{cyl}	2	–
Displacement	V_d	0.75	l
Bore diameter	B	85	mm
Stroke	S	66	mm
Connecting rod	l	115	mm
Compression ratio	ε	9	–
Idling speed	$\omega_{e,idle}$	1 200	rpm
Max. engine speed	$\omega_{e,max}$	6 000	rpm
Max. start engine speed	$\omega_{e,s,max}$	1 500	rpm
Number of IV per cyl.	N_{IV}	2	–
Number of EV per cyl.	N_{EV}	1	–
Number of CV per cyl.	N_{CV}	1	–
IV closing CA	ϕ_{IVC}	−114	°CA
EV opening CA	ϕ_{EVO}	114	°CA
IV acceleration	a_{IV}	0.19×10^{-5}	m/°CA^2
CV acceleration	a_{CV}	3.04×10^{-5}	m/°CA^2
CV diameter	d_{CV}	19.0	mm
Maximum valve lift	$y_{CV,max}$	4.0	mm
Initial engine position	ϕ_0	−95	°CA
Tank volume	V_t	30	l
Tank temperature	ϑ_t	50	°C

TABLE 3

Grid Γ used in the design example.

Variable	Lower bound	Upper bound	Step size
ϕ_{CVO}	−110°CA	50°CA	10°CA
ϕ_{CVC}	60°CA	220°CA	10°CA
d_{CV}	7 mm	21 mm	2 mm
p_t	7 bar	12 bar	1 bar

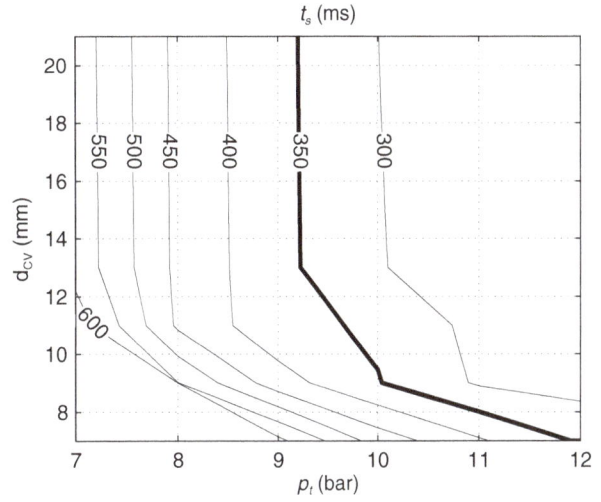

Figure 8

Minimum start time for various CV diameters and initial tank pressures.

time is chosen as $t_{s,max} = 350$ ms. These specifications are in accordance with the values used in [5] and [9].

Grid Definition

The grid Γ considered in the design example is given in Table 3.

Evaluation of the Process Model

In this step, the nonlinear process model f_{PM} is evaluated for Γ, which involves 13 872 model evaluations for the grid considered.

Determination of the Feasible Set and its Minimum Initial Tank Pressure

The analysis of the simulation results yields a minimum initial tank pressure of $\tilde{p}_{t,\Omega} \approx 9.2$ bar for $d_{CV} > 13$ mm.

Figure 8 visualizes this result. It shows the minimum start time achievable for fixed combinations of d_{CV} and p_t. The bold 350 ms line indicates the boundary of the feasible set in the p_t-d_{CV} subspace. All combinations of initial tank pressures and CV diameters with $t_s \leq 350$ ms are part of the feasible set.

On the test bench, CV with a diameter of $d_{CV} = 19$ mm are installed. This diameter is thus chosen for the design example and kept as a fixed value throughout the rest of the optimization procedure.

Definition of the Tank Pressure Range and the Weighting Function

Based on the data depicted in Figure 8 the tank pressure vector $\vec{p}_t$ can be defined. In this example, only one tank pressure close to the boundary of the feasible set is considered. Thus, no weighting is necessary:

TABLE 4

Optimal values of the design variables for $d_{CV} = 19$ mm and $p_t = 10$ bar

	Minimum search	PSO
ϕ^*_{CVO}	10°CA	13.7°CA
ϕ^*_{CVC}	110°CA	110.1°CA
m^*_a	10.15 g	10.13 g
t^*_s	350 ms	350 ms
N^*_{ps}	4	4

$$\vec{\mathbf{p}}_t = \{10 \text{ bar}\},\ w(p_t) = 1 \tag{10}$$

Minimum Search and Numerical Optimization

According to the procedure described in Section 2, the optimal CV design can be found by a minimum search on the feasible set. Since the tank pressure considered is on the grid, no further simulations are necessary. Table 4 lists the values for the design variables found by the minimum search and the PSO, respectively, for $d_{CV} = 19$ mm and $p_t = 10$ bar. The resulting valve timings found with the two methods are very close to each other. The resulting start time is 350 ms. Thus, the maximum start time is fully exploited, and the solution lies on the boundary of the feasible set. The number of power strokes required to start the engine is $N_{ps} = 4$.

Figure 9 shows the air consumption for the relevant part of $\{\phi_{CVO}, \phi_{CVC}\} \in \Gamma$ and a tank pressure of 10 bar.

3.1 Discussion

The optimization procedure yields a CV opening angle after TDC. This result is advantageous because the first engine revolution would become critical if the CV was to be opened before TDC and the initial tank pressure was large, i.e. significantly larger than the in-cylinder pressure at TDC. In that case, a negative torque would be produced by the injection of pressurized air before the piston passes TDC. If this torque exceeded the torque produced by the ES, the engine could not be started pneumatically.

Figure 9 clearly shows that a later closing of the CV is unfavorable with respect to the air consumption. The reason for this fact is that the opening angle of the EV is at $\phi = 114$°CA. A design where $\phi_{CVC} > 114$°CA implies that pressurized air flows from the CV directly into the exhaust manifold. Such a design increases the air consumption without producing significantly higher torques.

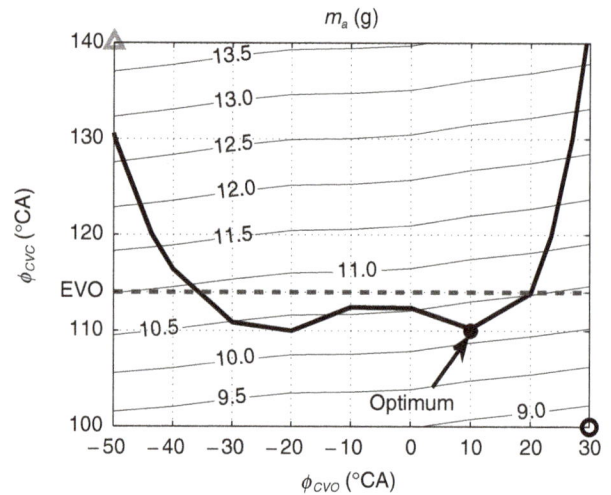

Figure 9

Consumption of pressurized air for various CV opening and closing angles with $d_{CV} = 19$ mm and $p_t = 10$ bar. The black bold line denotes the boundary of the feasible set. The label EVO denotes the EV opening angle. Black circle: minimum air consumption. Gray triangle: maximum air consumption.

Sensitivity Analysis

Based on the results depicted in Figures 8-10, a sensitivity analysis of the parameters in the set Γ can be performed.

Figure 8 shows that for small CV diameters the minimum tank pressure to satisfy $t_s \leq t_{s,max}$ increases significantly. For small values of d_{CV} less air can be transferred through the CVs due to the flow restriction. Thus, the torque is lower. The increased flow restriction needs to be compensated by a higher density, i.e. a higher tank pressure. For $d_{CV} > 14$ mm, the start time is almost constant for a fixed tank pressure. Larger values of d_{CV} might be expected to result in more air being transferred and hence in reduced start times. However, there is a counteracting effect. For large valve diameters, the pressure difference between tank and cylinder decreases rapidly once the CV is opened. This effect results in a reduction of the mass flow rate.

Figure 10 shows the start time for various combinations of ϕ_{CVO} and ϕ_{CVC} for fixed values of $p_t = 10$ bar and $d_{CV} = 19$ mm. The maximum start time on the grid shown is equal to 370 ms, and it is indicated by a gray triangle. The minimum start time is 341 ms and is indicated by a black circle. Thus, the maximum start time is 8% higher than the minimum start time, which indicates a rather small sensitivity. Around TDC, the start time t_s is almost constant for varying CV opening angles.

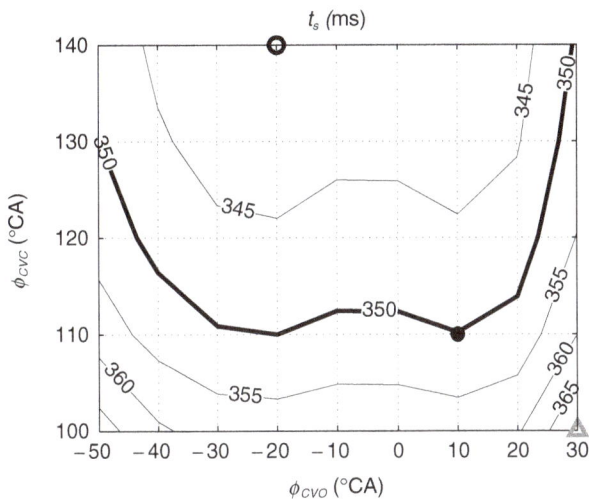

Figure 10

Start times for various CV opening and closing angles for $p_t = 10$ bar and $d_{CV} = 19$ mm. The bold black line denotes the boundary of the feasible set where all combinations above the line are feasible. Black circle: minimum start time. Gray triangle: maximum start time.

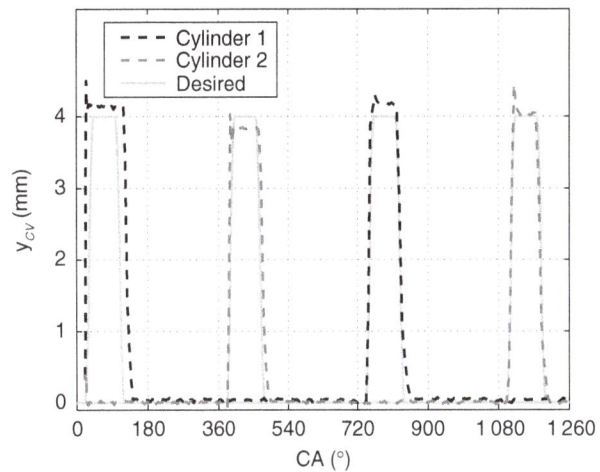

Figure 11

Desired and emulated CV lift profiles.

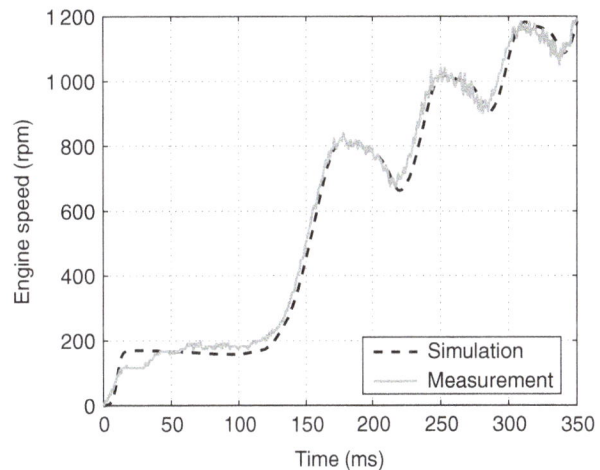

Figure 12

Measured and simulated pneumatic engine start.

On the other hand, the start time can be reduced by a later closing of the CV.

The air consumption depicted in Figure 9 shows a rather high sensitivity to the CV timing. In the CV timing intervals depicted, the minimum air consumption is 8.7 g (black circle). The maximum air consumption amounts to 13.8 g (gray triangle), which is 58% larger. Analogously to the start time, around TDC there is a small variation of m_a in terms of the CV opening angle. However, the gradient of the air consumption as a function of the CV closing angle points in the opposite direction of the gradient of the start time.

These results allow the conclusion that a later closing of the CV implies shorter start times but an increased air consumption. Comparing the time optimal and air optimal designs within the grid considered indicated by the black circles in Figures 9 and 10 leads to the statement that accepting an increase in the start time by 8% results in an air saving of 36%.

3.2 Experiments

In order to verify the quality of the model and the CV design found, the pneumatic start is implemented on a test bench engine with the optimal CV timings obtained from the design example. The CV are actuated by a fully variable valve system which allows the emulation of a wide variety of CV lift profiles. Figure 11 shows the desired and the emulated CV lift profiles of both cylinders for the engine start shown in Figure 12.

Figure 12 shows the engine speed trajectories for a measured and a simulated pneumatic start. The limit of $t_s = t_{s,max} = 350$ ms is fully exploited, which corresponds to the result predicted by simulation and presented in Table 4. The pressure drop in the air tank over the entire pneumatic engine start phase is approximately 300 mbar. The corresponding air mass consumption is 10.5 g, which agrees with the value predicted by the process model. The discrepancy is 3%.

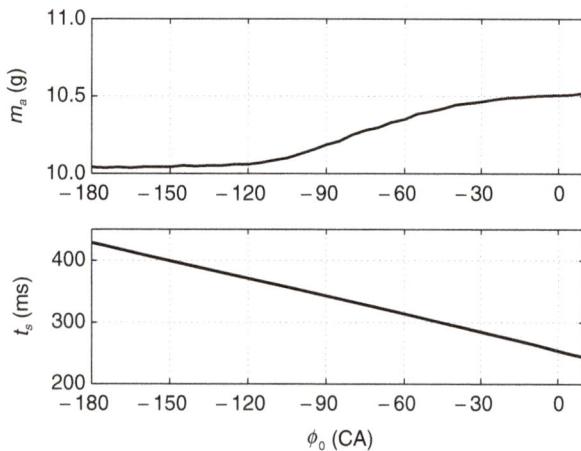

Figure 13

Air consumption and start time for various initial engine positions ϕ_0 for $\phi^*_{CVO} = 10°\text{CA}$, $\phi^*_{CVC} = 110°\text{CA}$, $d_{CV} = 19\,\text{mm}$ and $p_t = 10\,\text{bar}$.

Only four power strokes are required to reach the desired final engine speed $\omega_{e,s} = \omega_{e,idle} = 1\,200\,\text{rpm}$. During the first third of the start time, the engine is driven only by the ES until ϕ_{CVO} is reached. This duration can only be reduced by a more powerful ES.

The good agreement between measurement and simulation data confirms the quality of the process model.

3.3 System Analysis

3.3.1 Influence of the Initial Engine Position

The throttle strategy during the engine shutdown presented in Section 1.3 showed reliable results on the test bench. In this section, the importance of the initial engine position is highlighted by simulating the engine start for various initial positions. Figure 13 shows the resulting air consumption and start time data as a function of the initial engine position ϕ_0. If the start is initiated closer to the CV opening position the start time is shorter because it takes less time until pressurized air is injected and, thus, a high torque is applied. Furthermore, the first CV opening lasts longer since the engine speed is low. Hence, more air is injected, which yields a higher torque. The influence of the start position on the start time is significant. If the engine is started from the BDC position, it takes 430 ms. The start from the TDC position takes just 259 ms, which is faster by 40%.

In contrast, the air consumption is not very sensitive to the initial engine position. For the start positions considered, an increase of just 5% can be observed. The air consumption is rather constant for start positions from $\phi = -180°\text{CA}$ to $\phi = -120°\text{CA}$. This can be explained by the open IV, through which air is blown out. The air consumption is highest if the initial position is just before the CV opening because in this case the engine speed and the cylinder pressure are rather low during the first air injection. Hence, the CV is open for a very long time and a lot of compressed air is injected.

In conclusion, a good positioning of the engine during the shutdown substantially influences the start time while the air consumption is affected only slightly.

3.3.2 Variation of the Initial Tank Pressure

During the operation of the engine, the tank pressure varies due to emptying and refilling of the air pressure tank. In this section, the start performance of camshaft driven CV with respect to the air used and the start duration is analyzed for various initial tank pressures. The analysis is conducted for the optimal CV design with $\phi^*_{CVO} = 10°\text{CA}$ and $\phi^*_{CVC} = 110°\text{CA}$ using the process model f_{PM}. Figure 14 shows the results. The first subplot shows the total amount of air used, m_a, and the amount of air used when the desired engine speed is reached $m_a(t_s)$. The second subplot shows the start time t_s and the time $\hat{t}_s$ when the valves are closed. The third subplot shows the number of power strokes N_{ps} required to reach the desired engine speed.

As stated in Equation (6), the amount of air used to start the engine is calculated as the integral of the air mass flow. Therefore, the same amount of air is used whether a larger mass flow occurs for a short time or whether a small mass flow exists for a longer time duration. For instance, the air consumptions for initial tank pressures of $p_t = 9.3\,\text{bar}$ and $p_t = 12\,\text{bar}$, respectively, are equal. However, the start times are different.

Obviously, a higher tank pressure leads to a higher air mass flow. The torque exerted is larger, which leads to a shorter start time t_s. The total amount of air m_a used for an engine start is a piecewise continuous function of the tank pressure. The steps occur when the number of power strokes changes. For an equal number of power strokes the amount of air used increases along with an increasing tank pressure mainly because the valve cannot be closed when the desired engine speed is reached. If a CV is available that can be deactivated immediately the air consumption can be reduced to $m_a(t_s)$. However, the piecewise continuity remains.

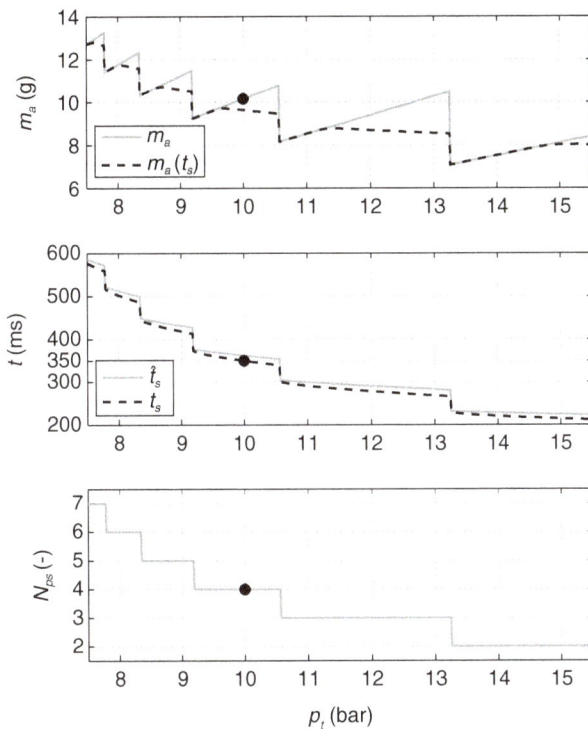

Figure 14

Air consumption, start time and number of power strokes for various tank pressures for $\phi^*_{CVO} = 10°$ CA, $\phi^*_{CVC} = 110°$ CA and $d_{CV} = 19$ mm. The black dot indicates the optimal result of the design example.

CONCLUSION

A general design methodology for camshaft driven charge valves used for pneumatic engine starts is presented. The design procedure is exemplified for a two-cylinder engine, and the results are verified experimentally. The discrepancy between the results predicted and those obtained on the testbench is 3%.

Pneumatic engine starts enable significantly reduced engine start times compared to those of conventionally started engines. In contrast to conventional engine starts, a large positive torque is produced already during the first expansion stroke. Another advantage of pneumatic engine starts is the fact that they are applicable on engines with port fuel injection as well as on those with direct injection. Starting the engine without burning any fuel also reduces the emission of hydrocarbons and carbon-monoxide which are caused by the incomplete combustion occurring at very low engine speeds.

In summary, pneumatic engine starts offer the possibility to implement stop/start strategies that satisfy the comfort demands of the driver without significantly

increasing the complexity and cost of the whole engine system.

OUTLOOK

The pneumatic engine start treated in this paper was investigated on a gasoline engine with port fuel injection. However, the possibility to inject pressurized air directly into the cylinders offers additional advantages for engines with direct gasoline injection. As mentioned e.g. in [6] and [7], the direct start using fuel injection into the stopped engine suffers from the fact that a successful start is guaranteed only for a limited range of engine rest positions. The injection of pressurized air can be used to extend the range of initial engine positions where the direct engine start is guaranteed to be successful. Additionally, the pneumatically assisted engine start with early direct fuel injection can help to further reduce the start time. The combination of the pneumatic start and the direct start is being considered in ongoing research.

REFERENCES

1 Dönitz C., Vasile I., Onder C., Guzzella L. (2009) Dynamic Programming for Hybrid Pneumatic Vehicles, *American Control Conference*, 3956-3963.

2 Silva C., Ross M., Farias T. (2009) Analysis and Simulation of "Low-cost" Strategies to Reduce Fuel Consumption and Emissions in Conventional Gasoline Light-duty Vehicles, *Energy Conversion and Management*, 215-222.

3 Zülch C.D. (2007) Konzepte für einen sicheren Direktstart von Ottomotoren, *PhD Thesis*, University of Stuttgart.

4 Rau A. (2009) Analyse und Optimierung des Ottomotorischen Starts und Stopps für eine Start Stopp Automatik, *PhD Thesis*, Technical University of Clausthal.

5 Fesefeldt T., Müller S. (2009) Optimization and Comparison of Quick and Hybrid Start, *SAE Technical Paper* 2009-01-1340.

6 Kulzer A., Laubender J., Lauff U., Mössner D., Sieber U. (2006) Direct Start – From Model to Demo Vehicle, *MTZ Worldwide*. **67**, 9, 12-15.

7 Kramer U. (2005) Potentialanalyse des Direktstarts für den Einsatz in einem Stopp-Start-System an einem Ottomotor mit strahlgeführter Benzin-Direkteinspritzung unter besonderer Berücksichtigung des Motorauslaufvorgangs, *PhD Thesis*, University of Duisburg-Essen.

8 Ueda K., Kaihara K., Kurose K., Ando H. (2001) Idling Stop System Coupled with Quick Start Features of Gasoline Direct Injection, *SAE Technical Paper* 2001-01-0545.

9 Vasile I., Dönitz C., Voser C., Vetterli J., Onder C., Guzzella L. (2009) Rapid Start of Hybrid Pneumatic Engines, *Proceedings of the IFAC Workshop on Engine and Powertrain Control, Simulation and Modeling, E-COSM'09*, IFP Energies nouvelles, Rueil-Malmaison, France, 30 Nov.-02 Dec, pp. 123-130.

10 Moser M., Voser C., Onder C., Guzzella L. (2012) Design Methodology of Camshaft Driven Charge Valves for Pneumatic Engine Starts, *Proceedings of the IFAC Workshop on Engine and Powertrain Control, Simulation and Modeling, E-COSM'12*, IFP Energies nouvelles, France, 23-25 Oct., pp. 33-40.

11 Ebbesen S., Kiwitz P., Guzzella L. (2012) A Generic Particle Swarm Optimization Matlab Function, *Proceedings of the American Control Conference*, Montréal, Canada, 27-29 June, pp. 1519-1524.

12 Guzzella L., Onder C. (2010) *Introduction to Modeling and Control of Internal Combustion Engine Systems*, Springer, 2nd ed.

13 Pischinger R., Krassnig G., Taucar G., Sams T. (1989) *Thermodynamik der Verbrennungskraftmaschine*, Springer, Wien, New York.

APPENDIX A: PROCESS MODEL

The most important relations used in the process model f_{PM} are described in this appendix. The engine cylinders are modeled as receivers with variable volume. Every cylinder $i = \{1, ..., N_{cyl}\}$ has its crank angle position ϕ_i. For $\phi_i = 0°$ CA the piston is located at the TDC after the compression stroke.

Mass Balance

The air mass of every cylinder $m_{cyl,i}$ is determined by the mass flows $\dot{m}_{k,i}$ through each valve type $k = \{IV, EV, CV\}$:

$$\frac{dm_{cyl,i}(t)}{dt} = \dot{m}_{IV,i} + \dot{m}_{CV,i} - \dot{m}_{EV,i} \tag{A1}$$

Blow-by is neglected. According to [12] the mass flow through the valves is modeled as a compressible flow restriction:

$$\dot{m}_{k,i} = c_d \cdot A \cdot \frac{p_{up}}{\sqrt{R_a \cdot \vartheta_{up}}} \cdot \psi\left(\frac{p_{up}}{p_{down}}\right) \tag{A2}$$

where c_d denotes the discharge coefficient, A is the maximal opening area of the valve, p_{up} and p_{down} correspond to the upstream and downstream pressures, respectively, ϑ_{up} is the upstream temperature, R_a is the ideal gas constant of air and $\psi(.)$ is the flow function. For the discharge coefficient c_d the relation of [13] is used, where it is defined as a function of the relative lift y_{CV}/d_{CV}. Figure A1 shows the relations for all engine valves.

The flow function $\psi(.)$ is defined by:

$$\psi\left(\frac{p_{up}}{p_{down}}\right) = \begin{cases} \sqrt{\kappa\left[\frac{2}{\kappa+1}\right]^{\frac{\kappa+1}{\kappa-1}}} & \text{for } p_{down} < p_{cr} \\ \left[\frac{p_{down}}{p_{up}}\right]^{\frac{1}{\kappa}}\sqrt{\frac{2\kappa}{\kappa-1}\left[1 - \left(\frac{p_{down}}{p_{up}}\right)^{\frac{\kappa-1}{\kappa}}\right]} & \text{for } p_{down} \geq p_{cr} \end{cases} \tag{A3}$$

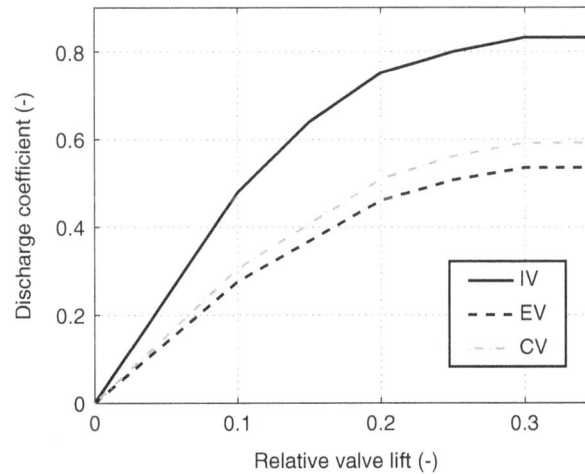

Figure A1

Discharge coefficient as a function of the relative valve lift. The relative valve lift is the ratio of the absolute valve lift and the valve diameter.

where κ is the ratio of the specific heats. The critical pressure p_{cr} at which the flow reaches sonic conditions in the narrowest part is given by:

$$p_{cr} = \left[\frac{2}{\kappa + 1}\right]^{\frac{\kappa}{\kappa - 1}} \cdot p_{up} \tag{A4}$$

Energy Balance

The internal energy balance of every cylinder is given by the enthalpy flows $\dot{H}_{k,i}$, the heat transfer and the instantaneous work done:

$$\frac{dU_i}{dt} = \dot{H}_{IV,i} + \dot{H}_{CV,i} - \dot{H}_{EV,i} - \dot{Q}_i - p_{cyl,i} \cdot \dot{V}_{cyl,i} \tag{A5}$$

where $\dot{Q}_i$ is the heat transfer to and from the walls. The cylinder volume $V_{cyl,i}$ depends on the cylinder's crank angle position

$$V_{cyl,i}(\phi_i) = V_{cyl,TDC} + A_{cyl} \cdot (l + r \cdot (1 - \cos \phi_i)) - A_{cyl} \cdot \left(\sqrt{l^2 - r^2 \cdot \sin^2 \phi_i}\right) \tag{A6}$$

where r is the crank radius, l is the length of the connecting rod and A_{cyl} is the piston area. The cylinder temperatures and pressures are calculated using the definition of the internal energy:

$$\vartheta_{cyl,i} = \frac{U_i}{m_{cyl,i} \cdot c_{v,a}} \tag{A7}$$

and the ideal gas equation:

$$p_{cyl,i} = \frac{m_{cyl,i} \cdot R_a \cdot \vartheta_{cyl,i}}{V_{cyl,i}} \tag{A8}$$

respectively, where $c_{v,a}$ is the specific heat of air at constant volume. The instantaneous torque T_i of each cylinder is defined as:

$$T_i = p_{cyl,i} \cdot A_{cyl} \cdot \left[r \cdot \sin \phi_i + \frac{r^2 \cdot \sin \phi_i \cdot \cos \phi_i}{\sqrt{l^2 - r^2 \cdot \sin^2 \phi_i}}\right] \tag{A9}$$

Conservation of Angular Momentum

The law of the conservation of angular momentum determines the engine's acceleration:

$$J_e \cdot \frac{d\omega_e}{dt} = \sum T_i - T_{fric}(\omega_e) + T_{ES}(\omega_e) \tag{A10}$$

where J_e is the engine's inertia and T_{fric} is the friction torque. The engine speed ω_e is the time derivative of any crank angle:

$$\frac{d\phi_i}{dt} = \omega_e \tag{A11}$$

APPENDIX B: CATALYST TEMPERATURE

This section discusses the influence of the pneumatic start on the catalyst temperature which is an important issue for the implementation.

During the pneumatic engine start, the pressurized air from the tank is expanded in the cylinder to produce a torque that accelerates the engine. This expansion decreases the temperature of the gas flowing through the catalyst. The cold gas might lower the catalyst temperature leading to an efficiency drop of the latter.

Due to the lack of a catalytic converter on the test bed no analysis has yet been performed. Here, several qualitative arguments are given on how the temperature of the catalyst is influenced. Firstly, the number of pneumatic power strokes to start the engine is only 2-6 as shown in Figure 14. Hence, the total amount of cold gas leaving the cylinders is very low. Secondly, the cold gas will be heated up by the hot exhaust pipes before it reaches the catalyst. If the cooling of the catalyst remains an issue, the number of pneumatic power strokes can be further reduced by an earlier switch to the combustion mode during the engine start.

APPENDIX C: INFLUENCE OF THE AIR TEMPERATURE IN THE TANK

During the operation of the engine also the air temperature in the tank can vary. However, according to the arguments mentioned in Section 2 the relative variation of the tank temperature is restricted to a significantly smaller range than the relative variation of the tank pressure. Hence, a variation of the tank temperature is not taken into account in the design methodology. To justify this simplification the optimal CV design is computed for various tank temperatures using the PSO algorithm. The results are shown in Figure A2. The two top plots show that the resulting optimal CV timings only vary within a very small range. For all design temperatures, the resulting start time is $t_s = 350$ ms. The bottom plot of Figure A2 shows that the air mass used increases with lower temperature due to the temperature dependence of the air density.

The dashed line shows the air consumption of the design given in Table 4 for various temperatures. This design has a slightly lower air consumption but the start time t_s is 1-4 ms above the limit $t_{s,max} = 350$ ms.

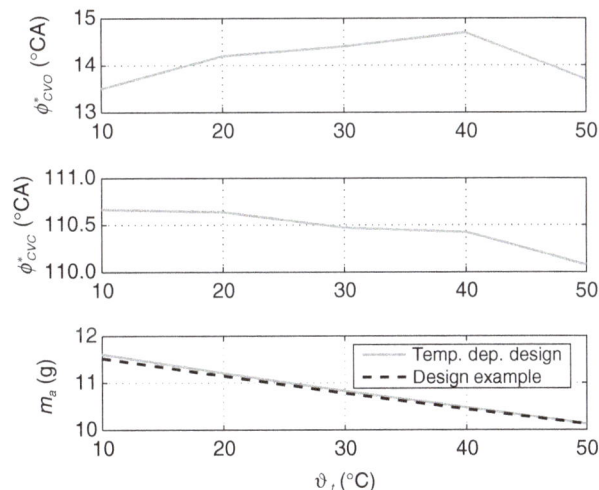

Figure A2

Optimal CV timings and resulting air consumption obtained for CV designs optimized at the respective tank temperature (temp. dep. design). The air consumption obtained with the design given in Table 4 is also shown (design example). This design yields a slightly lower air consumption but exceeds the maximum start time limit $t_{s,max}$ by 1-4 ms at temperatures below 50°C.

12

Development of Reactive Barrier Polymers against Corrosion for the Oil and Gas Industry: From Formulation to Qualification through the Development of Predictive Multiphysics Modeling

X. Lefebvre[1*], D. Pasquier[2], S. Gonzalez[2], T. Epsztein[3], M. Chirat[3] and F. Demanze[3]

[1] IFP Energies nouvelles, 1-4 avenue de Bois-Préau, 92852 Rueil-Malmaison - France
[2] IFP Energies nouvelles, Rond-point de l'échangeur de Solaize, BP 3, 69360 Solaize - France
[3] TECHNIP, rue Jean Huré, BP 7, 76580 Le Trait - France
e-mail: xavier.lefebvre@ifpenergiesnouvelles.fr

* Corresponding author

Abstract — *Corrosion is a key issue for operators in the oil and gas industry since production fluids contain some water and both CO_2 and H_2S acid gases. In this context, this paper illustrates the development of a reactive barrier polymer against corrosion by H_2S of offshore flexible pipes. The role of this reactive material, called anti-H_2S material, is to avoid H_2S reaching the structural steel layers of the flexible pipe during the whole service life of the structure, usually 20 years, and hence to place the steel layers in a sweet service environment. Placed between the existing pressure sheath and the steel layers, the anti-H_2S material has the ability to neutralize H_2S during its diffusion within the material. The neutralization is ensured by an irreversible chemical reaction on reactive components that are dispersed in the material. The raw material selection is based on both accurate requirements for their use in a flexible pipe and expected performances in a sour service environment over a long period of time. Some laboratory qualifications and experimental techniques are used to qualify the behavior of the material and build the material database. A dedicated multiphysics model is developed based on the coupling of permeation mechanisms and gas-solid reactions. Qualification of both the material and the model is performed thanks to middle-scale and full-scale tests conducted in representative sour service conditions.*

Résumé — **Développement de matériaux barrières réactifs contre la corrosion pour l'industrie pétrolière : de la formulation à la qualification industrielle en passant par le développement de modèles multiphysiques prédictifs** — La corrosion liée à la présence de gaz acides tels que CO_2 et H_2S en présence d'eau dans les effluents pétroliers est un enjeu majeur pour les opérateurs pétroliers. C'est dans ce contexte que cet article décrit le développement d'un matériau barrière réactif permettant de limiter la corrosion par H_2S des flexibles offshore. Le rôle de ce matériau barrière, appelé matériau anti-H_2S, consiste à empêcher l'H_2S d'atteindre les éléments métalliques structuraux du flexible pendant toute sa durée de vie, généralement 20 ans. De cette manière, les éléments métalliques bénéficient d'un environnement « *sweet service* », peu agressif en terme de corrosion. Placé entre l'actuelle gaine de pression et les éléments métalliques, le matériau anti-H_2S a la faculté de pouvoir neutraliser l'H_2S pendant sa diffusion au sein même du matériau. La neutralisation est assurée par une réaction chimique irréversible sur des charges réactives dispersées au sein du matériau. Les matières premières constitutives de ce matériau sont sélectionnées selon leur performance en milieu fortement corrosif « *sour service* », tout en garantissant leur compatibilité avec les applications flexibles offshore. La qualification du matériau à l'échelle du laboratoire est décrite ainsi que les techniques qui nous permettent d'alimenter la base de données nécessaire à l'identification d'un modèle. Un modèle multiphysique dédié a été développé pour décrire les couplages entre les mécanismes de diffusion-réaction. La qualification du matériau et du modèle est effectuée sur la base de résultats d'essais conduits en conditions représentatives, et réalisées sur éléments de structure et sur structures réelles.

INTRODUCTION

Corrosion is a key issue for operators in the oil and gas industry. In production systems, the production fluid usually contains a hydrocarbon phase, some water, and varying concentrations of CO_2 and H_2S. Both CO_2 and H_2S are acid gases; therefore, when associated with water, they lead to the formation of a corrosive environment.

Corrosion mechanisms resulting from the presence of these two acid gases have their own characteristics [1-8] (Sulfide Stress Cracking – SSC, Hydrogen-Induced Cracking – HIC). In particular, the presence of wet H_2S promotes and exacerbates many types of environmental cracking. These types of cracking may become an integrity concern and hence require a specific metallurgical design or operational precautions. These wet H_2S service conditions are called sour service conditions, in contrast with sweet service conditions where no specific metallurgical design is normally required. Sour service conditions lead to high corrosion rates by SSC and environmental fracture due to hydrogen atom uptake in the steel (HIC).

Nevertheless, metallurgical or operational recommendations do not always result in competitive solutions. On one hand, specific steel grades are more expensive and their use results in non-optimum structure design due to the low mechanical properties of sour service steel grades compared with sweet service ones. On the other hand, operational precautions can be very restrictive and can be detrimental from a technical and economic point of view. That is the reason why the opportunity to use sweet service steel grades in sour service environments is a major challenge for the oil and gas production industry.

A corrosive environment is both H_2S and CO_2 content-dependent and the synergistic effects of these two species are a point of consideration [1-10]. Three domains are usually defined in the oil and gas industry: the H_2S sour regime domain, the $CO_2 + H_2S$ mixed regime domain, and the CO_2 sweet regime domain [8]. As a result, a H_2S-free environment is always considered as a sweet service environment. That is the reason why we focus on the neutralization of this species in production systems.

This work was carried out in a collaboration between *IFP Energies nouvelles* and *Technip*.

Two main types of pipes are used in the oil and gas production industry: flexible and rigid pipes. In rigid pipes, the production fluid is usually in contact with the structural steel, and although some limitations exist the thickness of the pipe and the metallurgical composition of the steel can usually be determined so that the pipe will resist sour service conditions.

Technip's flexible pipes are mainly used when flexibility is needed, for offshore production in particular. Flexibility is provided by the multilayer structure of the pipe. During manufacturing, laying and operation, each function of the flexible pipe is provided by an unbonded layer. Figure 1 presents a conventional flexible pipe made of the following layers:

– the first layer of a rough bore structure is a carcass (1) made of stainless steel, in direct contact with the transported fluid and sustaining the external pressure;
– the second layer is made of a leakproof continuous polymer layer and is called the pressure sheath (2);
– the third layer sustains internal pressure thanks to a steel vault wire (3) spiralled at short pitch. For high-pressure applications an additional layer called a spiral can be added between the first vault layer and the armor layers.
– the armor layers (4) are made of one or two pairs of helicoidally spiralled steel wires and designed to ensure tensile load resistance;
– the tightness of the annulus regarding the external environment is provided by an external sheath (5) made of polymer;
– if necessary, the thermal performances of the flexible pipe can be improved by a second annulus composed of syntactic foam tapes (6).

The polymer pressure sheath protects the carbon steel layers sustaining internal pressure and axial tension from any direct contact with the fluid transported in the bore. However, due to the intrinsic properties of polymer materials, active components such as water, carbon dioxide and hydrogen sulfide will permeate through the polymer pressure sheath over a long period of time [11, 12]. The design of

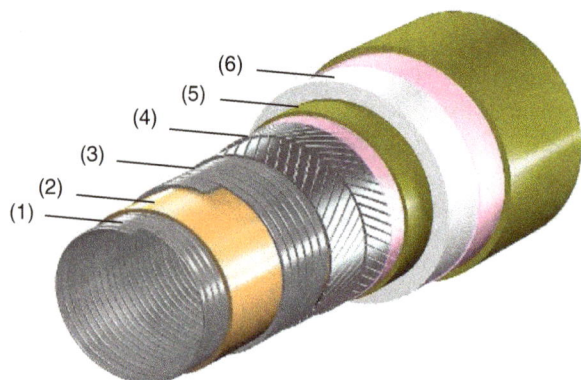

Figure 1

Layers of a conventional flexible pipe.

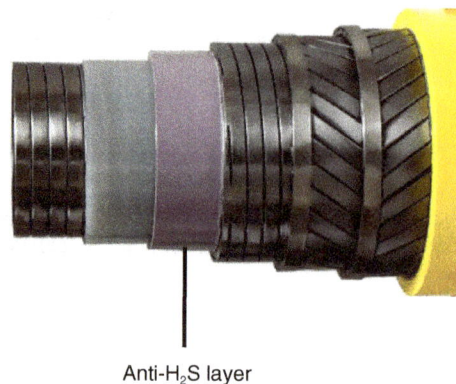

Figure 2

Position of the anti-H$_2$S layer in a flexible pipe.

every flexible pipe takes into account this permeation phenomenon using a dedicated fluid permeation model [13]. This model enables one to predict the partial pressure of H$_2$S in the annulus based on the operating conditions of the structure. This partial pressure of H$_2$S is then taken into account in order to select the appropriate steel grade for the vault and armor layers. Considering the permeation phenomenon with time in the pressure sheath, current operating conditions lead more and more frequently to sour service conditions in the annulus after several months of operation.

The objective of the anti-H$_2$S layer new material is to avoid H$_2$S from reaching the structural steel layers during the whole service life of the pipe and hence to place the steel layers in a sweet service environment.

We chose to develop a material based on a polymer matrix which can replace or be extruded on the existing polymer pressure sheath. The target of this new sheath is to neutralize H$_2$S during its diffusion in the sheath. The neutralization is ensured by chemical reactions on reactive components that are dispersed in the polymer matrix. This is a reactive barrier polymer material. To ensure a H$_2$S-free annulus, a layer of this material can be placed between the pressure sheath and the annulus (Fig. 2).

This paper illustrates the development of a barrier polymer material for flexible pipe applications: from the formulation of the reactive material, through the development of a dedicated multiphysics model, to qualification in sour service conditions.

The selection of raw materials is described in the first section of this paper. The selection is based on both accurate requirements for their use in a flexible pipe and expected performances in a sour service environment over a long period of time. In Section 2, the laboratory qualification of the material is described, as well as the experimental techniques used to qualify the behavior of the material and build the material database. In Section 3, the development of the dedicated multiphysics model is presented. Finally, the qualification of both the material and the model is performed thanks to middle-scale and full-scale tests conducted in representative sour service conditions.

1 SELECTION OF RAW MATERIALS AND FORMULATION

The raw material selection is based on both accurate requirements for their use in a flexible pipe and expected performances in a sour service environment.

The main components of the reactive material are the polymer matrix and the reactive components. The challenge consists of successfully developing the composite material, based on the mix of matrix and reactive components, knowing that it must meet several strict requirements to satisfy the application.

1.1 Polymer Matrix Selection

Requirements for the use of a polymer material as a sheath in a flexible pipe include several criteria, amongst which mechanical properties, durability in service conditions and transport properties are to be considered.

Each flexible pipe structure is designed for a given field, depending on several parameters such as temperature and pressure of the production fluid, H$_2$S and CO$_2$ contents, water depth, etc.

For mechanical and chemical stability considerations, three categories of thermoplastic polymers are commonly used for the pressure sheath: PolyEthylene (PE), PolyAmide (PA) and PolyVinylidene DiFluoride (PVDF). The selection of the polymer during the design of a flexible pipe is strongly dependent on the temperature of the production fluid.

Diffusion mechanisms are temperature-dependent and roughly obey Fick's law [14]. Permeability characterizes the ability of a chemical species (such as H_2O, H_2S, CH_4 and CO_2), to be transported through a polymer, and depends on the interaction between the polymer and the chemical species (the solute). Hence, it is an intrinsic property of the polymer – solute couple. The driving force for diffusion of solute in the polymer is the spatial difference in chemical activity of the solute in the polymer (Sect. 3).

Permeability is the product of the diffusion coefficient of the solute in the polymer, by the solubility of the solute in the polymer [11-14]. In a first approximation, for a given polymer – solute couple, both the diffusion coefficient and solubility value can be considered as temperature-dependent only. Their evolution as a function of temperature usually follows Arrhenius' law [11-14], but one must keep in mind that they can be pressure-dependent or depend on other solute contents [15-17]. In this case, mixture laws exist to describe the interaction between all solutes and the polymer [18-22].

A first generation of anti-H_2S material has been developed based on a polyethylene matrix to be used for temperatures up to 70°C. This paper will mainly be dedicated to the presentation of the qualification work performed on this first generation of anti-H_2S material.

1.2 Reactive Component Selection

The barrier function of the material is provided by the presence of reactive components added to the matrix. As previously said, we focused on the neutralization of H_2S in contact with the reactive fillers, during its diffusion within the composite material. The objective is to ensure a H_2S-free annulus during the service life of the flexible pipe (20 years generally). For this reason, we have to select highly reactive components so as to minimize the ratio of H_2S diffusion in the matrix to H_2S reaction with the active components.

One characteristic of transition metal oxides is their propensity to react with acid species. The catalytic activity of many metal oxides depends on both the strength and the amount of Lewis and Brønsted acid-base sites. Cationic metal sites act as Lewis acids while anionic oxygen sites act as Lewis bases. Concerning surface hydroxyl groups, they are able to serve as Brønsted acid or base sites as they are able to give up or accept a proton [23].

Metal oxide can exist in many forms (size, shape, etc.), so the selection of a metal oxide for our application is challenging: it is based on several technical parameters such as reactivity and the chemical nature of the reaction products,

but also toxicity, cost, reliability of supply, etc. Experimental characterization of some metal oxides and thermodynamic calculation will help the selection.

The chemical reaction taking place in the anti-H_2S material is a reaction of a gas species on a solid particle. Gas-Solid reactions are known to be particle size- and shape-dependent (Sect. 3.2). Because the metal oxide surface is directly accessible, surface reactions are quicker than volume reactions. Now, considering the metal oxide as a spherical particle, the smaller the particle, the higher the ratio of its surface to its volume, and so the higher the relative amount of rapid surface reactions. This notion is related to the specific surface of the particle (m^2/g), which is defined as the average ratio of the surface of particles to their weight.

So, to promote rapid reactions on surface particles, rather than reactions in the volume of the particle, the smallest particles may be selected.

Based on both experimental sulfidation test results and thermodynamic calculations, zinc oxide (ZnO) and iron (III) oxide (Fe_2O_3) were found to be among the best candidates for the application.

Chemical reactions involved in the composite material are:

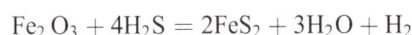

$$ZnO + H_2S = ZnS + H_2O$$

$$Fe_2O_3 + 4H_2S = 2FeS_2 + 3H_2O + H_2$$

The amount of metal oxide introduced into the material is determined according to the target performance while maintaining both appropriate mechanical properties of the material for the application, and good extrudability and processing properties.

1.3 Compounding

In order to qualify anti-H_2S compositions, a laboratory study was carried out. Lab-scale formulations were performed and composites characterized. Thermoplastic polymer/metal oxide composites were prepared by melt compounding using a *Thermo*® PTW 16/25 XL corotating twin screw extruder. A volumetric screw feeder is used for the polymer. A gravimetric screw feeder allowed bringing the metal oxide inside the polymer matrix.

Different parameters were studied, such as the nature and the grade of the thermoplastic matrix and the nature, grade and weight fraction of the metal oxide. Based on this study some formulations were selected toward their H_2S-barrier properties and their mechanical behavior. This paper deals with the material called PEZnO, composed of a thermoplastic matrix in PE compounded with ZnO and Fe_2O_3. Both oxides will react with H_2S. The reaction with ZnO will be responsible for the efficiency of the system, whereas Fe_2O_3, which gives its initial purple color to the material, will be used as a visual tracer of the reaction with H_2S.

2 LABORATORY-SCALE QUALIFICATION AND DATABASE ACQUISITION FOR MODELING

2.1 Objective of the Laboratory-Scale Qualification

In flexible pipeline applications, the aim of anti-H_2S materials is to ensure that no H_2S crosses the anti-H_2S sheath during the whole service life of the pipe. To achieve this goal, we need to understand the mechanisms that take place in the material in order to develop a predictive physical model.

In the presence of H_2S, anti-H_2S material is the place of a competition between two mechanisms: diffusion of H_2S in the polymer matrix, and chemical reaction of H_2S with metal oxide particles. These two mechanisms have opposite effects on H_2S advance in the composite material.

As illustrated in Figure 3, the competition between H_2S diffusion in the matrix and H_2S reaction with particles forms a reaction area within the anti-H_2S layer. This reaction area will progress inside the material. For a given diffusion coefficient of H_2S in the polymer matrix, the more reactive the particle, the smaller the reaction area.

To understand the behavior of such a material on the laboratory scale, we need to determine:
- the position of the H_2S front with time;
- the degree of conversion in the material: upstream of the reaction front to characterize the reactivity of the material,

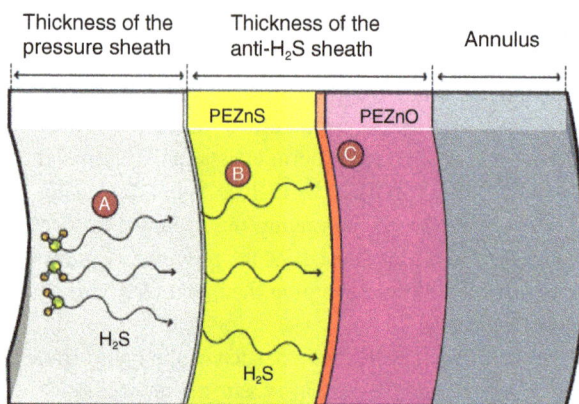

and downstream of the reaction front to ensure that no H_2S has passed and reacted beyond this point;
- the H_2S concentration in the downstream environment so as to determine, for given experimental parameters, the time after which H_2S will finally cross the material. This time is commonly called the breakthrough time.

Two main kinds of tests can be performed to produce data required to understand and characterize the behavior of anti-H_2S materials:
- breakthrough test: based on permeability test principles [12], this test consists of measuring the time after which H_2S is detected by Gas Chromatography (GC) downstream of a sample, while applying upstream an environment containing H_2S. This breakthrough time fully characterizes the behavior of the material since the test conditions are very similar to the application. Moreover, several analyses can be performed on the sample after the breakthrough test: degree of conversion measurements in the thickness, position of the reaction front at breakthrough time, etc. Nevertheless, this kind of test can be very long since the breakthrough time can be equal to several tens or even hundreds of days. So, we will favor the performing of this kind of test during the validation step on middle-scale and full-scale representative structures;
- immersion test: coupons are placed in a vessel in which they are immersed in an environment containing pure H_2S or a gas mixture including H_2S. Pressure and temperature are adapted to perform the tests in appropriate conditions to build the database. Typically, the H_2S partial pressure during the test is less than 6 bar, and the temperature is between 20°C and 60°C. Samples are collected after several intervals of time in order to follow the position of the H_2S front, degree of conversion, weight evolution, etc.

Mainly pure H_2S tests were performed to produce the database needed for the development and identification of the model. These conditions have the effect of accelerating the mechanisms in the material. Tests were performed at several temperatures representative of the application in order to be able, afterward, to differentiate the temperature dependence of each mechanism. The pertinence of this approach will be confirmed by small-scale and full-scale tests performed in more representative conditions of the application (low partial pressure of H_2S, for instance).

2.2 Analytical Techniques Used for Data Determination

To produce the database, we focus on H_2S front progression and measurements of the degree of conversion, both measured in the thickness of the material. To this end, immersion tests were performed in several conditions of temperature and H_2S partial pressures. During these tests, samples were

Figure 3

Progression of the reaction front in the anti-H_2S layer during time. Step A: H_2S transport through pressure sheath. Step B: H_2S transport through PEZnS (anti-H_2S material already reacted). Step C: H_2S reaction with ZnO (reaction area in red).

placed in a pressure vessel which was then filled with the test gas at the test temperature and pressure. To follow the mechanisms with time, samples were then removed at regular intervals of time, and subsequently cut in half along the diameter for analyses.

2.2.1 Techniques for H$_2$S Front Progression Measurement

Two main techniques were developed to follow the reaction front progression:

– optical microscopy: the PEZnO material contains Fe$_2$O$_3$, which is responsible for the change in color of the material after its reaction with H$_2$S. A microscope equipped with a measuring plate allows one to measure the position of the interface between the already reacted material (gray: FeS$_2$) and the non-reacted material (purple: Fe$_2$O$_3$). A *NIKON* Measurescope 10 is used. Its theoretical accuracy is 1 μm. The distance from the surface of the sample to this interface corresponds to the progression of H$_2$S within the material during the time of the test (Fig. 4);

– Electron Probe Micro-Analysis (EPMA): to verify that the change in color within the material does correspond exactly to H$_2$S progression in the material during the test, a *CAMECA* SX 100 EPMA is used to locate the position of the reaction front and to ensure that no ZnO has reacted after the interface. Figure 5 illustrates, for several immersion times in given immersion conditions, the good agreement obtained between optical and EPMA progression front measurements up to several millimeters. In addition, permeability tests and both middle-scale and full-scale tests performed in permeability configuration and monitored by gas chromatography never revealed any breakthrough of H$_2$S as long as the reaction front did not reach the downstream side of the material. *SRA* microGC are used on test units.

Optical microscopy measurements are performed to follow routine tests on coupons, whereas EPMA measurements are used to confirm front locations on chosen samples (middle-scale and full-scale structures, for instance). Taking

Figure 5

Comparison between optical microscopy and EPMA front progression measurements.

into account all sources of uncertainty, the expected accuracy of both these techniques is of the order of 0.1 mm.

2.2.2 Techniques for Degree of Conversion Measurement

Several methods have been developed to measure the degree of conversion of the material. Depending on the technique employed, this measurement can be performed on a global scale for the whole sample or on a more local scale at several points of the front progression:

– weight uptake: global information for the whole sample. The weight uptake relies on the chemical transformation of the metal oxide in sulfide compounds during operation. This measurement allows one, for instance, to follow the progress of a test or to ensure that a sample is completely converted into sulfur compounds;

– Thermo-Gravimetric Analysis (TGA): TGA can be used to measure the weight loss which characterizes the oxidation of ZnS (or FeS$_2$) in ZnO (or Fe$_2$O$_3$) at very high temperature in the presence of oxygen, and hence to know the relative content of ZnS in the material;

– Elemental Analysis (EA): EA is used to measure the sulfur content in samples. Knowing the initial composition of the material, this measurement can be correlated with the degree of conversion of the reaction. For instance, samples can be cut into successive thin strips (with a microtome) parallel to the front advance in order to know the evolution of the degree of conversion;

– EPMA: already used to measure the position of the reaction front within the material, this technique can rate the degree of conversion at every point of the analyzed

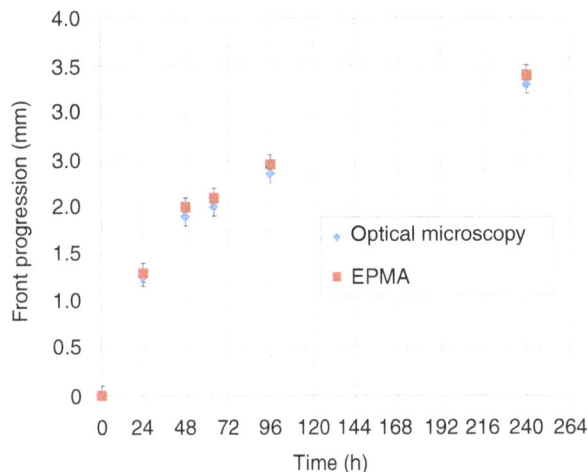

Figure 4

Anti-H$_2$S coupon cut in half along the diameter after a H$_2$S exposure test in immersion configuration. Gray area (border): H$_2$S progression area (the material has reacted with H$_2$S). Purple area (core): unreacted material (absence of H$_2$S).

surface. To do this, a dedicated methodology has been developed in order to compute the quantitative degree of conversion from Wavelength Dispersive Spectroscopy (WDS) spectra obtained on PEZnO samples (Sect. 3.3.2).

3 DEVELOPMENT OF A DEDICATED MULTIPHYSICS DIFFUSION-REACTION MODEL

At the time of an anti-H_2S flexible pipe design, the initial thickness of the anti-H_2S layer will be chosen in order to ensure that the reaction front does not cross the entire anti-H_2S layer to reach the annulus before the end of the lifetime of the pipe. As a consequence, the model will be developed to be able to predict the position of the reaction front, for all testing conditions.

The first step consists of identifying the right physics, based on the comprehension of the mechanisms that take place in the material and experimental characterizations described in Section 2. Then, parameter identification will be performed on the database.

3.1 Diffusion Mechanisms in the Matrix: The Driving Force of the Gas Advance in the Material

As for all polymeric materials, chemical species can permeate through the material. So, at the surface of the material in contact with H_2S, H_2S solubilizes and penetrates into the material then slowly diffuses across the material thickness to the other side.

3.1.1 Diffusion Coefficient, D

The driving force behind the progression of one species in the material is the gradient of the chemical potential of this particular species between the upstream and downstream sides of the material. The diffusion coefficient, D, stands for the intensity of this driving force and hence reflects the speed of the molecules under the effect of the chemical potential gradient between the upstream and downstream sides of the material. D is the kinetic component of the transport.

So, considering a one-way flow of particles along the x axis, the elementary flow of particles through a section, J [mol/(m²/s)], is linked to the local concentration gradient of gas through the diffusion coefficient D. This dependence is expressed by Fick's first law [14]:

$$J_x(x,t) = -D\frac{\partial c}{\partial x}(x,t)$$

Mass balance implies that the evolution of the gas concentration in a given section is equal to the amount of incoming molecules less the amount of outgoing molecules. This balance is expressed by Fick's second law [14]. For a unidirectional flow:

$$\frac{\partial c}{\partial t}(x,t) = -\frac{\partial J_x}{\partial x}(x,t) = \frac{\partial D}{\partial x}\frac{\partial c}{\partial x}(x,t) + D\frac{\partial^2 C}{\partial x^2}(x,t)$$

When the diffusion coefficient, D, can be considered as independent of the gas concentration, c, of the position of the molecule in the material, x, and of time, t, this expression leads to the fundamental equation of diffusion [14]:

$$\frac{\partial c}{\partial t}(x,t) = D\frac{\partial^2 C}{\partial x^2}(x,t)$$

The diffusion coefficient, D, is generally expressed in $m^2.s^{-1}$.

3.1.2 Solubility Coefficient, S

The solubility coefficient is the thermodynamic component of the transport mechanism in polymer: it characterizes the propensity of a given chemical species to solubilize into the polymer. So, at balance at a given pressure (in bar for instance), the solubility represents the amount of gas (in mol, for instance) inside a given volume of the material (one cubic meter, for instance). Consequently, the solubility coefficient can be expressed in $mol.m^{-3}.bar^{-1}$.

The local concentration of gas solubilized in the polymer, c, can be connected to the local partial pressure of the gas, P, by the following equation:

$$c = S(c) \cdot P$$

where $S(c)$ is the solubility coefficient, which depends on the temperature and can vary with c (or P).

If there is no interaction between the diffusing molecule and the polymer, the above equation can be linked to Henry's Law, which suggests, for a given temperature, a linear variation of the concentration, c, with the pressure, P [11, 12, 14]. In this case, the coefficient solubility is equal to the Henry constant.

Formally, the use of Henry's law requires the environment to meet several requirements: no synergistic effects between the gases, presence of only low condensable gas (N_2, O_2, H_2, CH_4, etc.), intermediate pressure (typically < 100 bar), and polymer above its glass transition temperature. For higher pressures or more condensable gas (CO_2, H_2S), it is necessary to take into account condensation and interactions between the gases of different chemical nature. As a consequence, in gas mixtures containing a few percent of H_2S in both CO_2 and CH_4, the partial pressure of H_2S is no longer representative of the activity of this gas. In this case, it is essential to use the fugacity of H_2S which is calculated based on the thermodynamic data of all gas species present in the environment [15-22].

3.2 Chemical Reaction on the Particles: An Obstacle to Gas Progression in the Material

The chemical reactions take place on metal oxide particles, themselves embedded in a polymer. As a result, the gas amount around each reactive particle is a function of the diffusion mechanism described in Section 3.1. Nevertheless, on the material scale, these two mechanisms are interdependent since the local concentration of H_2S depends on both H_2S transport by diffusion and its neutralization rate on particles. So, to model the behavior of such a material, *i.e.* the competition between diffusion and reaction, we need to develop a model in which these two physical mechanisms are coupled. To do that, the Finite Element Method (FEM) was selected especially because of the flexibility in boundary condition definition.

In the case of a chemical reaction of a fluid on a reactive solid, the properties of the solid can be a major point of consideration, because the global kinetics of the reaction can greatly depend on the morphology of the solid (size, shape, porosity rate, etc.). As reported in the literature among fluid-particle reaction theories [24-27], one model (and several variants) has been successfully used: the Shrinking Core Model (SCM), originally developed by Yagi and Kunii [24, 25]. On the scale of one particle (Fig. 6), this model suggests that the chemical reaction first occurs at the outer skin of the particle and then moves into the solid, leaving behind a completely converted and inert solid, commonly called "ash". So, simultaneously with the reaction front advance inside the particle, the unreacted core of the particle shrinks in size. On the scale of the particle, this mechanism leads

to typical evolutions of the degree of conversion, based on the combination of all individual mechanisms of this fluid-particle reaction.

The reaction rate is based on several mechanisms occurring successively at different levels within the particle, each mechanism having its own kinetics. Resistances to the reaction induced by each of these steps are often very different from each other. Since these steps occur successively, it is therefore possible to consider only the one(s) which control(s) the reaction rate. Indeed, the slowest mechanism imposes its rate on other mechanisms. This is the rate-limiting step. Formally, the gas-solid reaction can be divided into five successive steps (Fig. 7) [26, 27]:

- diffusion of H_2S through the polymer-particle interface: corresponds to the diffusion of H_2S across the boundary layer surrounding the particle. In the present case, this boundary layer is the interface between the polymer and the particle. This mechanism, also called transfer resistance, corresponds to the ability of the gas to get out of the polymer to enter the particle. The mass transfer coefficient between the polymer and the particle, k_D, is by definition equal to D_{PEZnS}/R_0 [26, 27]: the ratio of H_2S diffusion in the composite material divided by the radius of the particles;

- diffusion of H_2S in the already reacted part of the particle: corresponds to the penetration of H_2S into the particle and its diffusion through the ZnS layer to the reaction front,

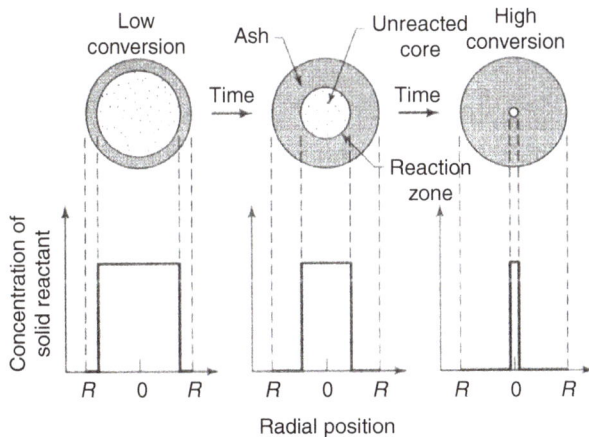

Figure 6

SCM - Evolution of the reaction front inside the reactive particle with time (from [26]).

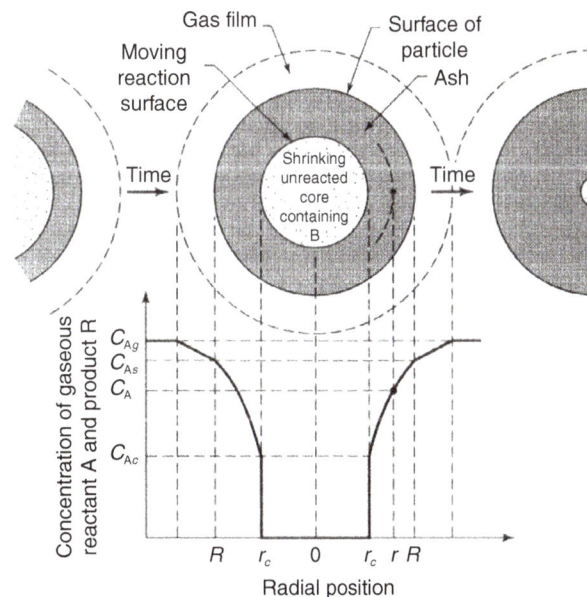

Figure 7

Representation of the successive steps of the SCM (from [26]).

that is to say, the surface of the unreacted core. This mechanism reveals the ability of H_2S to diffuse through the ZnS layer. D_{ZnS} is the diffusion coefficient of H_2S in ZnS, and R_0 is the mean radius of ZnO particles;

- chemical reaction at the surface of the unreacted core. During this step, H_2S is transformed into H_2O and ZnO is transformed into ZnS. The reaction rate constant is k_c;
- diffusion of reaction products (H_2O) in the "ash" back into the exterior of the particle. In our case, we consider that the diffusion of water into the ZnS does not limit the reaction kinetics. According to our experimental results, the presence of water would even tend to favor the global reaction;
- transfer of reaction products (H_2O) back into the polymer through the interface. In our case, we consider that the diffusion of water back into the external environment does not limit the reaction. As for the previous point, according to our experimental results the presence of water would even tend to favor the global reaction.

3.3 KilHyS Parameter Identification and Results

The model developed to describe the behavior of the reactive material was named KilHyS, for Kill Hydrogen Sulfide. This model is based on the mechanisms and parameters described in Sections 3.1 and 3.2.

KilHyS parameter identification is based on immersion test results. These tests were mainly performed with pure H_2S in order to accelerate the mechanisms in the material, but some gas mixture or multiphasic condition tests were also performed. The relevance of this approach is confirmed by larger-scale tests in more representative conditions of the application described in Section 4.

Tests were carried out at several temperatures representative of the application in order to be able to differentiate the temperature dependence of each mechanism.

At regular intervals of time (from a few days to more than one year), samples are removed from the pressure vessel and cut into two parts in order to identify both front progression and front shape by optical and electron microscopy.

3.3.1 Identification of Transport Parameters

Except when components promote an increase in tortuosity in the material (that is to say, a geometrical lengthening of the diffusion path), their incorporation in a polymer generally leads to an increase in the permeability of the composite material. This rise can be the consequence of an increase in the diffusion coefficient, D, and/or the solubility, S. Greater free volume and preferential path formation are potential causes of a composite permeability increase. Thus, transport parameters of the anti-H_2S material must not be determined on the matrix, but on the composite material.

3.3.2 Identification of the Gas-Solid Reaction Parameters

The main method used to identify the diffusion coefficient of H_2S in ZnS, D_{ZnS} and the reaction rate constant of H_2S with ZnO, k_c, is EPMA along the reaction front advance, in the thickness of the material. Knowing the transport parameters in the material, D_{PEZnS} and S_{PEZnS}, then the diffusion coefficient of H_2S in ZnS and the reaction rate constant of reaction of H_2S with ZnO can be identified thanks to degree of conversion measurements performed in the thickness of the specimen taken after several immersion times in H_2S.

As an example, Figure 8 represents the experimental degree of conversion (dots) measured by EPMA after 3 duration times in given test conditions and the corresponding KilHyS calculations (solid curves). The three specimens analyzed in this figure experienced different aging times in the same aging conditions. We can see good agreement on both the front position and front shape of the curves. EPMA precision is limited by the heterogeneity of the material on the scale of the electron beam. The expected accuracy on the determination of the degree of conversion is of the order of 10% of the measured value.

3.3.3 Identification Assessment

A total of more than 50 tests were performed to establish the database. These tests were performed in different conditions:

- temperatures ranging from 38°C to 71°C;
- H_2S partial pressures ranging from 26 mbar to 5.8 bar;
- nature of the fluid: pure H_2S, gas mixture (H_2S-CO_2-CH_4), multiphase conditions (aqueous + gas).

The results of these tests, in terms of front advances, along with the results of the KilHyS model predictions, are

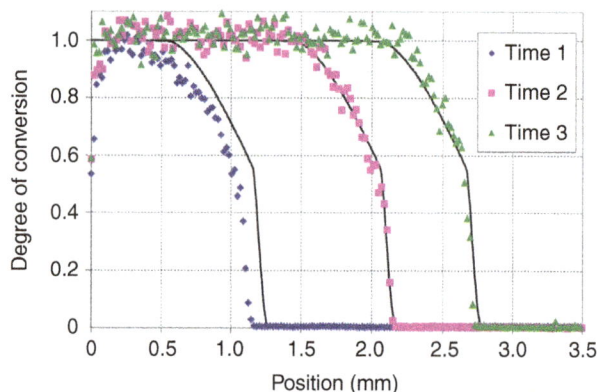

Figure 8

Experimental degrees of conversion (dots) and corresponding KilHyS calculations (solid curves) in the thickness of the material, from the surface to the core (Fig. 4).

presented in Figure 9. Very good agreement between the database and the model is obtained.

Even if the aim of the identification step is to obtain a good fit between experimental data and numerical simulation, the fact that experimental data and numerical simulations are consistent in various experimental conditions of temperature, the pressure and nature of the fluid, and resulting in the determination of physical parameters that follow some Arrhenius laws, promotes the idea that the diffusion-reaction mechanisms described in Sections 3.1 and 3.2, and present in the KilHyS model, correctly describe the experimental behavior of the anti-H_2S material. Described in the next section, the validation in representative environments of both the material and the model will confirm this choice of modeling.

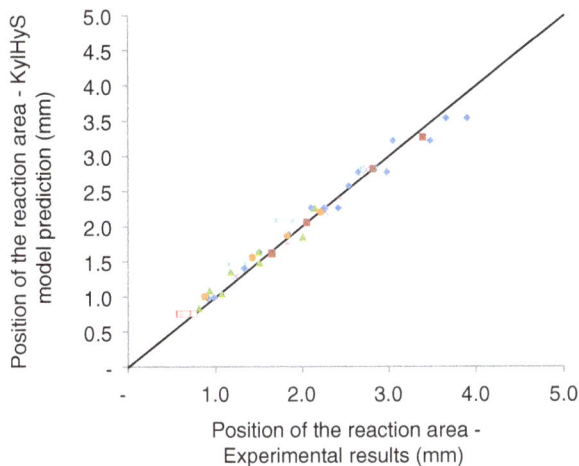

Figure 9

Assessment of the correlation between the experimental database and KilHyS prediction after the parameter identification. Each symbol stands for an experimental condition (H_2S partial pressure, temperature).

4 QUALIFICATION OF BOTH THE MATERIAL AND THE MODEL ON STRUCTURES TESTED IN REPRESENTATIVE ENVIRONMENTS

4.1 Tests on Two-Layer Middle-Scale Tubes in Representative Conditions

The main objectives of these tests on simplified structures are to validate:
- the anti-H_2S concept and model prediction in a two-layer axisymmetric configuration;
- model predictions in a system showing a temperature gradient;
- model predictions in more representative environments than during tests on material samples.

As presented in Figure 10, the two-layer configuration is composed of a tube of polyamide 11 (a pressure sheath material) surrounded by a tube of anti-H_2S material before sealing. The internal diameter is 3.5". A metallic piece with gaps, in contact with the outer surface of the anti-H_2S tube, simulates the geometry of the pressure vault. Test durations vary between 1 month and 4 months.

The progression of the reaction area in the anti-H_2S tube is measured after cutting the tube in several cross sections as shown in Figure 10b. The measurement is performed by using the color change induced by the reaction with H_2S and EPMA. The change in color is illustrated in Figure 11. The gray area corresponds to the bore side where the material has reacted with H_2S. The purple area corresponds to the annulus side where the material is still in its initial state (unreacted with H_2S).

Gas chromatography measurements [12] are also performed at regular intervals of time during these tests in order to monitor the composition of the gas permeating through the polymer materials and to verify that no H_2S is able to penetrate into the annulus.

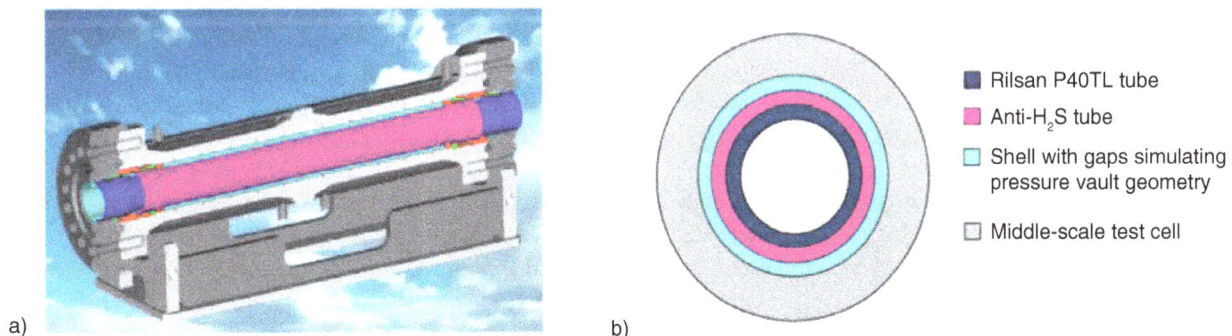

a)   b)

Figure 10

Two-layer middle-scale structure configuration.

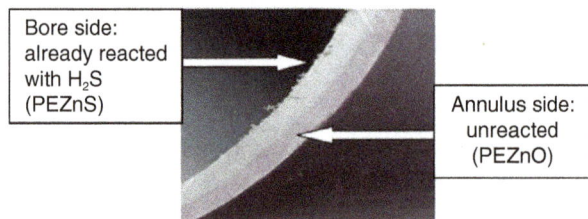

Figure 11

Change in color due to the reaction of H_2S in the anti-H_2S tube shown in Figure 10 (the total thickness of the tube is 4 mm).

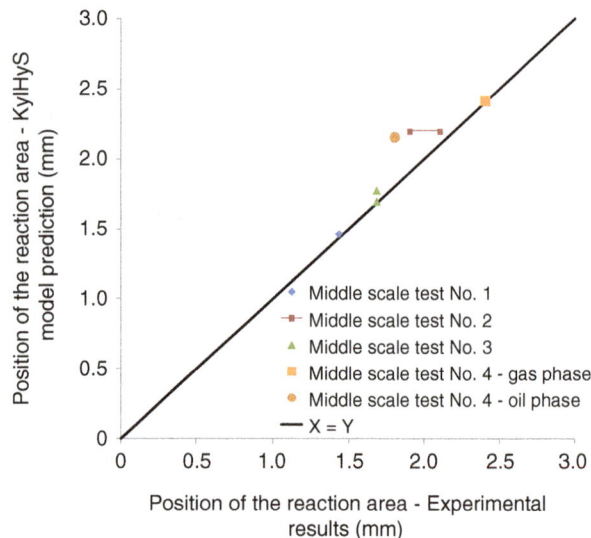

Figure 12

Assessment of the correlation between middle-scale experimental data and KilHyS predictions. The position of the reaction area corresponds to the front progression in the anti-H_2S layer.

As presented in Figure 12, four tests have already been performed, some of them being as long as 120 days. These tests were performed using either pure H_2S or gas mixture CH_4/H_2S, in single-phase (gas) or multiphase conditions (ASTM Gasoil + gas mixture). Out of all tests performed in this configuration, the partial pressure of H_2S is of the order of 5 bar, for a total pressure from 50 bar to 110 bar. The temperature in the bore is 70°C. Very good agreement between the experimental results and KilHyS model prediction is achieved. These results illustrate the ability of the model to describe the right physical mechanisms even in complex experimental conditions as long as each mechanism has been identified on samples.

We can note a lower progression of the experimental front in the oil phase than in the gas phase (test No. 4). We explain this observation by the delay induced by the solubilization of the gas in the oil phase before it can reach the polymer sheath, which is not taken into account by the model.

4.2 Tests on Full-Scale Flexible Pipes in Static and Dynamic Conditions

In addition to the tests performed in pressure vessels and on simplified bi-layer structures, full-scale tests on representative 4.8" flexible pipes structures are ongoing in order to validate the concept on real flexible pipes. The 8 meter long structures were manufactured using *Technip* Le Trait (France) manufacturing unit industrial equipment. The overall duration of these tests is 2 years of effective testing with sampling of anti-H_2S samples of the structure every 6 months.

These full-scale tests are performed on seven representative structures: one dynamic prototype, one static prototype, and five samples of structures. The objectives of these tests are to demonstrate:
- the efficiency of the material on a full-scale structure in both dynamic and static configurations;
- the mechanical performance of the anti-H_2S material both before reaction and after reaction with H_2S.

Each test structure contains a PEZnO anti-H_2S layer. The structures are exposed to a gas mixture containing both CO_2 and H_2S in CH_4. The molar composition of the gas mixture is: $H_2S = 1.5\%$, $CO_2 = 5.0\%$, $CH_4 = 93.5\%$. The pressure of the gas mixture in the bore is 200 bar, the temperature in the bore is 80°C.

The testing devices are located at *IFP Energies nouvelles* in Solaize near Lyon (France), where two units are designed to receive prototypes up to 10 meter long and test them in severe conditions. One unit is dedicated to testing static structures, whereas the second is dedicated to testing structures in dynamic conditions: a hydraulic actuator is used to perform representative mechanical solicitation on the structure on demand (in bending, tension, etc.). In the present case, as shown in Figure 13, the dynamic prototype is slung up in bending between a fixed trolley and a mobile trolley so that the anti-H_2S layer of the structure exhibits a given radius of curvature. Curvature cycling is then performed around this mean value. The surface of the prototype is cooled by closed-loop refrigerated water (not visible in Fig. 13).

Not all the prototypes will be tested during the full 2 years. Every 6 months of effective testing, a sample of the structure, as shown in Figure 14, will be removed from the test area in order to be thoroughly dissected. This way it will be possible to follow the position of the reaction area with aging time.

To date:
- 450 000 cycles have been performed on the dynamic prototype;

Figure 13

8-meter-long flexible pipe tested in cyclic dynamic bending in a H$_2$S-compliant unit (*IFPEN*, Solaize).

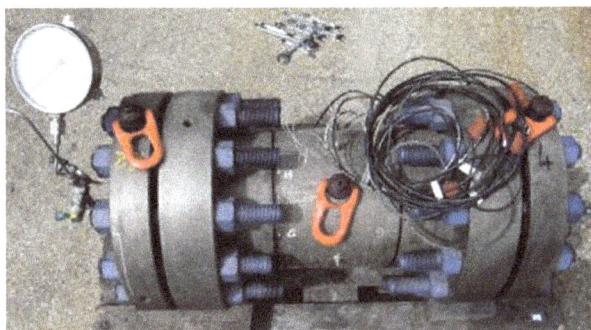

Figure 14

Sample of structure removed every 6 months of effective chemical aging.

– three samples of structure have been taken from the test area in order to be dissected. Two of the three samples have been dissected. The third dissection is ongoing. Observations performed on the anti-H$_2$S layer of the dissected sample confirm the predicted chemical behavior of the material, and consequently the efficiency of the anti-H$_2$S solution;

– no H$_2$S has been detected in the annulus of the samples.

CONCLUSION

As a result of a collaboration between *IFPEN* and *Technip*, a barrier material against H$_2$S has been successfully developed for large-scale and severe industrial applications. From the development of the material (selection of raw materials, compounding, characterization) to its qualification in a representative environment, multiphysics modeling based on the coupling of Fick diffusion in the polymer and gas-solid reactions enables the interpretation of experimental data

and the assessment of the results' relevance. The parameter identification was carried out on the results of immersion tests performed on anti-H$_2$S material samples in several conditions. Then, the robustness of the model was confirmed through the results of middle-scale tests performed on two-layer tubes with a thermal gradient in representative conditions (gas mixture with gasoil, for instance).

Full-scale tests with representative 4.8" flexible pipe structures are ongoing in static and dynamic configurations. To date, the progression of H$_2$S in the anti-H$_2$S layer is consistent with the KilHyS model predictions.

Since the qualification in static conditions is nearly completed, this new anti-H$_2$S layer will soon be available for static flexible flowline application.

ACKNOWLEDGMENTS

The authors would like to acknowledge *Technip*'s Product Engineering Department in Le Trait and both the Applied Chemistry and Physical Chemistry Division and the Applied Mechanics Division of *IFP Energies nouvelles* for their contribution to this development.

REFERENCES

1 *Guidelines on materials requirements for carbon and low alloy steels for H$_2$S-containing environments in oil & gas production* (1995) London, UK, EFC Pub. 16, The Institute of Materials.

2 Kermani M.B. (1991) Proc. Conf. Corrosion '91, Cincinnati, OH, USA, NACE, Paper 21.

3 Kane R.D., Horvath R.J., Cayard M.S. (1996) *Wet H$_2$S cracking of carbon steels and weldments*, NACE International, Houston, TX, USA.

4 NACE Technical Report 1F192 (1993 Revision), *Use of Corrosion Resistant Alloys in Oilfield Environments*, NACE, Houston, TX, USA, 1993.

5　Kermani M.B. (2000) Material Optimisation for Oil and Gas Sour Production, *Corrosion conference*, Paper No 156.

6　Craig B.D. (1993) Sour Gas Design Considerations, *SPE Monograph No. 15*.

7　Crolet J.L., Maisonneuve G. (2000) Construction of a Universal Scale of Severity for Hydrogen Cracking, *NACE Annual Conference, Corrosion 2000*, 26-31 March, Orlando, Florida, USA, NACE-00127.

8　Sulfide Stress Cracking Resistant Metallic Materials for Oilfield Equipment, NACE MR0175-2002/ISO 15156.

9　Kermani M.B., Martin J.W., Esaklul K.A. (2006) Materials Design Strategy: Effects of H_2S/CO_2 corrosion on materials selection, NACE International, *Corrosion conference, Corrosion 2006*, 12-16 March, San Diego, California, USA, NACE-06121.

10　Plennevaux C., Kittel J., Frégonèse M., Normand B., Ropital F., Grosjean F., Cassagne T. (2013) Contribution of CO_2 on hydrogen evolution and hydrogen permeation in low alloy steels exposed to H_2S environment, *Electrochem. Comm.* **26**, 17-20.

11　Klopffer M.H., Flaconnèche B. (2001) Transport properties of gases in polymers: Bibliographic review, *Oil & Gas Science and Technology* **56**, 3, 223-244.

12　Flaconnèche B., Martin J., Klopffer M.H. (2001) Transport properties of gases in polymers: Experimental methods, *Oil & Gas Science and Technology* **56**, 3, 245-259.

13　Aubry J.C., Saas J.N., Taravel-Condat C., Benjelloun-Dabaghi Z., De Hemptinne J.C. (2002) Moldi (Tm): A fluid permeation model to calculate the annulus composition in flexible pipes, *Oil & Gas Science and Technology* **57**, 2, 177-192.

14　Crank J. (1979) *The Mathematics of Diffusion*, Oxford University Press.

15　Peng D., Robinson D.B. (1976) New 2-Constant Equation of State, *Industrial & Engineering Chemistry Fundamentals* **15**, 1, 59-64.

16　Soave G. (1972) Equilibrium constants from a modified Redlich-Kwong equation of state, *Chemical Engineering Science* **27**, 6, 1197-1203.

17　Redlich O., Kwong J.N.S. (1949) On the Thermodynamics of Solutions. V. An Equation of State. Fugacities of Gaseous Solutions, *Chemical Reviews* **44**, 1, 233-244.

18　Nath S.K., de Pablo J.J. (1999) Solubility of small molecules and their mixtures in polyethylene, *Journal of Physical Chemistry B* **103**, 18, 3539-3544.

19　Nath S.K., Banaszak B.J., de Pablo J.J. (2001) Simulation of ternary mixtures of ethylene, 1-hexene, and polyethylene, *Macromolecules* **34**, 22, 7841-7848.

20　Raharjo R.D., Freeman B.D., Paul D.R., Sarti G.C., Sanders E.S. (2007) Pure and mixed gas CH_4 and n-C_4H_{10} permeability and diffusivity in poly (dimethylsiloxane), *Journal of Membrane Science* **306**, 1-2, 75-92.

21　Raharjo R.D., Freeman B.D., Sanders E.S. (2007) Pure and mixed gas CH_4 and n-C_4H_{10} sorption and dilation in poly (dimethylsiloxane), *Journal of Membrane Science* **292**, 1-2, 45-61.

22　Memari P., Lachet V., Klopffer M.H., Flaconneche B., Rousseau B. (2012) Gas mixture solubilities in polyethylene below its melting temperature: Experimental and molecular simulation studies, *Journal of Membrane Science* **390**, 194-200.

23　Marcilly C. (2005) *Acido-Basic Catalysis*, Editions Technip.

24　Yagi S., Kunii D. (1955) Studies on combustion of carbon particles in flames and fluidized beds, in *Proceedings of the 5th Symposium (International) on Combustion*, Hottel H.C. (ed), Reinhold, New York, USA, pp. 231-244.

25　Yagi S., Kunii D. (1961) Fluidized-Solids Reactors with Continuous Solids Feed. I: Residence Time of Particles in Fluidized Beds, *Chemical Engineering Science* **16**, 3-4, 364-391.

26　Levenspiel O. (1999) *Chemical Reaction Engineering*, 3rd edn., John Wiley & Sons, Inc., New York, USA, pp. 566-588.

27　Houzelot J.-L. (2000) Réacteurs chimiques polyphasés : couplage réaction/diffusion, *Techniques de l'Ingénieur, Traité Génie des Procédés*, 5J4012, J4012.1-J4013.2.

Biofuels Barrier Properties of Polyamide 6 and High Density Polyethylene

L.-A. Fillot*, S. Ghiringhelli, C. Prebet and S. Rossi

LPMA (UMR 5268 CNRS/Solvay), Advanced Polymer Material Department, R&I Centre Lyon, SOLVAY, 69192 Saint-Fons - France
e-mail: louise-anne.fillot@solvay.com

* Corresponding author

Abstract — *In this paper, a comparison of the biofuels barrier properties of PolyAmide 6 (PA6) and High Density PolyEthylene (HDPE) is presented. Model fuels were prepared as mixtures of toluene, isooctane and ethanol, the ethanol volume fraction varying between 0% and 100%. Barrier properties were determined at 40°C by gravimetric techniques or gas chromatography measurements, and it was shown that polyamide 6 permeability is lower than that of polyethylene on a wide range of ethanol contents up to 85% of ethanol (E85) in the biofuel, permeability of PA6 being 100 times lower than that of HDPE for low ethanol content fuels (E5, E0). The time-lags were also compared, and on the whole range of ethanol contents, HDPE permeation kinetics appears to be much faster than that of PA6, the time lag for a 1 mm thick specimens in presence of E10 being 50 days for PA6 and 0.5 days for HDPE. The compositions of the solvent fluxes were analyzed by FID (Flame Ionization Detector) gas chromatography, and it turned out that the solvent flux was mainly made up of ethanol (minimum 95%) in the case of PA6, whereas in the case of HDPE, solvent flux was mainly made up of hydrocarbons. The implication of this difference in the solvent flux composition is discussed in the present article, and a side effect called the "fuel exhaustion process" is presented. The influence of the sample thickness was then studied, and for the different biofuels compositions, the pervaporation kinetics of polyamide 6 appeared to evolve with the square of the thickness, a long transitory regime being highlighted in the case of PA6. This result implies that the time needed to characterize the steady state permeability of thick PA6 parts such as fuel tanks can be very long (one year or more), this duration being far superior to the Euros 5 or Euro 6 standard emission measurements time scale. The influence of temperature on the permeability was finally assessed, and the activation energy that is the signature of the temperature dependence of the barrier property turned out to be similar for the different biofuels compositions.*

Résumé — Propriétés barrière aux bio essences du polyamide 6 (PA6) et du polyéthylène haute densité (PEHD) — Dans cet article, les propriétés barrière aux bio essences du PolyAmide 6 (PA6) sont comparées à celles du PolyEthylène Haute Densité (PEHD). Des essences modèles sont préparées en faisant varier entre 0 et 100 % la fraction volumique d'éthanol dans des mélanges ternaires éthanol/toluène/isooctane. Les propriétés barrière aux bio essences sont caractérisées à 40 °C par gravimétrie ou par des mesures de chromatographie en phase gaz, et il est montré que la perméabilité du PA6 est plus faible que celle du PEHD sur une grande gamme de compositions d'essences contenant jusque 85% d'éthanol, le ratio de performance atteignant un facteur 100 pour les essences faiblement alcoolisées (E5, E0). Les temps d'induction ont aussi été comparés, et pour toutes les compositions d'essences évaluées, il apparait que la cinétique de pervaporation du PEHD est bien plus rapide que celle du PA6, le temps d'induction pour un échantillon de 1 mm d'épaisseur en présence de E10 étant 50 jours dans le cas du PA6 contre 0,5 jour dans le cas du PEHD. La composition du flux de solvant a ensuite été analysée par chromatographie en phase gaz avec détecteur FID (*Flame Ionization Detector*), et il s'avère que le flux de solvant est dans le cas du PA6 très majoritairement composé d'éthanol (minimum 95 %), tandis que dans le cas du PEHD, le flux de solvant est composé majoritairement d'hydrocarbures. L'implication de cette différence de composition du flux de solvant est discutée dans le présent article, et un effet collatéral appelé processus d'appauvrissement de l'essence est présenté. L'influence de l'épaisseur de l'échantillon est ensuite étudiée, et pour toutes les compositions d'essence, il apparait que la cinétique de pervaporation du PA6 évolue avec le carré de l'épaisseur de l'échantillon, un long régime transitoire étant observé dans le cas du PA6. Ce résultat implique que le temps nécessaire pour caractériser la perméabilité en régime permanent de pièces massives en PA6 telles que des réservoirs d'essence peut être extrêmement long (une année ou plus), cette durée étant bien supérieure à l'échelle de temps associée aux mesures d'émissions réalisées dans le cadre de normes telles que Euro 5 ou Euro 6. L'influence de la température sur la perméabilité du PA6 est finalement étudiée, et il apparait que la valeur d'énergie d'activation qui traduit la dépendance de la perméabilité avec la température dépend peu du taux d'éthanol dans l'essence.

INTRODUCTION

In the last ten years, petrol price hike and environmental issues strongly encourage to reduce greenhouse gas emissions, an increasing interest being set on fuels derived from the biomass (biofuels), and notably fuels containing alcohol. In parallel, in order to lighten cars, metal is progressively replaced by plastic, but polymer parts such as fuel tanks must fulfill strict specifications in term of barrier properties. The emission standards are defined in a series of directives such as the Euro 5 (since 2009) or the Euro 6 (starting from 2014) directives in Europa, or American programs of the Californian Air Resources Board (CARB) such as the Partial Zero Emission Vehicles (PZEV). Evaporative emission permeation tests such as the SHED or the mini-SHED tests (Sealed Housing Evaporative Determination) are performed by manufacturers in order to evaluate and homologate the performance of fuel systems. In the SHED configuration, a whole vehicle which has rolled 3 000 km and which contains a fuel tank filled at 40% with E5 (5% ethanol, 95% gasoline) is placed in a chamber in which a specific temperature variation between 23°C and 31°C is applied during 24 hours. During this one day cycle, the total emissions are measured with a Flame Ionization Detector (FID), and in the case of Euro 6, this total emission should not exceed 2 g/day per vehicle, the fuel system being estimated to be responsible for around 80% of the observed evaporations [1].

The most commonly used polymer based solution for fuel systems was initially High Density PolyEthylene (HDPE), but as regulations became more and more drastic, a substantial improvement of the barrier properties of fuel systems was made by adding a fluorination post-treatment to HDPE, or by developing multilayer structures comprising a layer of a high barrier material such as Ethylene Vinyl Alcohol copolymer (EVOH) sandwiched between layers of polyethylene or polyethylene/polyamide blends. But the fluorination solution is not environmentally-friendly, and the co-extrusion of multilayers plastic fuel tanks requires specific and expensive processing equipment. As PolyAmide 6 (PA6) exhibits intrinsically a high barrier property to biofuels, its use in fuel tanks designed for a monolayer strategy is very relevant.

In this paper, the barrier properties of PA6 are compared to the barrier properties of HDPE in presence of fuel containing various ethanol contents. In the last 30 years, research and development of alcohol fuels has taken center stage as a way to reduce vehicle tailpipe emissions and the reliance to fossil fuels. And for the past several decades, a variety of automotive manufacturers have been designing and manufacturing vehicles to run on fuels made of 85% ethanol and 15% gasoline (E85), the fuel E10 (10% ethanol, 90% gasoline) being the fuel commonly supplied in French gas stations. In contrast with polyethylene, polyamide is a polar polymer due to the presence of amide groups in its structure. A direct consequence of this polarity difference between polyamide and polyethylene is that polyamide is expected to present excellent barrier properties to non-polar solvents such as hydrocarbons, whereas polyethylene is expected to present excellent barrier properties to polar solvents such as ethanol or water. The behavior of these polymers in presence of hydrocarbons or some polar/apolar solvents mixtures [1-3], and the strategy to add polyamide in polyolefins to increase the barrier performance to hydrocarbons have been widely studied in the literature [4-12], but to our knowledge, few comparative studies of the behavior of these polymers in presence of toluene/isooctane/ethanol ternary mixtures have been conducted.

As permeability of polymers to solvents depends on both solvent solubility and solvent diffusion kinetics, sorption experiments on PA6 and HDPE were conducted in this work in addition of pervaporation experiments. Gravimetric tests were used as well as FID gas chromatography analysis which allowed quantifying the different molecules present in the solvent flux. The obtained results were then discussed regarding the actual methodology and the expectations of the Euros 6 emission standards. Beyond the comparison of the biofuels barrier properties of PA6 and HDPE, the goal of this article is to shed some light on the environmental parameters that affects significantly the biofuels barrier properties of PA6 and HDPE (fuel composition, sample thickness, temperature) and to highlight the consequences of the differences observed between PA6 and HDPE.

1 EXPERIMENTAL

1.1 Materials

Polyamide 6 (PA6) was a commercial grade supplied by *Solvay* with a number average molecular weight determined by size exclusion chromatography of 27 600 g/mol. HDPE was a commercial grade supplied by *Polymeri Europa* with a number average molecular weight determined by size exclusion chromatography of 21 300 g/mol. It should be kept in mind that the polymers evaluated in this work were virgin PA6 or HDPE grades, and not formulations specifically developed for fuel systems.

1.1.1 Injection Molded Plates

Injection molded plates of PA6 and HDPE of dimensions 100×100 with a thickness of 0.8 mm were realized using a *Demag* injection molding machine exhibiting a clamping force of 80 tons. PA6 and HDPE were processed in similar processing conditions: the injection speed was 170 mm/s, the melt temperature was set at 270°C and the mold temperature at 80°C. It should be pointed out that these processing conditions and namely the processing temperature are for HDPE a bit outside

its conventional processing window, but the integrity of the HDPE polymer turned out to be little affected. Indeed, thermogravimetric analyses showed only a 2.5% weight loss after maintaining the polymer 30 minutes at 290°C, high temperature size exclusion chromatography measurements performed on HDPE before and after processing showed only minor changes on weight average molecular mass (Mw: −15%) and Polydispersity Index (PI: + 50%), and capillary rheometry measurements on HDPE before and after processing showed no significant viscosity change.

1.1.2 Cast Extruded Films

Extruded cast films of PA6 were realized using a co-rotating twin screw *Leistritz* extruder of diameter 34 mm and ratio length in diameter L/D of 35 associated to a film set-up composed of a specific flat die of 300 mm width, followed by two chill-roll set at 125°C, and series of rolls that finally allow to wind the film. The gap of the die and the speed of the rolls were changed in order to generate different film thicknesses of about 50, 125, 200, 500, and 800 μm. The extruder throughput was 7.8 kg/h, the melt temperature set at 255°C, and the screw speed set at 250 rpm.

1.2 Experimental Techniques

1.2.1 Characterization of Samples Crystallinity

The crystallinity of polyamide 6 samples was characterized at several scales with Differential Scanning Calorimetry (DSC), Wide Angle X-ray Scattering (WAXS) and optical microscopy. HDPE was only characterized by DSC. The crystalline fraction was determined by DSC with a TA Q2000 calorimeter. Samples (between 7 and 15 mg) were put in non-hermetic aluminium pans and heated from 25°C to 300°C at a heating rate of 10°C/min. The crystalline fraction (X_c in $\%_{weight}$) was then determined from Equation (1):

$$X_c = \left(\frac{\Delta H_m}{\Delta H_m^{100\%}} \right) * 100 \qquad (1)$$

where ΔH_m is the measured melting enthalpy, and $\Delta H_m^{100\%}$ the reference melting enthalpy associated to a theoretical 100% crystalline polymer (taken as 191 J/g for PA6 and 293 J/g for HDPE [12]).

The crystalline structure of PA6 was characterized by WAXS at room temperature (23°C), using a *Brucker* D8 Advance diffractometer. The source consisted in a ceramic tube with a copper anode generating CuKα

TABLE 1
Composition of the different solvent mixtures studied in this work (volumetric proportions)

Mixture	E0	E5	E10	E22	E40	E60	E85	E100
% ethanol (vol)	0	5	10	22	40	60	85	100
% toluene (vol)	50	47.5	45	39	30	20	7.5	0
% isooctane (vol)	50	47.5	45	39	30	20	7.5	0

radiation, and the measurements were performed in the reflection mode.

The spherulite structure in PA6 was characterized by polarized optical microscopy. Specimens were first included in epoxy resin and placed in an oven at 70°C overnight. From these prepared specimens, sections of 2 μm thick were cut at room temperature by using a *Reichert* Ultracut S microtome equipped with a diamond knife. The sections were analyzed by using a *Leica* light polarization microscope equipped with video camera *Leica* DFC420C.

1.2.2 Solvent Mixtures Preparation

Commercial fuels were assimilated to ternary mixtures of ethanol, toluene and isooctane. These mixtures were prepared with different ethanol volume fractions, the list of the prepared mixtures and their composition being shown in Table 1. Analytical-grade solvents were used and ethanol was kept anhydrous with a desiccant.

1.2.3 Solvent Sorption Assessment by Gravimetry

Samples of dimensions 50 × 50 were cut in extruded films or in injection molded plates to study their solvent intake by gravimetry. Before immersion in the solvent, PA6 samples were dried overnight at 110°C under vacuum. The sorption device consisted in a crystallizing dish filled with the solvent or the solvent mixture and regulated in temperature thanks to a stir plate set at 40°C. Samples were immerged in the solvent and were regularly weighed during time with a balance precision of 0.1 mg until the sorption equilibrium was reached. The solvent weight fraction in the polymer was then plotted as a function of time.

1.2.4 Pervaporation Assessment by Gravimetry

Samples of dimensions 50 × 50 were cut in extruded films or in injection molded plates to study their barrier property to biofuel at 40°C by gravimetry. Prior testing, the thickness of the samples was carefully measured, and in the case of PA6, samples were dried overnight at

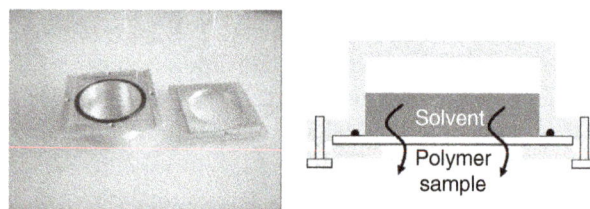

Figure 1

Picture of a permeation cup and configuration of the gravimetric pervaporation experiment.

110°C under vacuum. Polymer samples were then fixed to aluminum permeation cups filled with the solvent or the solvent mixture and comprising a *Viton* sealing joint, a picture of the permeation cup being shown in Figure 1.

During the pervaporation experiment, the polymer surface was directly in contact with the solvent (no air space between the polymer and the solvent) and the exchange surface between the polymer and the environment consisted in a disc of radius equal to 2 cm (exchange surface = 12.56 cm²). The permeation cup was then introduced in an oven regulated in temperature (40°C) and in hygrometry (20%), and the set (permeation cup + polymer sample + solvent) was weighed during time with a balance precision of 0.1 mg. The weight loss divided by the exchange surface and multiplied by the sample thickness (normalized weight loss) was then plotted as a function of time. The derivative of this normalized weight loss as a function of time was then calculated, the obtained value being called hereafter "reduced Flux – J.L" because it corresponds to a solvent flux (J) related to specimen thickness (L). During the pervaporation experiment, the reduced flux firstly remains close to 0 during a certain time period (induction period), and then it increases progressively until it reaches the stationary state. The value of the reduced flux at the stationary state was used for defining the permeability of the sample. Another parameter called "time-lag" was also determined: it corresponds to the end of the induction period and was determined

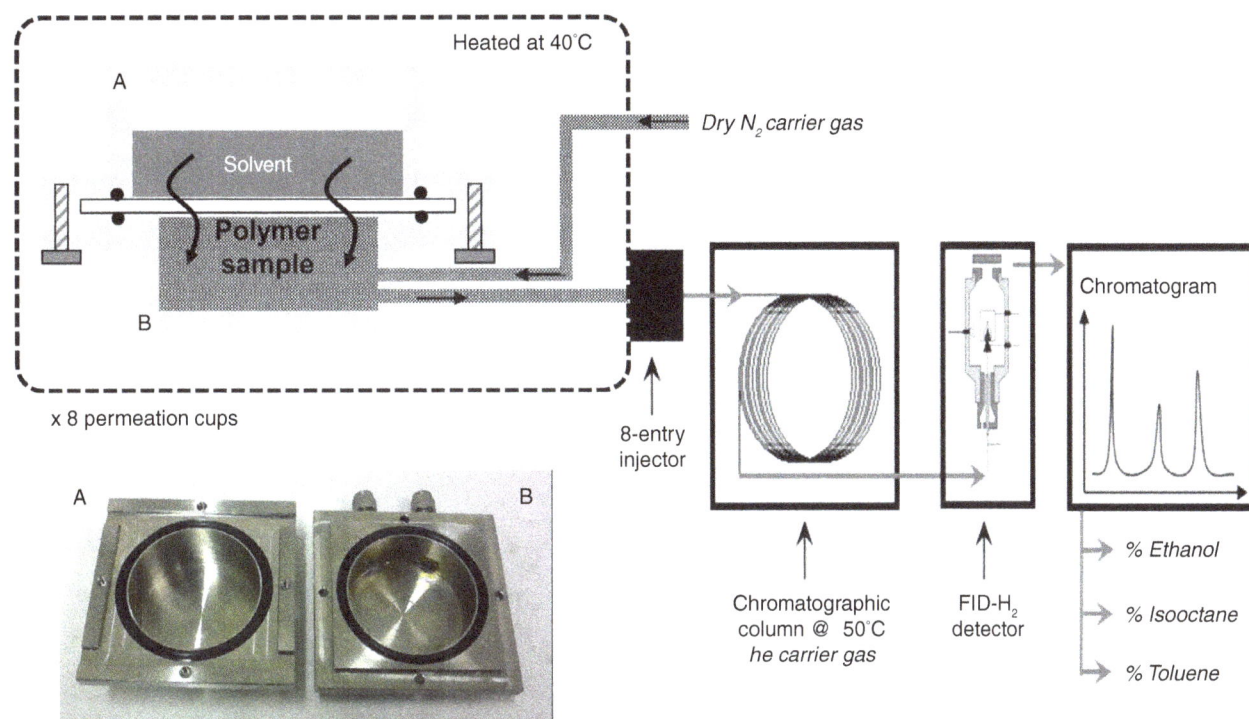

Figure 2

Set-up for analyzing the solvent flux by FID gas chromatography.

as the intersection of the tangent of the normalized weight loss at the stationary state and the *x*-axis (*Fig. 6*).

1.2.5 Pervaporation Assessment by FID Gaz Chromatography

In order to analyze the composition of the solvent flux permeating through the polymer sample, a variant of the permeation cup described in the previous paragraph was connected to an *Agilent* G1530N Gas Chromatograph (GC) equipped with a FID. The GC was equipped with a *Heliflex* AT-1 non-polar column, a FID detector, and an 8-entry 1:50 splitless injector (injection volume = 50 μL) with a fixed split ratio of 1:1. The carrier gas from the sample to the injector was dry nitrogen with a flux set at 15 mL/min, and from the injector to the chromatographic column, helium with a flux set at 20 mL/min was used as the carrier gas. Both the injector and the column were heated at 50°C. The solvent vapors were separated in the chromatographic column and quantified by the FID detector. The result of the experiment was a chromatogram evidencing until 3 peaks associated to ethanol, toluene and isooctane, the surface of the peaks being proportional to their concentration *via* a proper calibration constant. Thanks to the selection valve,

8 permeation cups containing the polymer and the solvent could be connected to the GC-setup, each sample being analyzed once in one hour. The permeation cups were placed in an oven set at 40°C, the connection between the sample and the GC-FID being also heated at 40°C. A schematization of the experimental setup is shown in Figure 2.

2 RESULTS & DISCUSSION

2.1 Characterization of the Crystallinity

It is generally accepted that diffusion of small molecules in semi-crystalline polymers mainly occurs at the solid state in the amorphous phase of the polymer, the accessibility of crystalline domains to solvents being very limited [13, 14]. A direct consequence of this restricted accessibility is that permeability depends on crystalline fraction [15], and in addition, for a given crystalline fraction, it has been shown that the crystalline structure of PA6 also affects the barrier properties of polyamide 6, notably in presence of ethanol or ethanol-toluene mixtures [16]. In order to interpret the permeability results

TABLE 2
Crystalline fraction of the different materials determined by DSC

Polymer	PA6						HDPE
Process	Film cast extrusion					Injection molding	
Thickness	50 μm	125 μm	200 μm	500 μm	800 μm	1 mm	1 mm
Xc (%) – (+/−3%)	31	30	30	30	32	34	65

obtained on materials processed in different conditions (for instance film casting *versus* injection molding), it is thus of great importance to accurately characterize the crystalline morphology of the materials.

Table 2 firstly shows the crystalline fraction of the different specimens determined by DSC. Crystalline fraction of HDPE (65%) is much more important than the one of PA6 (30-35%). The crystalline fraction in PA6 films does not seem to depend on film thickness, and the mean value of 31% obtained for films is close to the one obtained for the PA6 injection molded plate (34%).

The crystalline structure of the 50 μm PA6 film and the 1 mm PA6 injection molded plate were then characterized by WAXS, the diffraction patterns being shown in Figure 3.

The 50 μm extruded film and the 1 mm injection molded plates both exhibit one main intense peak at 2 theta = 21° as well as one minor peak at 2 theta = 10°, suggesting the presence in both samples of mainly the gamma phase. The broadening of the main peak for the injection molded sample suggests the additional presence of a small fraction of either the alpha phase, or the metastable beta phase [17].

The spherulite structure was finally assessed by polarized optical microscopy, the obtained pictures of the extruded films and the injection molded plate being shown in Figure 4.

Well defined spherulites with an average diameter of 5-10 μm are observed on the films of different thicknesses as well as on the injection molded plate. But contrary to the extruded films, the injection molded plate exhibits a 10 μm skin free of spherulites, suggesting that the cooling condition in the injection molding process is more severe (*i.e.* fast cooling), than the cooling condition in the film casting extrusion process. This is consistent with the temperature of the cooling devices (80°C for the mold used in injection-molding process and 125°C for the Chill-Roll used in film cast extrusion)

To conclude on the crystalline structure of the studied materials, it seems that the crystalline fraction and the crystalline structure (*i.e.* crystalline lattice) are similar in the films presenting different thicknesses and in the

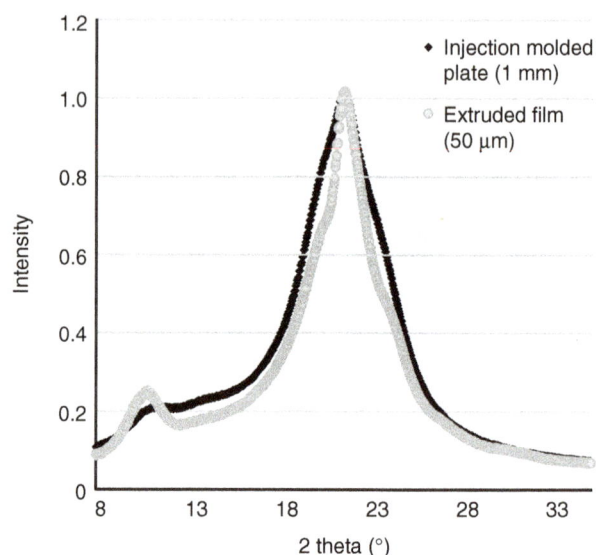

Figure 3

X-ray Diffraction patterns of the 50 μm PA6 film and the 1 mm PA6 injection molded plate at room temperature.

injection molded plate (X_c ~32%, presence of mainly the gamma phase). But at the scale of the spherulites, a 10 μm thick skin deprived of spherulites is only seen in the injection molded plate. Concerning HDPE, it should be kept in mind that its crystalline fraction (65%) is far more important than the one of PA6.

2.2 Analysis of Sorption and Permeation Data in Presence of E10

Figure 5 shows the sorption curves obtained by gravimetry of 1 mm thick PA6 and HDPE plates immersed in E10 at 40°C, the dash lines representing only guides for the eye.

The amount of E10 sorbed at the thermodynamic equilibrium in PA6 and HDPE is very close (7.5-8%), but if we consider that solvents penetrate only the amorphous phase, it turns out that the amount of E10 in the amorphous phase of HDPE (23%) is higher than the one

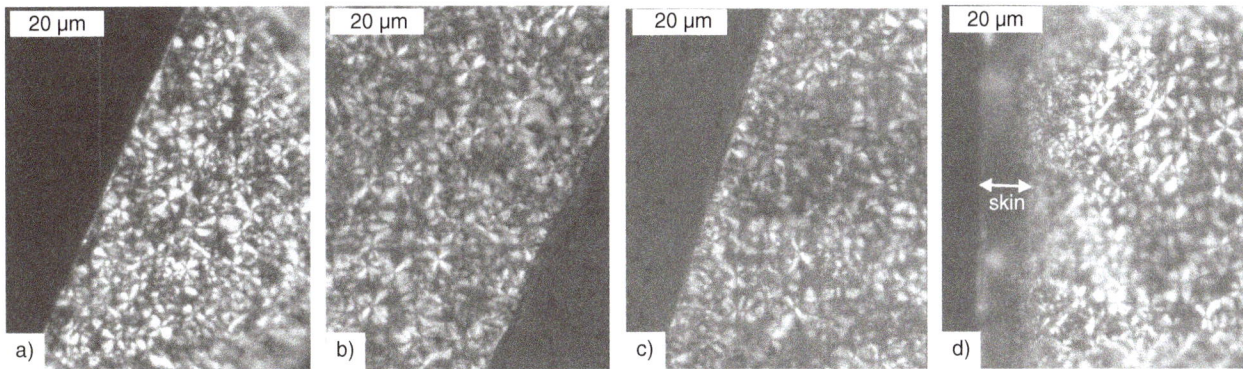

Figure 4

Polarized optical micrographs of a) 50 μm film, b) 200 μm film, c) 800 μm film, d) 1 mm injection molded plate.

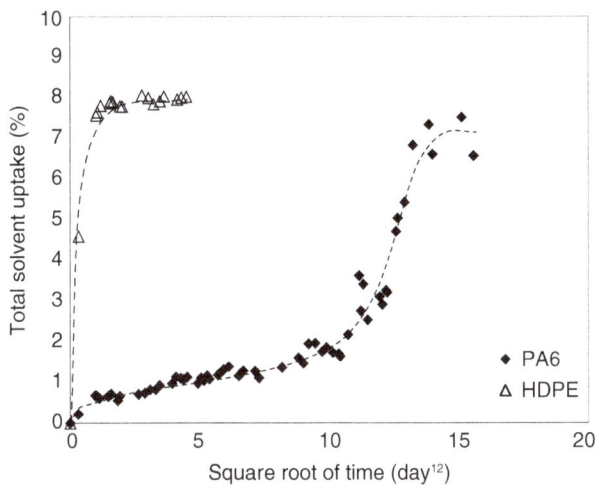

Figure 5

E10 uptake as a function of the square root of time for 1 mm thick PA6 and HDPE plates at 40°C.

in the amorphous phase of PA6 (12.3%), the equation used for calculating solvent intake in the amorphous phase being defined as:

$$G_{amorphous\,phase} = \frac{G_{total}}{(1 - X_c)} \qquad (2)$$

with $G_{amorphous\,phase}$ the solvent uptake in the amorphous phase, G_{tot} the measured solvent uptake and X_c the crystalline fraction of the polymer.

Figure 5 also shows that the sorption kinetics in HDPE is much faster than that of PA6. In the case of 1 mm thick HDPE sample, swelling is completed after 1 day whereas in the case of 1 mm thick PA6, 200 days

are needed to reach the sorption equilibrium. The evolution of solvent uptake with the square root of time is not linear in the case of PA6, which suggests a non fickian behavior. This kind of diffusion called abnormal or Case-II diffusion is generally observed when the solvent is able to modify the molecular mobility of the polymer during the sorption experiment, this modification consisting in decreasing the glass temperature of the polymer until a value equal or inferior to the testing temperature [18-21]. In the first stages of abnormal diffusion, the solvent is absorbed following a fickian diffusion (governed by the solvent concentration gradient), but at some point (corresponding roughly in our case to the inflexion point observed on the PA6 sorption curve of Fig. 5), the concentration of the polymer will be such, that locally the presence of the solvent will allow the relaxation of the polymer at the testing temperature. In this zone, the diffusion of the solvent is accelerated since the polymer is plasticized, and thus there is the appearance of a plasticization front. The diffusion is no longer governed by the solvent concentration gradient, but rather by the ability of the solvent to modify the molecular mobility of the polymer. We showed in a previous work [22] that the solvents present in E10, and mainly ethanol, strongly impacts the molecular mobility of the amorphous phase of PA6 (Tg of PA6 swollen with E10 = −30°C). Consequently, at a testing temperature of 40°C, PA6 is expected to become rubbery during the sorption experiment. The non fickian behavior of PA6/ethanol systems has been described and discussed in the literature, on the basis of heuristic modified diffusion-relaxation models [19, 23] or on the basis of a thermodynamical approach [24].

Figure 6 shows then the pervaporation curves obtained by gravimetry of 1 mm thick HDPE and PA6 plates put in contact with fuel E10 at 40°C. In the case of HDPE, pervaporation starts quasi immediately, whereas in the case of PA6, no significant weight loss is observed before 40-50 days (time-lag $\sim$50 days). The small initial weight gain observed on the PA6 curve should be attributed to water intake, initially dried polyamide progressively equilibrating at the current hygrometry level of the pervaporation test (20% in the present case).

In order to precisely determine the E10 permeabilities of PA6 and HDPE, the derivative of the normalized weight loss as a function of time (Reduced Flux J.L.) is plotted in Figure 7. In the case of HDPE, the reduced flux at the stationary state is about 65 g.mm/m^2.day whereas it is only 2 g.mm/m^2.day in the case of PA6.

It should be also pointed out that 60 days are necessary for PA6 to reach a stationary regime whereas less than one day is required for HDPE. It is known that permeability (P) is driven by the solubility (S) and the diffusion coefficient (D) of the diffusing species in the polymer, the relationship $P = SD$ being valid in the case of fickian diffusion. But even if the diffusion is not fickian, it is clear that P depends on both S and D. In the present case, it is shown that HDPE and PA6 exhibits similar E10 solubilities, but diffusion kinetics is much faster in the case of HDPE. This fast diffusion kinetics in HDPE may thus be at the origin of the poor E10 barrier property of HDPE in comparison with the one of PA6. It should be underlined that PA6 E10 permeability is 35 times lower than the one of HDPE, despite a crystalline fraction in PA6 divided by almost 2.

2.3 Effect of Ethanol Content

The influence of ethanol content in the fuel on the barrier properties of HDPE and PA6 was then studied. Different ethanol/toluene/isooctane mixtures containing different ethanol fractions were prepared according to the protocol described in the experimental section, and HDPE and PA6 samples were immerged in these mixtures at 40°C until the sorption equilibrium was reached. The solvent uptake at the equilibrium measured by gravimetry for both polymers is plotted in Figure 8 as a function of the ethanol content in the fuel. In the case of PA6, solvent uptake at the equilibrium increases with ethanol content, whereas the opposite is observed for

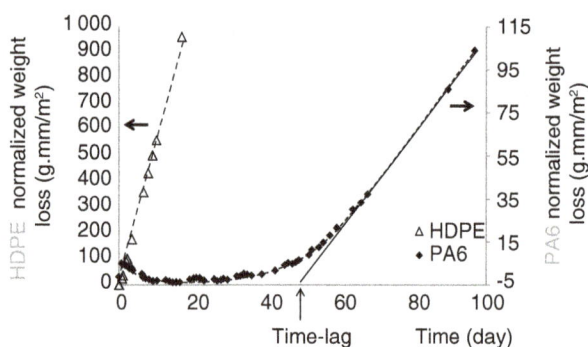

Figure 6

Weight loss as a function of time for 1 mm thick PA6 and HDPE plates in presence of E10 at 40°C (HDPE: left axis, PA6: right axis).

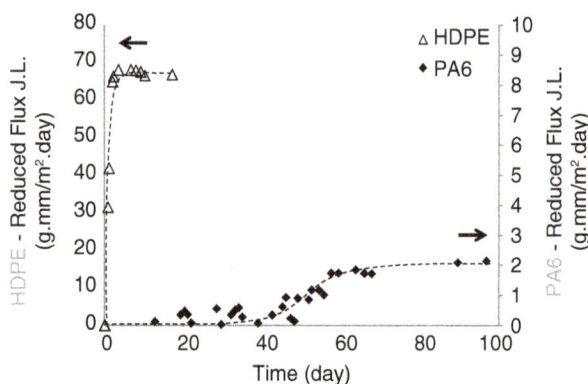

Figure 7

Reduced Flux (J.L.) as a function of time for 1 mm thick PA6 and HDPE plates in presence of E10 at 40°C (HDPE: left axis, PA6: right axis).

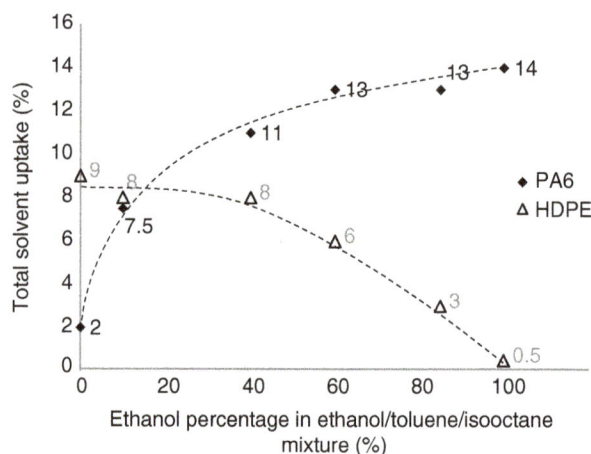

Figure 8

Solvent uptake at the equilibrium as a function of the ethanol content in fuel for PA6 and HDPE immerged at 40°C in ethanol/toluene/isooctane mixtures.

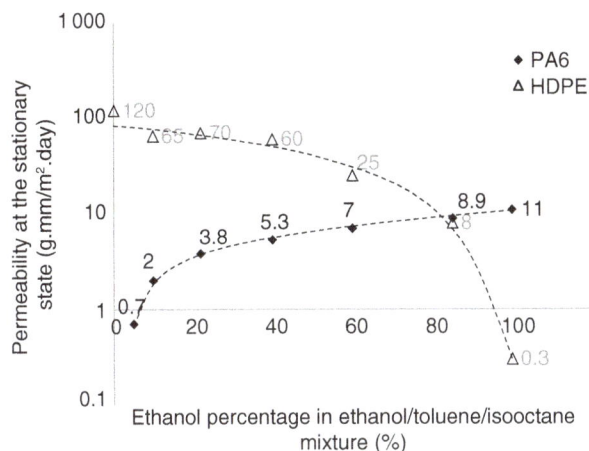

Figure 9

Steady state permeability as a function of the ethanol content in fuel for PA6 and HDPE put in contact with ethanol/toluene/isooctane mixtures at 40°C.

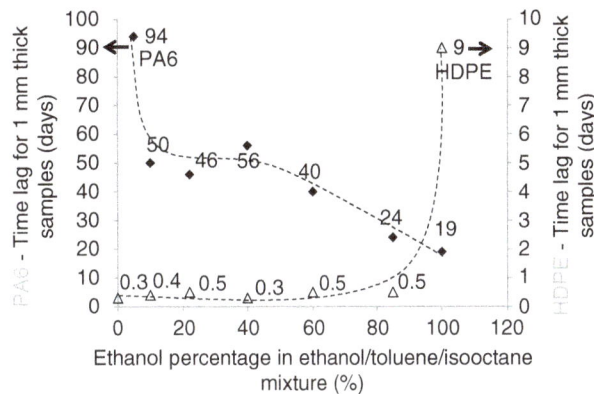

Figure 10

Time-lag values as a function of the ethanol content in fuel for 1 mm thick PA6 and HDPE plates put in contact with ethanol/toluene/isooctane mixtures at 40°C.

HDPE, a similar solvent uptake for PA6 and HDPE being observed for the composition E10, as already described in the previous paragraph. The difference of polarity between PA6 and HDPE is naturally at the origin of these two opposite trends, the presence of polar amide groups in polyamide favoring the interaction with polar molecules such as ethanol. Above 10% of ethanol in the fuel, fuel uptake is larger in PA6 than in HDPE.

The influence of ethanol content in fuel on the steady state permeability measured by gravimetry is then shown in Figure 9, the values obtained for HDPE being close to data found in the literature (69 g.mm/m^2.day in presence of E10 and 90 g.mm/m^2.day in presence of E0 according to Nulman et al. [1]). Similarly to the equilibrium solvent intake, permeability of PA6 increases with ethanol content whereas it decreases in the case of HDPE. But permeability values of HDPE are much higher than the ones of PA6 until 85% of ethanol in the fuel, composition for which permeability of HDPE is close to that of PA6 (8-9 g.mm/m^2.day). In the case of low ethanol content fuels (E5, E0), permeability of PA6 turns out to be at least 100 times lower than that of HDPE. It should be also pointed out that the evolution of the permeability of PA6 or HDPE with ethanol content is monotonic, no maximum suggesting a "positive synergy" being observed as it is the case for PA12, PVDF or EVOH in presence of methanol/toluene mixture [2].

The influence of ethanol content in the fuel on the time-lag measured during gravimetric pervaporation tests on 1 mm thick PA6 and HDPE samples is then shown in Figure 10. On the whole fuel composition range (0% to 100% ethanol), the time-lag is much higher for PA6 than for HDPE, suggesting slower diffusion kinetics for PA6 on the whole fuel composition range. The slow diffusion kinetics of E10 in PA6 between 10% and 85% of ethanol in the fuel seems to counterbalance the higher solubility observed for PA6 in Figure 8 (because the permeability of PA6 is lower than the one of HDPE on that fuel composition range). But beyond 85% of ethanol in the fuel, it seems that the slower diffusion kinetics of PA6 does not counterbalance the higher solubility of polyamide, a poorer barrier property being observed in the case of PA6 in presence of E100.

In the case of PA6, three regimes can be distinguished: time-lag firstly strongly decreases between 5 and 10% of ethanol in the mixture (we could not determine the time-lag of PA6 in presence of E0 in a reasonable time scale), then a plateau is observed between 10 and 60% of ethanol in the mixture, and finally, time-lag slowly decreases between 60% and 100% of ethanol in the mixture. The strong decrease of time-lag between 0 and 10% associated to a strong increase of the solubility on that fuel composition range lead to a strong increase of permeability, and between 10% and 100%, the presence of a plateau or a moderate decrease of the time lag associated to a moderate increase of solubility lead to a moderate increase of the permeability.

2.4 Effect of Sample Thickness

The influence of sample thickness on the barrier properties of PA6 was then studied thanks to the extruded films

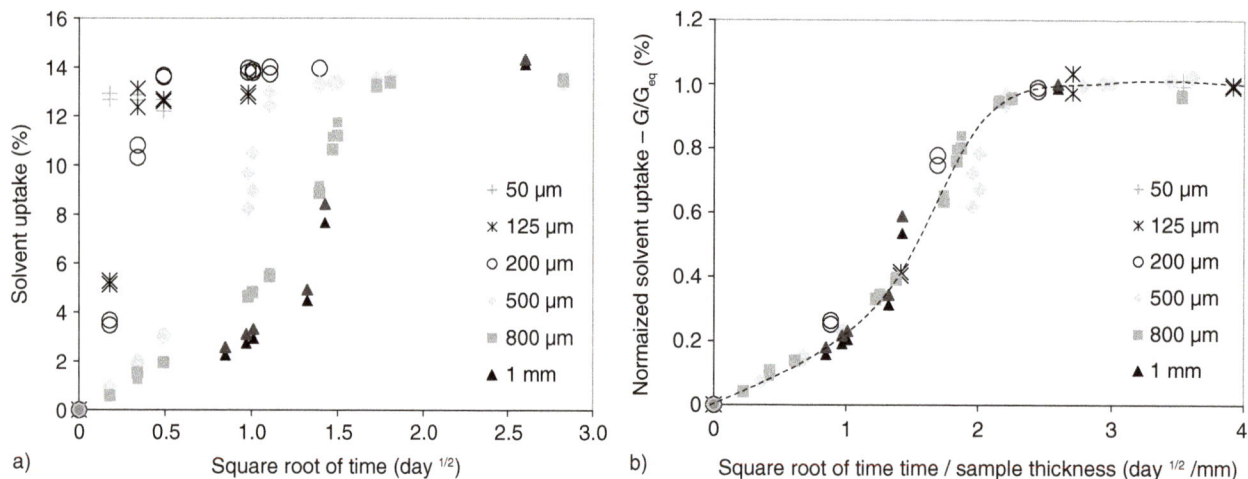

Figure 11

Ethanol sorption at 40°C of PA6 samples of various thicknesses a) solvent uptake as a function of square root of time b) normalized curve: solvent uptake divided by the uptake at the equilibrium as a function of the square root of time divided by sample thickness.

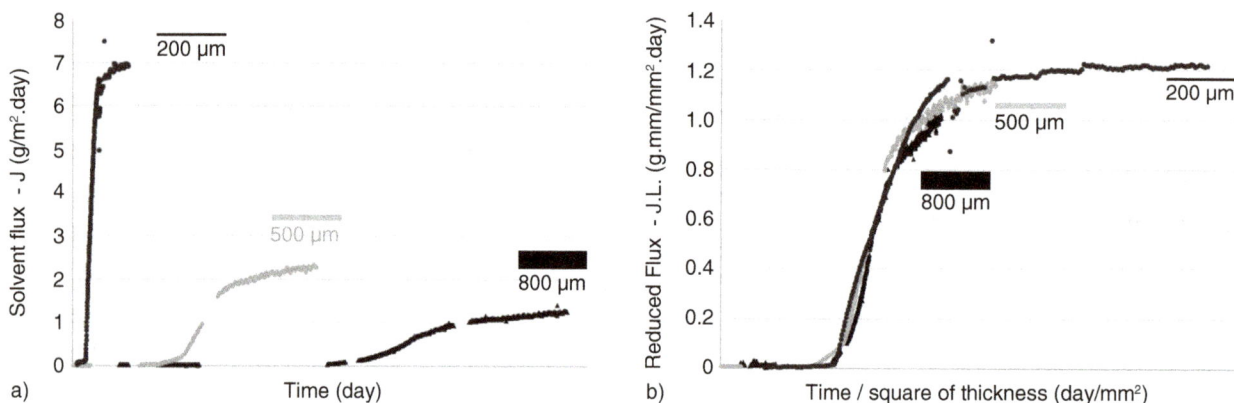

Figure 12

E10 pervaporation of PA6 samples presenting various thicknesses at 40°C a) solvent flux as a function of time b) reduced flux as a function of the time divided by the square of the thickness.

and the injection molded plate presenting different thicknesses varying from 50 μm to 1 mm. The influence of sample thickness on the ethanol sorption of PA6 measured by gravimetry at 40°C is firstly shown in Figure 11.

Figure 11a shows that solvent uptake at the equilibrium varies little with sample thickness, which is consistent with the fact that crystalline fraction as well as crystalline lattice nature are similar in the different samples. Then, the thinner the sample, the faster the sorption kinetics. By normalizing the x axis (square root of time) by the sample thickness (*Fig. 11b*), a mastercurve is obtained; suggesting that ethanol diffusion kinetics in

PA6 evolves with the square of the thickness. Since ethanol diffusion in PA6 is not fickian (as attested by the sigmoidal sorption curves), this result was not expected. The presence of a 10 μm skin deprived of spherulites in the injection molded sample does not seem to affect neither the solvent uptake nor the diffusion kinetics.

The influence of sample thickness on the permeability of PA6 in presence of E10 at 40°C was then studied on the basis of results obtained with FID gas chromatography on films presenting 3 different thicknesses of 200, 500 and 800 μm. Figure 12a shows that the thinner the sample, the higher the solvent flux and the faster the

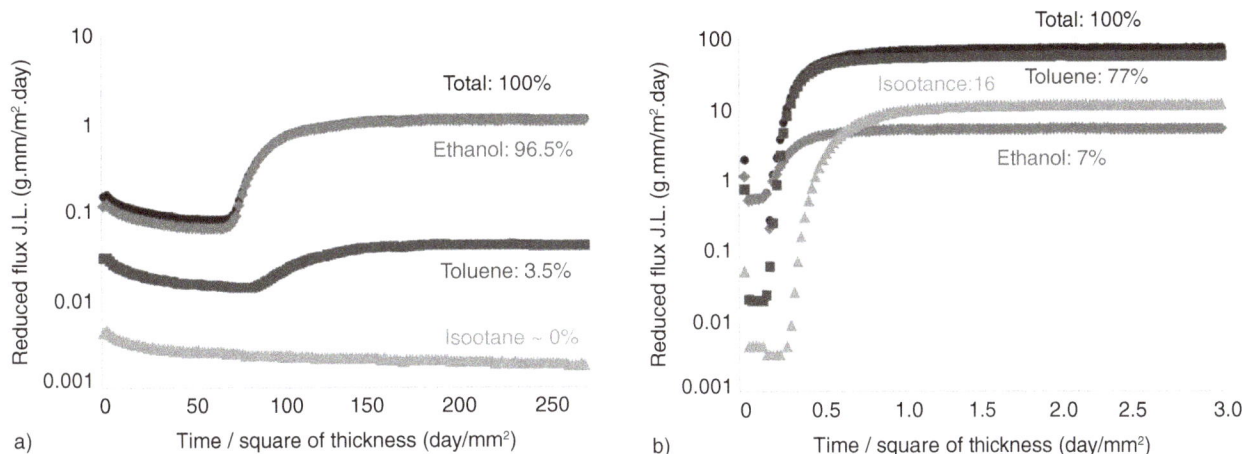

Figure 13

Separate solvents contribution to the reduced flux measured by FID gas chromatography at 40°C in presence of E10 as a function of the normalized time a) PA6, b) HDPE.

pervaporation kinetics. By normalizing the y-axis (solvent flux) by sample thickness and the x-axis (time) by the square of the sample thickness, a mastercurve is obtained. Similarly to sorption kinetics, permeation kinetics turns out to evolve with the square of the sample thickness. To summarize, it is important to underline that despite the pronounced plasticizing effect of ethanol on PA, and a resulting anomalous diffusion behaviour attested by sigmoidal sorption curves, sorption or permeation kinetics of biofuels in PA6 turns out to evolve with the square of the sample thickness (observed also in the case of PA6 in presence of fuels containing 5% or 85% of ethanol – results not shown here).

The validation of the evolution of pervaporation kinetics with the square of sample thickness allows extrapolating what would be the pervaporation kinetics of thick parts such as fuel tanks. For 1 mm thick samples, the time lag is about 75 days for extruded films (Fig. 12b, result obtained by FID chromatography) and 40 days for the injection molded plate (Fig. 7, result obtained by gravimetry). The difference between these results could originate from differences linked to polyamide processing (extrusion *versus* injection molding), pervaporation measurement (FID chromatography *versus* gravimetry), or the hygrometry level associated to the experiment (dry condition *versus* HR20). If we consider the minimum value of 40 days for the 1 mm injection molded plate, a simple calculation shows that the time lag for a 3 mm part will be $40 \times 3^2 = 360$ days, *i.e.* about 1 year. But a pervaporation test that lasts 1 year would not be long enough to assess the stationary state

permeability. Results on 1 mm injection molded plate show that the stationary state permeability is obtained after 60 days, *i.e.* after a relatively long transient regime. For a 3 mm thick part, the extrapolation indicates that the stationary state permeability would be reached after $60 \times 3^2 = 18$ months. The results obtained on films (time lag: 75 days, steady state: 200 days), imply an extrapolation for 3 mm thick parts of almost 2 years for the time lag and 5 years for the stationary state. Beyond the differences between films characterized by FID chromatography and injection molded plates characterized by gravimetry, these results show that the test duration necessary to assess the stationary state permeability of polyamides is very long regarding the time scale of the actual standards for assessing the fuel emissions of vehicles. Indeed, Euro 5 standard implies a measurement on a vehicle after 3 000 km of rolling, the time needed to complete such a distance being of the order of magnitude of a few months. As a consequence, if a SHED evaluation is performed on a PA6 fuel tank according to the actual standards, the measured emission could be close to zero, because the PA6 tank is likely to be still in its "time-lag" period. This is not the case for HDPE which exhibits much faster pervaporation kinetics, and for which a stationary state permeability should be obtained after 3 000 km of rolling (stationary state permeability of HDPE for 1 mm thick sample obtained in less than one day). This issue of underestimating the permeability of polymers because of a too short period of exposition of the tank to solvents is well known by automotive manufacturers who often apply a soaking procedure of the

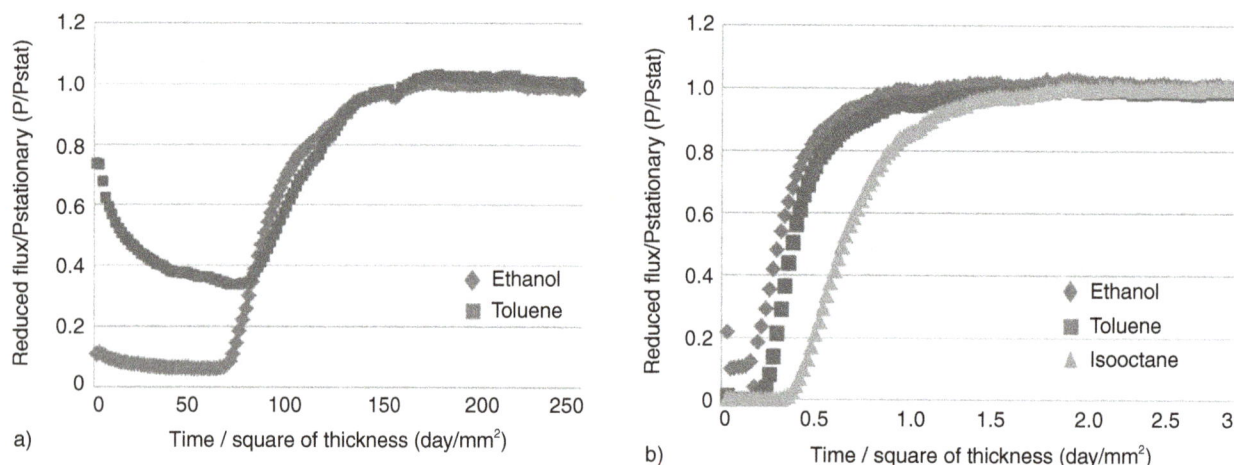

Figure 14

Reduced flux over permeability at the stationary state for each solvent flux measured by FID gas chromatography at 40°C in presence of E10 a) PA6, b) HDPE.

tanks of 20-25 weeks before the SHED measurement. But it turns out that this extended soaking procedure remains short in regards of fuel diffusion kinetics in PA6.

In order to accelerate the pervaporation experiment, thin specimens should be preferred, but in order to avoid changing polyamide processing for the obtention of these thin specimens (for example extruding films instead of blow molding fuels tanks), which could alter the PA microstructure and further its barrier properties, one possibility is to thin out the part. A good correlation was indeed obtained between machined samples for which the thickness was reduced from 1 mm to 0.3 mm, and non-machined samples obtained directly from the injection molding process (results not shown here).

2.5 Analysis of the Solvent Flux by FID Gas Chromatography

The composition of solvent flux was then studied by FID gas chromatography for PA6 and HDPE put in contact with fuels containing various ethanol contents at 40°C. Figure 13 firstly shows the results obtained for PA6 (200 µm thick extruded film) and HDPE (1 mm thick injection molded plate) in presence of E10. In the case of PA6 + E10, the species contained in the solvent flux are mainly ethanol (96.5%), a little amount of toluene (3.5%) and no detectable amount of isooctane. In the case of HDPE + E10, the solvent flux is composed of 77% of toluene, 16% of isooctane and 7% of ethanol. The slight decrease of the fluxes of ethanol, toluene and isooctane observed for PA6 at the very beginning of the permeation experiment could be attributed to

residual solvent traces which are present on the downstream part of the permeation cell, and which are progressively evacuated.

In order to compare the solvent diffusion kinetics, the ratio of the reduced flux on the permeability at the stationary state was plotted in Figure 14 for each solvent. A quite unexpected result is that both in the cases of PA6 and HDPE, diffusion kinetics of ethanol and toluene turn out to be relatively close, permeation of both solvents starting quite simultaneously (but it should be more accurately confirmed in the case of PA6 because the issue of residual traces of toluene hide the beginning of toluene permeation). Isooctane diffusion kinetics in HDPE appears to be slower than those of ethanol and toluene, isooctane starting to permeate slightly later. And as already discussed in Section 2.4., the diffusion timescales in PA6 and in HDPE are very different, solvents diffusion being much faster in HDPE than in PA6.

Figure 15 shows then the ethanol percentage in the pervaporation solvent flux measured at 40°C by FID gas chromatography as a function of ethanol content in the fuel for the two samples described previously. On the whole fuel composition range, ethanol is by far the main species present in the solvent flux of PA6. On the contrary, hydrocarbons are the main components of the solvent flux of HDPE, 30% of ethanol being nevertheless measured in the solvent flux of HDPE in contact with E85.

These results point out two important facts. Firstly, as ethanol is less toxic than aromatic hydrocarbons such as toluene, the use of polyamide should be preferable from the Health and Safety point of view. Indeed, if we take

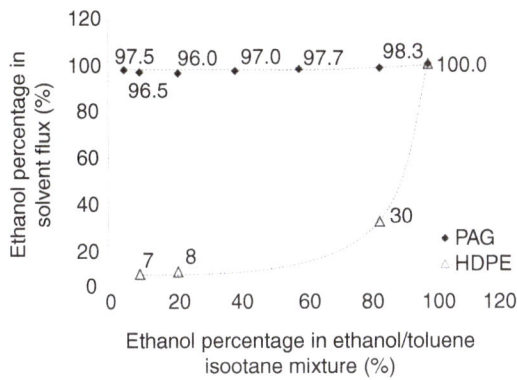

Figure 15

Ethanol percentage in the pervaporation solvent flux as a function of ethanol content in ethanol/toluene/isooctane mixtures – comparison of PA6 (200 μm extruded film) and HDPE (1 mm injection molded plate).

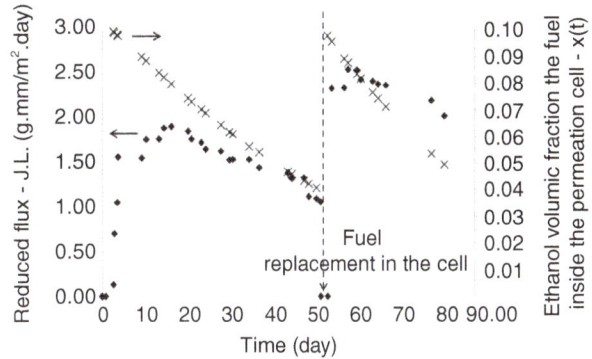

Figure 16

Illustration of the influence of fuel composition evolution during the experiment on the permeability value measured by gravimetry (result obtained on PA6 200 μm thick film in presence of E10 at 40°C).

into account both the vapors composition and the lower pervaporation value of PA6 (2 g.mm/m^2.day for PA6 instead of 65 g.mm/m^2.day for HDPE in presence of E10), it turns out that the hydrocarbons content in the permeant species is 900 times lower for PA6 + E10 than for HDPE + E10. Secondly, the vapors composition difference between PA6 and HDPE has to be kept in mind if the two materials are characterized by applicative assessment methods such as the SHED test. Indeed, SHED experiments involve the use of a FID detector but no chromatographic column that allows the separation of the species. SHED tests are generally calibrated with propane, which implies that all the molecules detected by the SHED FID detector are "seen" as propane molecules. But ethanol and hydrocarbons should not be counted up the same because the response coefficient of ethanol is much lower than that of hydrocarbons. If this difference is not taken into account (as it is the case for the SHED test), this will lead to an underestimation of the PA6 permeability. The relationship between the permeability measured by SHED and the permeability measured by gravimetry or by FID gas chromatography (with species separation and appropriate calibration) could be basically given by Equation (3):

$$P_{SHED} = P_{real} \left(\frac{x_{etoh}}{f} + x_{hyd} \right) \qquad (3)$$

with P_{SHED} the permeability measured by SHED experiment, P_{real} the permeability measured by gravimetry or

by FID gaz chromatography (with species separation and appropriate calibration), x_{etoh} the ethanol content in the permeant species, x_{hyd} the hydrocarbons content in the permeant species, and f the ratio between the response coefficient of hydrocarbons and the response coefficient of ethanol determined during the calibration of the FID detector (in our case, 5). In presence of E10, the simple numerical application of Equation (1) predicts that the permeability measured by SHED will be 4 times lower than that measured by gravimetry for PA6, whereas it will be almost unchanged in the case of HDPE.

2.6 Influence of Fuel Composition Evolution

It was shown in the previous paragraph that there is a preferential diffusion of polar molecules in the case of PA6 or apolar molecules in the case of HDPE, and a consequence of this preferential diffusion can be an evolution of the fuel composition present in the permeation cell (or in the fuel tank). As pervaporation is strongly dependent on the fuel composition (namely the ethanol content), this can lead to a variation of the permeability value during the experiment.

Figure 16 shows on its left axis the reduced flux measured by gravimetry of a 200 μm thick PA6 film in presence of E10 at 40°C as a function of time.

The reduced flux seems to stabilize around 10 days to a value close to 2 g.mm/m^2.day, but afterwards, reduced flux progressively decreases with time. The evolution of the ethanol content in the permeation cell was then calculated according to Equation (4):

$$x(t) = x_0 - \frac{f \cdot \Delta m(t)}{d_{etoh} \cdot V_0} \qquad (4)$$

with $x(t)$ the volume fraction of ethanol inside the permeation cell at a given time t, x_0 the initial volumic fraction of ethanol inside the permeation cell ($x_0 = 0.10$ in the case of E10), f the weight fraction of ethanol present in the solvent flux ($f = 0.965$ in the case of PA6 + E10, according to the results presented in *Sect. 4*), $\Delta m(t)$ the total weight loss at a given time t, d_{etoh} the density of ethanol at the temperature of the experiment (taken as 0.79 g/cm³), and V_0 the initial volume of solvent in the permeation cell (in our case $V_0 = 14$ cm³). This parameter $x(t)$ is plotted as a function of time on the right axis of Figure 16 and it can be seen that after 50 days, the ethanol content in the fuel inside the permeation cell decreases from 10% to about 4%, this decrease of the ethanol content being accompanied by a decrease of the reduced flux from 2 g.mm/m².day to almost 1 g.mm/m².day. In the current experiment, at 55 days, the fuel in the permeation cell was entirely replaced by a new preparation containing the right specific composition of fuel E10. Figure 16 shows that this fuel replacement induces an immediate increase of the reduced flux, reaching a value close but slightly higher than that of the reduced flux after 10 days. Then, reduced flux decreases again while the ethanol content in the permeation cell decreases.

This result obtained on a 200 μm thick PA6 film illustrates what could be called the fuel exhaustion process. In the case of polyamide, it is the ethanol fraction decrease that leads to a decrease of the measured permeability, and in the case of polyethylene it is the hydrocarbons fraction decrease that leads to a decrease of the measured permeability. In order to avoid this phenomenon, it is important to have an initial volume of fuel inside the permeation cell important in comparison with the measured weight loss, and in every case, attention should be paid to attest the possible evolution of the fuel composition. In case of doubt, a simple way to proceed is to replace the fuel in the permeation cell and to see if this replacement induces an increase of the reduced flux.

From an applicative point of view, we can wonder in what extent this evolution of fuel composition induced by preferential diffusion can be an issue in real size fuel tanks. In order to estimate the fuel composition evolution kinetics in a real fuel tank, it is possible to calculate the time t_i (in days) needed to deplete the component i with a ratio x according to Equation (5):

$$t_i(x) = \frac{1000 \, V \, \varphi_{tank} \varphi_i \, d_i \, x}{\left(\frac{P \, S_{exc}}{e}\right) f_i} \qquad (5)$$

with V the volume of the tank in L, ϕ_{tank} the filling rate of the tank, ϕ_i the volumic fraction of component i in the fuel, d_i the density of the component i in g/cm³, P the permeability of the polymer, S_{exc} the exchange surface in the fuel tank in m², e the thickness of the tank in mm, f_i the weight fraction of component i in the solvent flux and x the ratio of the loss amount of component i on the initial amount of component i ($x = 1$ for complete component depletion).

A very unfavorable situation corresponds to a large tank (important exchange surface) almost empty. If we consider the following hypothesis: 80 L cubic fuel tank (exchange surface 1 m²), tank thickness of 3 mm, filled up to 5% with E10 ($\phi_{tank} = 0.05$), permeability PA6 of 2 g.mm/m².day, ethanol representing 96.5% of the permeant species ($f_i = 0.965$), we obtain according to Equation 5 that the time needed to completely deplete ethanol in fuel (in other words transforming E10 into E0) is 16 months for PA6. In the case of HDPE (permeability HDPE = 65 g.mm/m².day, ethanol representing 7% of the permeant species), the time needed to deplete hydrocarbons (in other words transforming E10 into E100) is 4.5 months.

The ratio of the time needed to deplete ethanol (i) in a PA6 tank on the time needed to deplete hydrocarbons (j) in an HDPE tank is then given in Equation (6):

$$\frac{t_{i-PA6}}{t_{j-HDPE}} = \left(\frac{P_{HDPE}}{P_{PA6}}\right) \left(\frac{\varphi_i}{1 - \varphi_i}\right) \left(\frac{d_i}{d_j}\right) \left(\frac{f_j}{f_i}\right) \qquad (6)$$

This ratio mainly depends on the barrier performances of the polymers in presence of a specific fuel composition. But whatever the size of the tank, the filling rate and the tank thickness, it turns out that in presence of E10 ($P_{HDPE} = 65$ g.mm/².day, $P_{PA6} = 2$ g.mm/².day, $\varphi_i = 0.1$, $d_i = 0.79$, $d_j = 0.78$, $f_j = 0.93$, $f_i = 0.965$), the fuel exhaustion process of HDPE is 3.5 faster than the fuel exhaustion process of PA6, and this despite the fact that E10 contains much more hydrocarbons than ethanol, which is favorable for HDPE.

2.7 Influence of the Temperature on PA Permeability to Biofuels

The influence of temperature on the barrier properties of PA6 to biofuels presenting various ethanol contents was then studied by varying between 30°C and 50°C the temperature of the FID gas chromatography pervaporation experiment. Figure 17a shows the evolution of the stationary state permeability of 200 μm thick PA6 films with the temperature for 4 fuel compositions containing 10% (E10), 22% (E22), 85% (E85) or 100% (E100) of

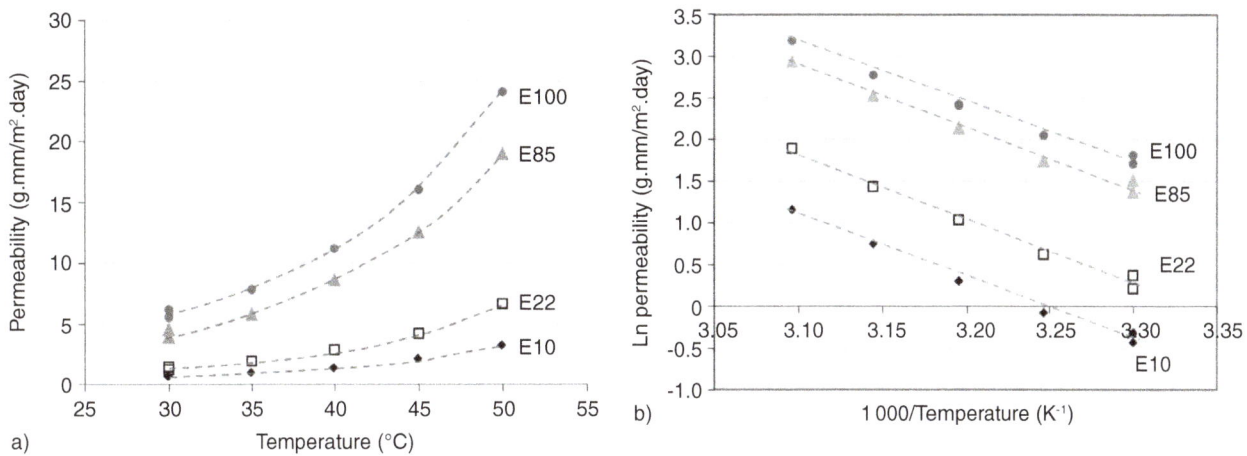

a)

b)

Figure 17

Influence of the temperature on the permeability of PA6 measured by FID gas chromatography for different fuel compositions containing 10, 22, 85 or 100% of ethanol a) plot of temperature as a function of time, b) Logarithm of the permeability as a function of 1 000/temperature.

TABLE 3

Apparent activation energies for the permeation process of different polymers in presence with fuels containing different ethanol contents – PA6: experimental data of the current work, HDPE* and PA12*: data from [1]

Fuel	E0	E10	E22	E85	E100
E_a (kJ/mol) – PA6	x	61.5	64.1	60.1	56.8
E_a (kJ/mol) – HDPE*	53.7	52.3	x	x	x
E_a (kJ/mol) – PA12*	72.2	50	x	x	x

ethanol. In accordance with the results presented in Section 2, the permeability of PA6 increases as ethanol content increases, and it is also evidenced that permeability increases as temperature increases.

It is well established that fickian diffusion of small-size molecules in polymers is a thermally activated process [25], and a large number of data in the literature suggests that the permeability depends on temperature *via* an Arrhenius's law defined in Equation (7) [26]:

$$P(T) = P_0 \exp\left(\frac{-E_a}{RT}\right) \quad (7)$$

with $P(T)$ the permeability at a temperature T, P_0 a pre-exponential factor representing the value of P for an infinite molecular agitation ($T \rightarrow \infty$), E_a the apparent activation energy for the permeation process, R the gas constant equal to 8.314 J/mol.K, and T the temperature.

The linear evolution of the natural logarithm of the permeability with the inverse of the temperature plotted

in Figure 17b suggests that the permeability of PA6 to biofuels follows effectively an Arrhenius law. The activation energies for the different fuels can be deduced from the slopes of the straight lines plotted in Figure 17b, the calculated values being given in Table 3.

The values of the apparent activation energies for the pervaporation process of PA6 is close to 60 ± 4 kJ/mol and it seems that the ethanol content in the fuel little affects the activation energy values. This value for PA6 is consistent with data from the literature describing the evolution of the permeability with the temperature of various polymers in presence of fuel C (isooctane/toluene mixture) or fuels CE10 (equivalent to the E10 denomination in the current work), the data for PA12 and HDPE reported in the paper of Nulman *et al.* [1] being reported in Table 3. These results are in agreement with an oft quoted rule of thumb stating that permeability of solvents through polymeric materials decreases (increases) by about 10% for a 1°C decrease (increase) in temperature.

CONCLUSION

This article showed that until 85% of ethanol in the bio-fuel, the barrier performance of polyamide 6 is much better than that of HDPE, the very slow diffusion kinetics in PA6 in comparison with HDPE being at the origin of the smaller permeability values of PA6. In addition, FID gas chromatography showed that more than 95% of the solvent flux that permeates through PA6 is composed of ethanol. A direct consequence, which is important to consider from an Health and Safety point of view, is that hydrocarbons content (including toxic aromatics components) is 900 times lower in the solvent flux of PA6 than in that of HDPE in presence of E10. Secondly, as SHED test does not allow any speciation of the solvent molecules, ethanol in this test configuration is counted as a propane molecule, and at the end, this should lead to an under-estimation of the permeability of PA6, little variation being expected for the permeation value of HDPE. A third side effect associated to the solvent flux composition is the fuel exhaustion process, which consists in a decrease of the reduced flux due to the evolution of the fuel in the tank or in the permeation cup, the fuel evolution being itself caused by the preferential diffusion of one of the solvent present in the mixture. This phenomenon should be taken in consideration when assessing the permeability of thin samples, the fuel composition evolution kinetics in thick tanks being a very slow process. But it should be pointed out that in presence of E10, the fuel exhaustion process is 3.5 times faster for HDPE than for PA6. The influence of temperature on PA6 biofuels barrier properties was also studied, and it turned out that the dependence of permeability with temperature varies little with the ethanol content in the fuel, the mean apparent activation energy obtained for PA6 (60 kJ/mol) being of the same order of magnitude than the values found in the literature for other polymers such as PA12 or HDPE. Concerning the diffusion processes, it was shown that in the case of PA6, despite a non fickian behavior due to the fact that ethanol strongly increases the molecular mobility of PA6 amorphous phase ("plasticizing" effect), solvents diffusion kinetics evolve with the square of the thickness: indeed, mastercurves were obtained if the solvent uptake (sorption experiment) is plotted as a function of the square root of time divided by the thickness, or if the solvent flux (pervaporation experiment) is plotted as a function of the time divided by the square of the thickness. This allows predicting the permeability of thick parts by extrapolating the value obtained on thin samples. To conclude, it emerges from this work that PA6 based formulations can be suitable for fuel tanks applications, and beyond their interesting intrinsic barrier performance, it should be underlined that in comparison with multi-layer structures, mono-layer constructions also offer a greater potential for accurate models and simulation tools. Indeed, from an easily accessible permeability measurement at the lab scale and an accurate knowledge of the fuel tank geometry, it is possible to predict the permeability of the fuel tank in specific conditions similar to the lab scale ones, and on the basis of the data presented in this article (influence of ethanol content, temperature, fuel composition evolution), permeability of parts subjected to various environments could also be accurately extrapolated.

ACKNOWLEDGMENTS

The work presented here is related to a research program conducted in the LPMA (*Solvay*-CNRS joint lab) on polymer-solvents interactions and diffusion mechanisms. We acknowledge scientific discussions with Didier Long, Paul Sotta, Agustin Rios de Anda and Erik Lange. The authors are grateful to Duramat project for the financial support. They are also very grateful to Vincent Curtil and Fabrice Chavand from *Solvay* facilities for the processing of the samples, and to Erwann Jeanneau from the "Centre Henri Longchambon" for the X-Ray diffraction measurements.

REFERENCES

1 Nulman M., Olejnik A., Samus M., Fead E., Rossi G. (2001) Fuel Permeation Performance of Polymeric Materials, *SAE Technical Paper* 2001-01-1999.

2 Gagnard C., Germain Y., Keraudren P., Barrière B. (2003) Permeability of Semicrystalline Polymers to Toluene/Methanol Mixture, *Journal of Applied Polymer Science* **90**, 2727-2733.

3 Berlanga-Labari C., Albistur-Goni A., Barado-Pardo I., Gutierrez-Pinado M., Fernandez-Carrasquilla J. (2011) Compatibility study of high density polyethylene with bioethanol-gasoline blends, *Materials and Design* **32**, 441-446.

4 Gonzalez-Nunez R., Padilla H., De Kee D., Favis B.D. (2001) Barrier properties of polyamide-6/high density polyethylene blends, *Polymer Bulletin* **46**, 323-330.

5 Yeh J.T., Chao C.C., Chen C.H. (2000) Effects of Processing Conditions on the Barrier Properties of Polyethylene (PE)/Modified Polyamide (MPA) and Modified Polyethylene 5MPE)/Polyamide (PA) Blends, *Journal of Applied Polymer Science* **79**, 1997-2008.

6 Subramanian P.M. (1985) Permeability Barriers by Controlled Morphology of Polymer blends, *Polymer Engineering and Science* **25**, 483-487.

7 Yeh J.T., Fan-Chiang C.C. (1997) The Barrier, Impact, Morphology, and Rheological Properties of Modified Polyamides and their Corresponding Polyethylene-Modified Polyamide Blends, *Journal of Applied Polymer Science* **66**, 2517-2527.

8 Yeh J.T., Jyan C.F., Yang S.S., Chou S. (1999) Influence of Compatibilization and Viscosity Ratio on the Barrier and Impact Properties of Blends of a Modified Polyamide-6 and Polyethylene, *Polymer Engineering and Science* **39**, 1952-1961.

9 Yeh J.T., Huang S.S., Yao W.H. (2002) Gasoline Permeation Resistance of Containers of Polyethylene, Polyethylene/ Modified Polyamide and Polyethylene/Blends of Modified Polyamide and Ethylene Vinyl Alcohol, *Macromolecular Materials and Engineering* **287**, 532-538.

10 Yeh J.T., Chen C.H., Shyu W.D. (2001) Gasoline Permeation Resistance of the As-Blow-Molded and Annealed Polyethylene, Polyethylene/Polyamide, and Polyethylene/ Modified Polyamide Bottles, *Journal of Applied Polymer Science* **81**, 2827-2837.

11 Yeh J.T., Fan-Chiang C.C. (1996) Permeation Mechanisms of Xylene in Blow-Molded Bottles of Pure Polyethylene, Polyethylene/Polyamide and Polyethylene/Modified Polyamide Blends, *Journal of Polymer Research* **3**, 211-219.

12 Yeh J.T., Chang S.S., Yao H.T., Chen K.N., Jou W.S. (2000) The permeation resistance of polyethylene, polyethylene/polyamide and polyethylene/modified polyamide blown tubes against unleaded gasoline, *Journal of Materials Science* **35**, 1321-1330.

13 Puffr R., Sebenda J. (1967) On the Structure and Properties of Polyamides. XXVII. The Mechanism of Water Sorption in Polyamides, *Journal of Polymer Science: Part C* **16**, 79-93.

14 Murthy N.S., Stamm M., Sibilia J.P., Krimm S. (1989) Structural Changes Accompanying Hydration in Nylon 6, *Macromolecules* **22**, 1261-1267.

15 Nilsson P.G., Lindman B. (1983) Water self-diffusion in nonionic surfactant solutions, hydratation and obstruction effects, *Journal of Physical Chemistry* **87**, 4756-4761.

16 Sabard M., Gouanvé F., Espuche E., Fulchiron R., Seytre G., Fillot L.A., Trouillet-Fonti L. Influence of montmorillonite and film processing conditions on the morphology of polyamide 6: effect on ethanol and toluene barrier properties, accepted in *Journal of Membrane Science*.

17 Penel-Pierron L., Depecker C., Seguela R., Lefebvre J.M. (2001) Structural and Mechanical Behavior of Nylon 6 Films. Part I. Identification and Stability of the Crystalline Phases, *Journal of Polymer Science. Part B: Polymer Physics* **39**, 484-495.

18 Long F.A., Richman D. (1960) Concentration Gradients for Diffusion of Vapors in Glassy Polymers and their Relationship to Time Dependent Diffusion Phenomena, *Journal of American Chemical Society* **82**, 513-519.

19 Berens A.R., Hopfenberg H.B. (1978) Diffusion and relaxation in glassy polymer powders: 2. Separation of diffusion and relaxation parameters, *Polymer* **19**, 489-496.

20 Hui C.Y., Wu K.C., Lasky R.C., Kramer E.J. (1987) Case-II Diffusion in Polymers. I. Transient Swelling, *Journal of Applied Physics* **61**, 5129-5136.

21 Hui C.Y., Wu K.C., Lasky R.C., Kramer E.J. (1987) Case-II Diffusion in Polymers. II. Steady State Front Motion, *Journal of Applied Physics* **61**, 5137-5149.

22 Rios A., Fillot L.A., Rossi S., Long D., Sotta P. (2011) Influence of the sorption of polar and non-polar solvents on the glass transition temperature of polyamide 6,6 amorphous phase, *Polymer Engineering and Science* **51**, 11, 2129-2135.

23 Sabard M., Espuche E., Gouanvé F., Fulchiron R., Seytre G., Fillot L.A., Trouillet-Fonti L. (2012) Influence of Film Processing Conditions on the Morphology of Polyamide 6: Consequences on Water and Ethanol Sorption Properties, *Journal of Membrane Science* **415-416**, 670-680.

24 Lange E., Masnada E., Rios A., Fillot L.A., Sotta P., Long D. Solvent diffusion in glassy polymers: a macroscopic approach, (in preparation).

25 Barrer R.M. (1937) Nature of the Diffusion process in Rubber, *Nature* **140**, 106-107.

26 Rogers C.E. (1985) Permeation of Gases and Vapours in Polymers, in *Polymer Permeability*, Comyn J. (ed.), Elsevier Applied science, pp. 11-73.

Permissions

List of Contributors

Jing Zhao
Laboratory of Macromolecular Physical Chemistry, LCPM
FRE CNRS-UL 3564, ENSIC, Université de Lorraine, 1 rue
Grandville, BP 20451, 54001 Nancy Cedex - France

Charbel Kanaan
Laboratory of Macromolecular Physical Chemistry, LCPM
FRE CNRS-UL 3564, ENSIC, Université de Lorraine, 1 rue
Grandville, BP 20451, 54001 Nancy Cedex - France

Robert Clément
Laboratory of Macromolecular Physical Chemistry, LCPM
FRE CNRS-UL 3564, ENSIC, Université de Lorraine, 1 rue
Grandville, BP 20451, 54001 Nancy Cedex - France

Benoît Brulé
Arkema – CERDATO, rue du Grand Hamel, 27470
Serquigny - France

Henri Lenda
Laboratory of Macromolecular Physical Chemistry, LCPM
FRE CNRS-UL 3564, ENSIC, Université de Lorraine, 1 rue
Grandville, BP 20451, 54001 Nancy Cedex - France

Anne Jonquières
Laboratory of Macromolecular Physical Chemistry, LCPM
FRE CNRS-UL 3564, ENSIC, Université de Lorraine, 1 rue
Grandville, BP 20451, 54001 Nancy Cedex - France

N. Cavina
DIN, University of Bologna, Viale Risorgimento 2, 40136
Bologna - Italy

E. Corti
DIN, University of Bologna, Viale Risorgimento 2, 40136
Bologna - Italy

F. Marcigliano
Ferrari SpA, Via Abetone Inferiore 4, 41053 Maranello
(MO) - Italy

D. Olivi
DIN, University of Bologna, Viale Risorgimento 2, 40136
Bologna - Italy

L. Poggio
Ferrari SpA, Via Abetone Inferiore 4, 41053 Maranello
(MO) - Italy

Giorgio Rizzoni
Center for Automotive Research and Department of
Mechanical and Aerospace Engineering,
The Ohio State University Columbus, OH 43212 - USA

Simona Onori
Automotive Engineering Department, Clemson University,
Greenville, SC 29607 - USA

P.-Y. Le Gac
IFREMER Centre de Bretagne, Marine Structures
Laboratory, 29280 Plouzané - France

P. Davies
IFREMER Centre de Bretagne, Marine Structures
Laboratory, 29280 Plouzané - France

D. Choqueuse
IFREMER Centre de Bretagne, Marine Structures
Laboratory, 29280 Plouzané - France

Olivier Grondin
IFP Energies nouvelles, 1-4 avenue de Bois-Préau, 92852
Rueil-Malmaison Cedex - France

Laurent Thibault
IFP Energies nouvelles, 1-4 avenue de Bois-Préau, 92852
Rueil-Malmaison Cedex - France

Carole Quérel
IFP Energies nouvelles, 1-4 avenue de Bois-Préau, 92852
Rueil-Malmaison Cedex - France

A. Karvountzis-Kontakiotis
Aristotle University of Thessaloniki, Laboratory of
Applied Thermodynamics (LAT), GR54125, POB 458,
Thessaloniki, Greece

L. Ntziachristos
Aristotle University of Thessaloniki, Laboratory of
Applied Thermodynamics (LAT), GR54125, POB 458,
Thessaloniki, Greece

Ivan Arsie
Dept. of Industrial Engineering, University of Salerno,
Fisciano (SA) 84084 - Italy

Rocco Di Leo
Dept. of Industrial Engineering, University of Salerno,
Fisciano (SA) 84084 - Italy

Cesare Pianese
Dept. of Industrial Engineering, University of Salerno,
Fisciano (SA) 84084 - Italy

Matteo De Cesare
Magneti Marelli Powertrain, Bologna 40134 - Italy

Tomas Nilsson
Department of Electrical Engineering, Linköping University - Sweden

Anders Fröberg
Volvo Construction Equipment, Eskilstuna - Sweden

Jan Åslund
Department of Electrical Engineering, Linköping University - Sweden

Lai Fengpeng
School of Energy Resources, China University of Geosciences, Beijing 100083 - China

Li Zhiping
School of Energy Resources, China University of Geosciences, Beijing 100083 - China

Li Zhifeng
China Huadian Engineering Co., Ltd, Beijing 100035 - China

Yang Zhihao
School of Energy Resources, China University of Geosciences, Beijing 100083 - China

Fu Yingkun
School of Energy Resources, China University of Geosciences, Beijing 100083 - China

Frank Willems
Eindhoven University of Technology, Faculty of Mechanical Engineering, P.O. Box 513, 5600 MB Eindhoven - The Netherlands
TNO Automotive, Steenovenweg 1, 5708 HN Helmond - The Netherlands

Frank Kupper
TNO Automotive, Steenovenweg 1, 5708 HN Helmond - The Netherlands

George Rascanu
Eindhoven University of Technology, Faculty of Mechanical Engineering, P.O. Box 513, 5600 MB Eindhoven - The Netherlands

Emanuel Feru
Eindhoven University of Technology, Faculty of Mechanical Engineering, P.O. Box 513, 5600 MB Eindhoven - The Netherlands

Michael M. Moser
Institute for Dynamic Systems and Control, ETH Zurich, Sonneggstrasse 3, 8092 Zurich Switzerland

Christoph Voser
Institute for Dynamic Systems and Control, ETH Zurich, Sonneggstrasse 3, 8092 Zurich Switzerland

Christopher H. Onder
Institute for Dynamic Systems and Control, ETH Zurich, Sonneggstrasse 3, 8092 Zurich Switzerland

Lino Guzzella
Institute for Dynamic Systems and Control, ETH Zurich, Sonneggstrasse 3, 8092 Zurich Switzerland

X. Lefebvre
IFP Energies nouvelles, 1-4 avenue de Bois-Préau, 92852 Rueil-Malmaison - France

D. Pasquier
IFP Energies nouvelles, Rond-point de l'échangeur de Solaize, BP 3, 69360 Solaize - France

S. Gonzalez
IFP Energies nouvelles, Rond-point de l'échangeur de Solaize, BP 3, 69360 Solaize - France

T. Epsztein
TECHNIP, rue Jean Huré, BP 7, 76580 Le Trait - France

M. Chirat
TECHNIP, rue Jean Huré, BP 7, 76580 Le Trait - France

F. Demanze
TECHNIP, rue Jean Huré, BP 7, 76580 Le Trait - France

L.-A. Fillot
LPMA (UMR 5268 CNRS/Solvay), Advanced Polymer Material Department, R&I Centre Lyon, SOLVAY, 69192 Saint-Fons - France

S. Ghiringhelli
LPMA (UMR 5268 CNRS/Solvay), Advanced Polymer Material Department, R&I Centre Lyon, SOLVAY, 69192 Saint-Fons - France

C. Prebet
LPMA (UMR 5268 CNRS/Solvay), Advanced Polymer Material Department, R&I Centre Lyon, SOLVAY, 69192 Saint-Fons - France

S. Rossi
LPMA (UMR 5268 CNRS/Solvay), Advanced Polymer Material Department, R&I Centre Lyon, SOLVAY, 69192 Saint-Fons - France